COMMUNITY LIBRARY

D0754425

DISUNITED
NATIONS

DIS UNITED NATIONS

The Scramble for Power in an Ungoverned World

PETER ZEIHAN

An Imprint of HarperCollins*Publishers*

HARPER
BUSINESS

DISUNITED NATIONS. Copyright © 2020 by Peter Zeihan. All rights reserved. Printed in the United States of America. No part of this book may be used or reproduced in any manner whatsoever without written permission except in the case of brief quotations embodied in critical articles and reviews. For information, address HarperCollins Publishers, 195 Broadway, New York, NY 10007.

HarperCollins books may be purchased for educational, business, or sales promotional use. For information, please email the Special Markets Department at SPsales@harpercollins.com.

FIRST EDITION

Designed by Bonni Leon-Berman

Library of Congress Cataloging-in-Publication Data has been applied for.

ISBN 978-0-06-291368-5

20 21 22 23 24   LSC   10 9 8 7 6 5 4 3 2 1

To the mentors who advised me that while
it is important to be bold and brash,
the smartest thing I could ever do is
find the smartest person in the room and just . . .
listen.

Candice Young
Robert Pringle
Susan Eisenhower
Matthew Baker

CONTENTS

INTRODUCTION: MOMENTS OF TRANSITION

At the end of the last millennium I found myself in a moment of transition.

Just after Christmas 1999, I drop-kicked my relationship, my job, and my life in the nation's capital, loaded up everything I owned into a rickety SUV, and took off for a fresh start in Austin, Texas. On January 10, 2000, I was a shiny new staffer at a place called Stratfor.com, at the time a media- and geopolitics-analysis house. My new boss was . . . a piece of work. Matt Baker was a wiry ball of over-caffeinated angst and passion, stuffed into a sharp-edged personality that oozed detailed knowledges (not opinions, *knowledges*) as regards Europe and Russia and China and Turkey and so on.

As my recent work in DC had involved quite a bit of Europe and Russia and China and Turkey, clearly I had things to contribute. Matt and I clashed openly, vitriolically, and often. In doing so we developed a deep mutual respect and became good friends.

One night over too many adult beverages we discussed moments of transition of the less personal, and more global, type.

Matt's core thesis was that the entire fabric of the international system was based on the American alliance network, which in turn depended upon a nearly feudal mix of American security commitments and deference to American desires. We back you up in your backyard where it matters to you, and in exchange, you back us up where it matters to us. Challenges from the allies had started to pop up after the Berlin Wall fell, triply so as many of the allies sensed the end of history with the 1990s Russian collapse, and disdained the screwiness of American foreign policy in places like Kosovo and Iraq. Matt's contention was that as the fear of nuclear Armageddon faded from memory into history, the Americans would have a harder time holding the allies together. In his mind 2020 would look a lot like 2003 with countries resisting American power, but the Americans would continue to muster suitable motivation to (successfully) pressure everyone into maintaining some version of the status quo.

In other words, everyone would keep working with America, because the most powerful country in history would want to keep it that way. It wasn't going to be quite as chaotic as herding cats, but it would be close.

In contrast, I felt that the alliance network, bereft of the security demands of the Cold War, had slid into a new role. Instead of defending the allies on the plains of Northern Europe or in the seas of East Asia, it instead was spreading security to the global commons. This was becoming an end unto itself that was superseding the old logic of alliance. The root of economic growth is physical security, and if the Cold War ended *and* the alliance could hold, the nearly automatic result would be a transformational boom global in scope that was both economic and technological. In my mind, 2020 would look a lot like 1950, albeit without the whole fear-of-nuclear-war thing. The spreading wealth, by force of information and capital flow, would grind down what less modern, less democratic pockets of resistance remained.

In other words, the world order would keep going, because unraveling it could mean unwinding decades of economic penetration and deny billions access to hamburgers and cell phones.

Who would want that? It wasn't quite going to be *Star Trek*, but it would be close.

I was Stratfor's unofficial economist. Matt was the gun guy. Most of our friction came from arguing over the power of the checkbook versus the power of Smith & Wesson. Yet so did our greatest collaborations.

After a bunch of backing and forthing and what-ifing and puzzling over oddly colored mixers, Matt asked point-blank, "So what happens to my 'suitable motivation' and your 'transformational boom' when the Americans change their minds about their alliance? No alliance system can last without a common threat." As the night's seventh drink soaked in, we both *hmmm*ed as we contemplated the dark, horrifying possibility that we might *both* be wrong. I'm pretty sure we cracked the code a few minutes later, but our mutual blackouts robbed us of the pertinent knowledge. Instead, as often was the case with Matt, I was left with more questions than answers.

Ever since, both at Stratfor and beyond, my professional life has been about building up enough understandings to close out that long-ago conversation. *Disunited Nations*, nearly twenty years in the making, is my best shot.

Disunited Nations is about what happens when major powers decide they are better off competing instead of cooperating. It is a book about what happens when the global Order isn't just falling apart but when many leaders feel their country will be better off tearing it down. We're going to look at the rise of Trump and leaders like him. We're going to think through Saudi Arabia and Iran's competition to rule (or misrule) the Middle East. We're going to look at how we match farmers to hungry mouths, minerals to manufacturing, oil to gas tanks.

Through these stories, we're going to keep two big ideas in mind.

The first is that geography might not be destiny, but it is damn close. It is the biggest factor in determining how we act and how we live and fundamentally who we are. Live in a desert and *bam!* you're going to fight to protect what little you have. Live on a coast and *bam!* you're going to eat a lot of foreign food. Live in a dense urban area and *bam!* you're probably not going to have an issue with Tongans, Thais, Tunisians, or transvestites. Live in the mountains and *bam!* you're going to be a bit . . . persnickety when folks from other regions roll through. As we were geographers at heart, in this Matt and I were in complete agreement.

Most of us consistently misread economies and conflicts because we don't take geography into account. It's so clear that we are not like the next town over, and city folk are not like country folk, but when we start trying to explain the world, geography often slips our mind. We misinterpret what's happening in the news, and think China is holding on to Hong Kong out of stubbornness or the fights about the American-Mexican border are only about race. Geography shapes everything. Including *us.* What's been different in recent decades is that geography has been suspended somewhat, enabling deep global economic interconnections. We've come to see those connections as a great strength; they are turning into weakness before our very eyes.

The second big idea is that *Disunited Nations* is being published now in 2020, not a few years from now, because the world has run out of time. That moment of transition when the Order will come crashing down is almost upon us.

It may not seem this way to Americans who have been engaged in some degree of warfare continuously since 1999 and had decades of duck-and-cover drills before that, but the world since 1946 is *as calm as the world has ever been.* In creating their anti-Soviet Cold War alliance, the Americans by hook, crook, carrot, and stick brought every significant power of the past five centuries together under a single banner: Norway, Sweden, Spain, Portugal, France, Germany, the Netherlands, Italy, Greece, Turkey, Iran, Japan, China—all of them and more allied in various degrees of formality against the Soviet Union. If they were to be fighting the Soviets, it wouldn't be particularly productive if they were also fighting one another. The American alliance didn't so much end history as freeze it in place.

Most Americans are broadly familiar with the European side of the equation—after centuries of conflict within the Continent the guns fell nearly silent, courtesy of the American security system—but the impact on Asia was even bigger. China and India did not have a *single* century in the past two millennia when they were not at war within themselves or under assault from a foreign power. Instead, from the American-enforced end of British colonial rule in India in the late 1940s and China's joining of the American-led alliance against the Soviets in the 1970s, the two have had the most security and wealth in their long histories.

As the decades rolled on, the Soviets ran out of gas. The American network expanded deep into the nonaligned countries of the developing world and the former Soviet Empire. But the Americans never adapted their overall strategies to a post–Cold War world, and so never built a case at home or abroad for their alliance in a world without the Soviet Union. It turns out—both to the detriment of Matt's and my own pet theories—the 1990s and 2000s were less a moment of transition and more a world on autopilot, with an old security strategy enabling some gangbusters growth. That gangbusters growth is what most of us think of as normal. It is not. It is nothing more than a moment in time made possible by some strategic inertia.

But now Matt's "suitable motivation" is over. The Americans *have* changed their mind about their alliance and *have* turned sharply more insular. There is no effort to ride herd. The W Bush administration abused the allies, the Obama administration ignored the allies, and the Trump administration insulted the allies. And so America's list of allies has shrunk from nearly everyone to the potentially useful to the obviously useful to the obviously loyal to those with little choice. In a world without America, the questions become: Who will still benefit from some lingering connection to the Americans? And who can go it alone?

My "transformational boom" is over as well. Without the global security the Americans guaranteed, global trade and global energy flows *cannot* continue. Seven decades' worth of global industrialization and modernization are not simply at risk, the very pillars of civilization are cracking. In a world without stability, the questions become: Who was most dependent upon the world that was and so will fall? And who was most restrained by the old Order and so will soar?

The world we know is collapsing. Entire countries are watching in horror as what makes them possible—global access, imported energy, foreign markets, American troops—slips through their fingers. For many, there just isn't enough access or energy or markets or security for them to maintain what they have, much less grow. In a world of want, the questions become: What do countries need to survive in a scrambled world? Who will shoot to get what they need? And who gets shot *at*?

Not all competitions and scarcities are created equal. Nearly all food is dependent upon global trade, whether in the form of imported inputs or the foodstuffs themselves. For decades, the world's experiences with famine have been crises of distribution, the inability to match foods with mouths. Global breakdown guts food supply itself. The security concerns of the past two decades were largely limited to terrorism, but the tools necessary to counter terror are radically different from those needed to protect agricultural supply chains. Fewer door-to-door manhunts, more beyond-the-horizon naval patrols. In a world of different scarcities and different tools, the questions become: Where will trade patterns hold and where will they collapse? Which ones are worth fighting over? Which tools will be brought to bear? Are we on the verge of a mess of overlapping and interlocking naval competitions for something as basic as the right to *eat*?

Both Matt and I were wrong, and for the same reasons. We

both suffered from a failure of imagination, believing in our respective, simplistic visions that the world of tomorrow would be some variation of the world we knew. The break is sharper. We stand at the end of the era that began with the Cold War. It'll be less like the messiness of the early 2000s or the raw potential of the 1950s, and more a disastrous combination of the battle royales and displacements of the 1870s against the economic backdrop of the 1930s. It. Will. Suck. A mad scramble for the scraps of the era just ending. Compared with the safety and wealth of the past several decades, it may seem like the literal end of the world. But the end of an era isn't the same as the end of history. Something new is coming. Something that, historically speaking, is far more "normal" than anything the Americans created. Just keep in mind that "normal" is far from synonymous with "comfortable," much less "favorable."

Disunited Nations is my effort to sketch out that "normal" future. To answer these questions and more. To identify the countries that will rise to dominate the human condition, and perhaps provide a glimpse as to what that condition will be, both in the time during and after this moment of transition. In the pages that follow, I'm going to argue that thinking the future will look more like the year 2000 than the year 1900 has negative effects in almost every sphere of our lives. On a grand scale, many of us are betting on the wrong horses. France will lead the new Europe, not Germany. We should be worried about Saudi Arabia, not Iran. We should be thinking about how to remedy mass starvation in China, not counter its economic and military clout. As we look at how each country has benefited from the Order, and what each brings to the table in a new world, these conclusions will seem obvious, rather than controversial.

In Chapters 1 through 4, we'll take a whirlwind tour of the past and present: the various epochs of history, combined with a quick and dirty review of how to run a global empire and what makes countries tick. Chapters 5 through 13 are dedicated to the major powers—both those we all *know* are the countries of the future (but in reality won't be able to hack it), as well as those we barely think about that will rise to rule their respective roosts.

Capping off each country-themed chapter is a bit of a report card. A cheat sheet, if you will. A few distilled lines of information on each country's guts and outlook to help readers filter the news and paranoia and fluff of the day to generate a more accurate appreciation for the country's potential and limitations. For those of you who like to share books with people who don't like to read, I've taken the liberty of refining the chapter's thoughts into a single, solitary word.

Finally, we'll end with a pair of chapters on the United States. The Americans might not be running things any longer, but it isn't like they're disappearing into a black hole. The United States will retain—by far—more power than any player, past or present. So it's important to understand both domestic limitations on Americans' capacity as well as the sorts of issues and allies they will still care about. In all cases, you can grab full, sharable versions of each and every graphic in this book at my website, www.Zeihan.com.

I knew Matt for only three years. In 2003 he was suddenly, horribly taken from us. I miss his passion. His brutal honesty. He would have *hated* the first draft of this book. He would have ripped it apart. I would have been so pissed at him. Then, we would have rebuilt it. Together. It would have been awesome!

This version is still pretty good.

DISUNITED
NATIONS

CHAPTER 1

THE ROAD SO FAR

There's no good way to launch into a book on the past and future without using a quote from *The Lord of the Rings*. Probably from an immortal elf matriarch. Something about how the world is changed. Since that's all heavily copyrighted, we're going to have to just jump right in.

Time erodes everything. Countries included. Surviving history requires a delicate balance of factors that most of the world lacks: border zones that are difficult to cross and an interior zone where it is easy to move people and goods and ideas around. Such a mix of crunchy and gooey is rare. Most locations are either so crunchy throughout that the locals don't get along, or so gooey on the edges that the neighbors' armies like to host block parties on your lawn. Historically speaking, most countries have been small, brittle, fragile, and, above all, short-lived. If countries cannot arise or thrive or survive, history tends to stand still.

But that is not true everywhere. There's a handful of countries whose geography is in balance. These countries defy time, and so have dominated much of human history. Let's start at the beginning.

THE FIRST AGE: EMPIRE

It all comes down to a pair of concepts we all instinctually grasp but spend little time pondering.

The first is *continuity*: the idea that the positive things that make your life today possible—health, shelter, clean water, food, education, clothing, a functioning government, and so on—will still be around tomorrow. No brigands will steal your cow; no dryad will kidnap your children; no horde will descend from the distant horizon and burn down your local Walmart. The modern equivalent? Flip a switch and the lights turn on. Each and every time. Historically speaking, continuity is a rare, precious thing. Few countries boast the sort of crusty borders that enable them to be protected from outside threats for more than a few decades at a time. For most, going a few years is about all they can hope for.

Not all threats to continuity are from bullets or sharp sticks. Drought and flood can wreck a system just as easily. Or a rampaging, homegrown mob. Or a coup. Or particularly crappy leadership. And never underestimate the power of a good plague. Breaks in continuity shatter institutions, disrupt food production, wreck infrastructure, break educational coherence, and severely damage cultures. Recovery is, of course, possible, but in many ways every time a country suffers an invasion or civil war or coup or famine, it must start over.

The second concept, *economies of scale*, is even more straight-forward. Imagine you've been tasked to build a computer. All the necessary information and equipment has been provided for you to melt and purify sand into silicon and draw it into crystals and slice them into wafers and score them with acid and lay them with metal and assemble them into circuit boards and so on. How long would it take you to do it? How long would it take you to learn all the steps independently? A lifetime? Ten? And then, at the end of it all, you'd have one measly computer.*

Economies of scale are ultimately about specialization. Instead of you learning and carrying out every individual task, you need to learn only one—say, scoring the silicon wafer with acid. Other people take charge of all the other tasks, one each. Different people are better at different things, and matching people with their niche makes the entire system more productive and efficient. After your first month on the job, you've etched several thousand wafers, and you've gotten pretty good at it, both in terms of speed and quality. Most everyone else is having a similar experience. Collectively, the team is cranking out dozens of computers an hour. The system is scaled *up*. Production becomes cheaper. End-sales prices go down. And to sweeten the deal, your deep knowledge of that one facet lets you innovate at a faster pace, too.

National success requires achieving *both* continuity and economies of scale. Those big enough to have economies of scale rarely have good borders that enable continuity—think Russia. Those sufficiently isolated to have long continuities rarely have scale—think New Zealand.

This is not the case in every place in every time. Some locations *can* make it work. Some countries have the geographic viability required to make a go of things. The best of these locations do more than succeed as countries. They can reach out and absorb other, less-than-ideal lands. In doing so, they don't just become bigger; they swallow up resources, knowledge bases, and taxes, and they make them their own, creating even larger economies of scale. They anchor their important lands on more secure footings.

---

* Be honest with me here. How long did it take you to put together a piece of furniture that was designed, sourced, and fabricated by someone else, and then transported to you with pre-drilled holes and ready-made, picture-heavy instructions?

They become *empires*. Most of human history is the story of how this or that imperial center didn't simply come to be, but expanded to dominate our understanding of humanity itself.

Any number of things can wound an empire: climatic shifts that gut food production can crash even the most powerful of entities terrifyingly fast; internal political disputes can (literally) behead the system at the top; outlying territories can rebel, costing the imperial center an economic lifeline or a strategic bulwark. But by far the most common way an empire perishes is by way of another empire waltzing over and punching it in the face.

It added up to make the Imperial Age a *brutal* era. Anyone who was not from an imperial core didn't have the best life, typically being used as cannon fodder in incessant wars among clashing empires. And it all lasted a looooooooooooooooooooooong time. The world's first city—Uruk, in Mesopotamia, home to Gilgamesh—upgraded from a small settlement to something more permanent sometime in the middle of the fourth millennium BC. The world's first empire—Akkadia—erupted upon its neighbors a mere thousand years later.*

For over four *thousand* years, empire was the norm.

How empires interacted had a lot to do with the horse they rode in on. For a long time, quite literally. If we view history as a series of inventions, the history of empires looks like a long arc of new efficiencies in movement . . . and in death. From the Mongolian use of stirrups to move at blazing speeds across the steppe to the newfangled Portuguese ships that crossed oceans, technology brought the world's empires into ever-greater proximity with ever-bigger consequences.

Until technology became *so* efficient that wars between em-

---

* Ish. Dating the specifics of the ancient world is a tricky business. The Uruk people weren't simply our urban forebears, they were also the first recyclers. Archeologists have had a devil of a time figuring out the details of precisely when everything happened because the Urukians never left anything old just lying around, going so far as to deconstruct even their temples for use as building materials.

pires tore the world to shreds. The two technological families of deepwater navigation and industrialization enabled *all* the empires to engage one another *everywhere* at the *same* time. The resulting ultimate, inevitable, catastrophic, system-ending conflict was the most deadly, destructive war in history.

We know it more commonly as the Second World War. It set the stage for something fundamentally new.

THE SECOND AGE: ORDER

World War II didn't so much end the Imperial Age as eradicate it. When the dust settled, only two powers remained—the United States and the Soviet Union—and they immediately got themselves uncomfortable for a long, drawn-out struggle. The Soviet system was the final remaining empire, and it operated like all the empires before it. A single group of people—the Russians—called most of the shots, and everyone else, whether they be Estonian or Polish or Slovak or Bulgarian or Armenian or Uzbek or Tatar or Ingush, was there to provide strategic depth, captive markets, and, if need be, a deep pile of bodies to throw at the neighbors. Anyone who had a problem with that could suck on some bullets in a work camp. Very old school.

In a way the power that rivaled the Soviet Union—the United States—was kind of, sort of, an empire as well. The Americans had expanded from a small coastal base to dominate the middle third—the best third—of the North American continent. But a mix of factors made them something else entirely.

First, the Americans started life as the extension *of* an empire. The common struggle of the original thirteen colonies against the British system granted them the most important aspect of a common identity: a common cause. Almost by definition, empires lack that—the imperial center tends to suck the life out of colonies and protectorates.

Second, because the Americans began as a branch of the world's most powerful military system, at independence they enjoyed a massive military advantage over most of their immediate neighbors. Combine that advantage with their resistance to Old World diseases, and the natives of North America were all but extinguished, leaving ample room for the new Americans to expand into. Historical revisionists have reevaluated and are reevaluating* the United States' role in the destruction of the native cultures, but the fact remains that Old World peoples cannot even fathom how extraordinarily low the barriers were to the Americans' territorial and economic expansion. The natives' annihilation was so complete the United States faces little of the internal heartburn of undigested peoples of imperial systems.

Third, that expansion—both in population and geographic terms—muddled the differences among the coastal states of the early nation. Thirteen distinct identities merged in the interior into something new and more holistic. Within a generation of the end of the War of Independence, the Americans thought of themselves less as citizens of the various states and more as citizens of one of two regions: the North and the South.

Fourth, these two regions duked it out in what remains the deadliest conflict in American history in both relative and absolute terms. At war's end, the North was victorious, but rather than purge or subjugate the defeated South, the Northerners reintegrated their defeated brothers into the Union while simultaneously beginning the long, multifaceted process of incorporating nonwhites into the greater whole. This Reconstruction was awkward and incomplete and absorbed the bulk of the split country's attention for three decades, but despite its at-best half-victories on racial inclusion it succeeded in both maintaining and expanding America's continuity while simultaneously deepening its economies of scale. By 1900 the United States boasted more arable land, better borders, a better

---

* And in my opinion, should continue to reevaluate.

integrated population than any other single self-ruling power, and a larger population than any such power save the decidedly *un*egalitarian czarist Russia.

Finally, the Reconstruction success deepened America's cultural integration process, blurring the differences between North and South. To give an idea of just how successful this has become in contemporary times, citizens of the American states of West Virginia, Virginia, and Pennsylvania—states on opposite sides of the Civil War who suffered the most brutal fighting of that conflict—are now most likely to list their *ethnic* background as "American."

Such melding is critical. The United States is *not* an empire. There is no clear political or economic core. New York, California, Texas, and Florida all have very strong ideas about just who is actually in charge. There is no singular region that enjoys preferential treatment. American politics might often be divisive, bitter, loud, rude, annoying, childish, venomous, inane, flabbergasting, willfully ignorant, and unfathomably obtuse, but it is pretty clear it is not any Americans' destiny to serve as cannon fodder. Combine that shared identity with fantastically crunchy borders and a truly wonderful gooey center, and post-Reconstruction America isn't simply a fundamentally different sort of political beast; it is the most powerful country on Earth.

But "most powerful" does not mean "all-powerful." When the Americans emerged from Reconstruction, they discovered a world in the midst of a significant military-technological upgrade that brought the empires into even greater proximity. New ships could cross oceans in weeks instead of months. New artillery could reach miles. Aircraft would soon debut. America's strategic isolation was being intruded upon. Ultimately, reluctantly, the Americans felt compelled to join in the world wars, the apex of industrialized slaughter.

And they really didn't care for what came next. One tender-hearted Joseph Stalin had taken the disaster of a country he'd inherited from Vladimir Lenin's failed economic experiments and used the white-hot fires of brutality and battle to forge a war machine of unparalleled size and power. With an inhuman lust for casualties normally reserved for AI-driven killer robots in post-apocalyptic horror novels, Stalin fueled that machine with millions of Soviet soldiers, and then wielded it with sufficient skill and fury to beat back the Nazis inch by bloody inch. GIs had good reason to freak out a bit, as they were now facing off against a battle-hardened Soviet force that was literally raping its way through eastern Germany and the German capital.

The field of combat did not appeal; the Northern European Plain is flat and open. The German war machine had charged west and east across it with frightening speed—a speed the Soviets may not have been able to match in their westward assault, but that they more than made up for in sheer horror. It crossed the mind of more than one American military commander that the Americans had found themselves facing the same quandary of the Germans and French and Dutch and Poles and Russians before them: how the hell do you establish a defensive position on the vast wide opens of Northern Europe?

The potential ally list wasn't encouraging either. The Europeans—former imperial centers all—had a historyful of antagonisms and backstabbing stretching back to their misty beginnings. How could the US take a ragtag group of war-broken countries with an epic poem of mutual grievances and get them to work together against a threat as monolithically terrifying as the Red Army?

In Asia it was even messier. At war's end the Americans were not simply occupying Japan in totality; rather, the *way* the Americans had fought the war had created the greatest power vacuum in Asian history.

The American campaign didn't eject the Japanese from every inch of the territory the Japanese had captured. Instead the

Yanks hopped from island to island, only seizing sufficient bases to maintain their advance. This continued until the Americans could angle themselves into a position where they could concentrate their bombs—industrial in volume, reach, and devastation capacity—upon the Home Islands. Well over 90 percent of the territory the Japanese captured in the war the Americans didn't touch—including *all* the Japanese territory in China and Korea (where the Japanese were *winning* right up to the day the nukes fell). Overnight, the East Asian rim went from a near-singular imperial government to the complete surrender and withdrawal of that government's authority and its troops. Chaos reigned.

In the course of less than six months, the Americans went from having the world's simplest geography in terms of security and wealth, to occupying—and needing to protect—its most complex.

This didn't mean there weren't perks.

At war's end, the United States floated what was indisputably the most powerful navy in history, while simultaneously the assets of nearly all other historical naval powers—the Japanese, Russian, French, German, Dutch, and Italian—were busy serving as the foundation for new reefs. The single exception of significance, the British Navy, was reduced to acting as an American adjunct. That enabled the Americans to think forward to what pieces of their newly expanded geography of responsibility could be excellent naval bases.

But for every new redoubt, there was a quagmire-in-waiting. For every Great Britain, there was a Fulda Gap. For every Singapore, there was a DMZ. For every Okinawa, a Saigon. For every Diego Garcia, a Beirut. Defending such locations under any circumstances is difficult, but defending them *all*? At the *same time*? For the Americans, these weren't simply territories with which they had little to no familiarity; they were on different continents, necessitating supply lines the likes of which no strategic planner had ever considered viable. Right from the beginning, it was obvious there was no possible way the Americans could shoulder the burden themselves.

The Americans needed allies to help, but far-flung allies meant far-flung commitments—every potential ally brought in a new geography with new complications. The Americans needed a *world*ful of allies if they were to contain and beat back a power as large, determined, battle-hardened, and ethically unfettered as the Soviets. They needed those allies to be fully motivated, committed to the fight.

In that charming way the Americans have of oversimplifying things, the Americans cut the Gordian knot. Carpe-ing the hell out of the diem, the American president—one Franklin Delano Roosevelt—invited everyone to a ski resort in the New England town of Bretton Woods and bribed everyone into joining him in forging a new world.

The Americans pledged total physical security for anyone who joined their alliance, protecting them with tanks, troops, ships, and the still-under-development nuclear umbrella. The Americans used their absolute mastery of the seas to protect any ship from anywhere transporting any product to any location. And that was just the beginning:

- Face pirates? No problem. The American Navy would take them out. No more imperial raiding of other countries' commerce.
- Soviets trying to overthrow your government? No problem. The American Congress would vote you some more reconstruction funds. The Marshall Plan surged forth.
- Can't get the lights on? No problem. The Americans would ensure you could import coal and oil from anywhere in the world. The Americans would patrol—in force—the Persian Gulf.
- Can't jump-start your economy? No problem. The American market—the only market of size to survive the war—was now open to your exports. America's trade deficit was born.

- Don't trust the American security guarantee? No problem. The Americans would join battle with the Soviets—even at the times and places of the Soviets' choosing—to prove their reliability. Enter the Korean and Vietnam Wars.

If you side with the Americans against the Soviets, the Americans will use their military might to protect you, their economic might to subsidize your existence, and *for free* provide you with everything every empire throughout history had ever fought for. For the former empires, it meant more than having access to sea lanes and markets and resources; it meant having access to *all* sea lanes and *all* markets and *all* resources *all* the time as if they had decisively won a global war. For secondary powers, it was like having access to the potential of the British Empire at its peak, without the commensurate military outlays. For the colonies, it meant independence from their old masters and access to the same global market as the former empires, as well as the opportunity to chart their own political and economic continuities. It no longer mattered if your lands couldn't possibly sustain a growing population or if you had a wide-open border. For the first time in history, geography was put on hold. Major empire, lagging power, or colony, join the Americans' new alliance against the Soviets, and you were secure, not just from the new Soviet threat, but from the old threats as well.

What we think of as "free trade" is much more than "just" the regular exchange of goods and services. It is much more than getting the latest doodad via Amazon Prime. It is much more than globalism. It is a global *management* system. A global *network*. A global *alliance*. A global *order*.

The *first* global Order.

It worked . . . amazingly well! The Order provided the economic grounding for the entire American grand strategy throughout the Cold War. Without the Order, there would have been no NATO. With the Order, the defeated Axis powers were transformed from implacable foes to pliable allies. The empires that had so vexed American strategic policy during its first fifteen decades of existence—including that of the Americans' own former colonial master—were lashed to the American will. The Americans had their global alliance.

Turns out, the creation of history's largest and most powerful alliance was just the beginning.

The Americans were in many ways champions of stability. By enforcing America's own brand of global peace with some Tiffany-size karats and Teddy Roosevelt–size sticks, the Americans banished from memory the old imperial boxing matches that tended to leave no room for weaker states. Instead of the vicious circle of imperial clashes, the Order created a virtuous circle of political stability, security, and economic development. This helped push the technological descendants of the Industrial Revolution even further, while keeping most governments busy with the needs of their people. Famine and disease weren't exactly banished, but they certainly lost their grip on day-to-day life. The result has been the deepest educational penetration, infrastructure expansion, economic growth, and technological advancement in history.

One of the Order's most impressive features was its universality. The United States guaranteed the safety of the imports, exports, and supply lines of *everyone*. Even countries it economically competes with. Even countries it *likes to bomb*. As Detroit was hollowing out, German automotive exports were sacrosanct. As Midwest farmers were struggling with low grain prices, those pursuing Brazilian agricultural expansion found it easy to import American fertilizers and equipment. As the Americans were sparring diplomatically and militarily with the Iranian Ayatollah, the American Navy maintained ironclad naval safety for all commercial vessels at all times. Even when the Americans were actively prosecuting a war—as in Vietnam—they persisted in protecting local commerce, even that of the other side.*

The Soviet Union found itself often on the defensive, forced to act as the agent of chaos, seeking to overthrow states that were economically thriving, many for the first time *ever*. It was a losing proposition. Yet by *any* measure of the old Imperial Age, the Russians were doing fabulously well. The nuclear doctrine of mutually assured destruction combined with the Americans being in a different hemisphere all but guaranteed that the Soviets would not face the sorts of land invasions that had so wrecked their continuity every few decades. The result was a degree of development and peace throughout their lands that the Russian people had never known.† Continuity and economies of scale reached all-time highs, and they took the Russians further and higher than they had ever been.

Yet the Soviet rise was *nothing* compared with what the Americans mustered with their Order. Almost *all* the world's richest lands united into a single, integrated economic system guaranteed by the American Navy and nuclear arsenal. The Soviets never had a chance, and five short decades later, their system—besieged, outnumbered, outmaneuvered, and outspent—collapsed.

The Order may be the world that most of us know as familiar, but from the long view of history, the Order is one of the most bizarre periods ever.

It is nothing compared with what came next.

---

* Events such as the Cuban embargo were the exceptions that prove the rule. It took an immediate existential (and Soviet-originated) threat for the Americans to revoke access.

† I'm expressly noting that this was a golden age for the *Russians*. For non-Russians in the Soviet Empire, life more or less sucked just as much after World War II as before.

THE THIRD AGE:
ORDER WITHOUT BORDERS

The Soviets saw the writing on the wall, and in the mid-1980s they started negotiations toward a managed defeat. What was intended to be the first step turned out to be the first, last, and only. Soviet forces relaxed their control of their Central European satellite states in early 1989. By 1992, the Cold War and the Soviet Union were in the ash heap.

The Cold War victors had a bit of a party, and then something curious occurred.

The American policy of global Order was a strategic one, designed as a means of doing battle in a global conflict. With the Cold War over, it was time to rejigger that Order toward a new goal.

With the Soviet fall, American president George HW Bush sensed history calling. He used his unprecedented popularity in the aftermath of the fall of the Berlin Wall and victory in the First Iraq War to launch a national conversation on what's next. What do the American people want out of this new world? He openly discussed a New World Order, his personal goal being "a thousand points of light," a community of free nations striving to better the human condition in ways heretofore unimaginable. Bush's background—he had previously served as vice president, budget chief, party chief, ambassador, House representative, and intelligence guru—made him the right person with the right skill set and the right connections and the right disposition in the right place in the right job at the right time.

So of course the Americans voted him out of office, and all serious talk of moving the Order onto newer footing for the new age, more relevant for the challenges and opportunities of the post–Cold War era, ceased.

American leadership in the years since has been, in a word,

underwhelming. Bill Clinton found foreign policy boring and did his best to avoid it. George W Bush became embroiled exclusively in the Middle East. Barack Obama proved so insular he refused to have many meetings with, well, anyone—even allies within his own Democratic Party. Donald Trump's "America First" expressly calls for divorce from the global system.

None of the four picked up the challenge of George HW Bush to reform the Order and build a better world. Nor did any of the four provide the necessary guidance to American military, intelligence, and diplomatic staff as to what America's goals actually *are*.

With no clear grand strategy, the Americans lurched from crisis to crisis—Haiti to Bosnia to Yugoslavia to Afghanistan to Iraq to Yemen to Syria—with the country's political, economic, and military elite seeing power as something to be used in an endless march of tactical situations, rather than a tool for shaping the broader picture. For a mix of reasons political, personal, and institutional, there has been a lack of imagination all around.

Bereft of direction from the top, American strategic policy slipped into a thought-free rut. The US Navy *continued* to provide global maritime security. The American military *continued* to protect Cold War allies. The American economic system *continued* to be a sink for global exports. The American system *continued* to enable all the nuts and bolts of global energy and finance and agriculture and manufacturing.

The only change was, the Americans stopped asking for anything in exchange.

It made the world a *very* strange place.

The Order had already created the greatest half century of global economic growth in history. However, in the post–Cold War iteration of the Order, the parts of the world that had been left behind due to local political instability or geographic inaccessibility or simply for being on the wrong side of the Cold War contest flooded into the system as well. Continuities and economies of scale extended and added new players. The increased demand for the inputs of industrialization bolstered Sub-Saharan Africa, Latin America, and the Middle East. Increased demand for more and better food boosted Latin America and the former Soviet Union. Increased demand for energy helped the Middle East and the former Soviet Union. Increased demand for manufactures boosted Europe and the East Asian rim. All without the Americans asking for a thing in return for making the entire system possible. The Order had been so successful, most forgot—or never learned—that it was even there in the first place.

The sheer size of the new system boggles the mind.

Just before World War II got moving in 1935, the world's largest fully integrated economy—Germany—had about 62 million people. With the imposition of the American-led Order in the late 1940s, the fusion of the American system with the first batch of postwar allies created a system with some 400 million members. Absorbing the Axis powers and several formerly neutral states in the 1950s brought the figure to just shy of one billion. By 1980 population growth and the admission of China increased it to three billion. And at the time of this writing in early 2020, pretty much all the world's 7.8 billion people are beneficiaries. With each step, the continuity of the Order deepened and broadened. With each step, the possible economies of scale expanded. With each step, we came a bit closer to the modern world of computers, on-demand deliveries, widespread and varied food availability, and reliable electricity.

But this ever-more-wealthy, ever-more-connected, ever-more-advanced world lost sight of one central, inconvenient truth: American involvement in the Order isn't about—was *never* about—free trade and its subsequent effects as an *end*. Free trade was the *means*. Free trade was part of the *bribe*. The Americans never enmeshed their economy into the Order. If they had done so, it would have been as if they were establishing an empire with the allies being the new provinces.

Instead, the Americans in their splendid isolation more or

less kept their continent-size economy to their continent-size selves; geographic isolation and abundance are powerful forces of crunch and goo. Today the United States remains the least integrated major economy in the world. Which means that with the notable exception of Soviet containment and defeat, *none* of the Order's advances—in energy, agriculture, finance, manufacturing, disease prevention, education, lifespan, democracy, child mortality, and on and on and on—were part of the plan. They were merely side effects.

No, the Order wasn't about money for the Americans, and everyone would do well to remember that longevity and broad political support are not the same thing as sustainability or permanence. The Americans forged, operated, and subsidized the free trade Order so that they would have allies to help face down the Soviets. But the Cold War ended in November 1989 with the fall of the Berlin Wall. On New Year's Eve as 1991 ended, the Soviet flag was lowered for the last time. The world has since changed, and—belatedly—the Americans are changing with it. With the Order's strategic rationale gone, the Order has had to justify its continuing existence to the Americans.

It has not gone well.

THE AMERICANS IN THE THIRD AGE

There was no single event that directly translated into American disenchantment with the Bretton Woods Order. With any strategic grounding dissolving in 1992 right along with the Soviet Union itself, the American alliance network has long since become unmoored from all the Cold War realties that cemented the system in place. Everything continued on increasingly surreal autopilot for a while, but a decade after the Order's grand success, three specific major events in quick succession . . . clarified things:

On April 1, 2001, a Chinese fighter jet collided with an American EP-3 spy plane it was shadowing, crashing both aircraft. The issue quickly escalated into a bilateral snit. The Chinese asserted that the EP-3 was spying on China,* while the Americans asserted that such flights were not only routine but that, since the EP-3 was over international waters, the flight was legal under international law.†

Bad blood had been rising between the two countries for some time. While the Chinese and Americans had never truly been comfortable with each other, they were allies in the Order. The core of US President Richard Nixon's historic trip to China in 1972 was to bring the Chinese into the anti-Soviet alliance network. The normalization process was a monumental success, the Chinese turned monumentally against the Soviets, and that shift exposed the Soviets' entire southern flank to monumental pressure, which probably accelerated the Soviet collapse by more than a decade. It also ushered China into the global network of resources and trade, launching the Panda Boom.

But by 2001 the Chinese were growing tired of deferring to the Americans' increasingly nebulous post–Cold War strategic policy, particularly when tendrils of that policy could be seen from Chinese shores. When the EP-3 accident occurred, the Chinese leadership decided to take a stand. The Americans, furious that an Order ally would embrace a military standoff with the country that made their economic success possible, took more than a bit of umbrage. But before the issue could be resolved one way or another, another transformational event occurred.

One idle Tuesday morning—September 11, 2001—Islamic militants hijacked four American passenger jets, successfully crashing three of them into buildings iconic of American power. For the first time since the War of 1812, a hostile power had struck a major American city.

* True.
† Also true.

Within a month, American forces were hunting the perpetrators across the length and breadth of Afghanistan. The effort quickly degraded into a whack-a-mole problem. American forces would advance upon a known militant hideout. The militants would catch wind of their approach and vacate. The Americans needed more eyes and more ears and more guns in more places, and so they called upon their entire alliance network to join the hunt.

The response was discouraging. Most of America's allies demurred; many actively resisted.

While much has been said and written about the wisdom, effectiveness, and/or execution of the Global War on Terror, the thing is that from the American point of view, the United States was still providing the economic and strategic largesse of the Order, but most of the allies were no longer providing the strategic deference that was supposed to balance the equation. And so the thinking went in Washington, if no one else is upholding their end of the bargain, why should we?

On New Year's Day 2002, a dozen European countries jointly ditched their national currencies in favor of a new, common currency: the euro. There are any number of reasons why the euro adoption wasn't all that realistic, but it is undeniable that part of the euro's raison d'être was to pool European economic power so that Europe could provide a non-American pole of power in global affairs.

A lingering distaste for the new currency rose up in America. During the Imperial Age, the Europeans had been at each other's throats. Peace among them had been possible only because of American involvement in World War II, American financial support in the postwar rebuilding effort, American strategic overwatch during the Cold War, and the de facto American subsidization of their economies since 1945 via the Order. Without the Americans, there could not be a European Union, much less a euro.

And to thank the Americans, the Europeans decided to launch a common currency expressly designed to chip away at American preeminence in global trade and finance.

As the years ticked by, the Americans found themselves ever-more ticked off.

First, the Chinese started claiming airspace throughout their Western Pacific periphery. Then seas. Then islands. Using a mix of economic and military tools, they bit by bit impressed their will upon their region—and did so while continuing to fully exploit every aspect of the Order. The Chinese wanted to push the American Navy back from Asian shores while expecting that same American Navy to continue to make the global oceans safe for Chinese merchandise exports and for Chinese energy and raw-material imports. This fact was not lost on the American Navy.

Saudi Arabia's idea of social policy is to hype up their disaffected youth with radical Islam, give them some firearms and explosives, and export them to war zones to fight for Saudi interests. Should some of those groups self-radicalize, give themselves a new name—like, say, al Qaeda—and start selecting their own targets, such as matching skyscrapers in New York City, the Saudi reaction was typically to shrug, disavow, and form a new group. Saudi Arabia's resistance—to taking responsibility for the 9/11 attacks, to ceasing their radical-export strategy, to allowing the Americans to hunt for radicals on Saudi soil—fundamentally changed the way many Americans view their "allies."

The rise of the euro complicated European-American relations, particularly when the Europeans came to Washington hat in hand for help in dealing with the European Financial Crisis, a crisis that would have not been nearly as severe if not for the euro's creation.

This is hardly a comprehensive list. Many other Bretton Woods "allies" have taken the shine off the alliance:

- A homegrown crisis in Argentina prompted the Argentines to go hat in hand to Washington for a bailout *while* they were disavowing debts to US citizens.
- Australia attempted to use the G20 to muscle its way into global management.
- Brazil, whose development would have been impossible without scads of foreign funds, led a global effort to diversify global finance away from American institutions.
- India regularly torpedoed American efforts to tweak the Bretton Woods system via the World Trade Organization.
- Despite being two of the countries that benefit the most from the Order, China supports perennial Asian bugaboo North Korea against the Americans, while Brazil often provides economic and diplomatic cover to Venezuela and Cuba.
- It is difficult to overstate American frustration with allies within the Order who insist on using their newfound strategic cover and economic stability to pick fights with *other* Order members: China with Taiwan, Greece with Turkey, South Korea with Japan.
- Malaysia publicly condemns American economic leadership while privately being the world's second-most trade-dependent country.
- Not just China, but also France, Israel, and South Korea regularly engage in espionage against American corporations in order to steal everything from technology to business plans to negotiating guidance.
- After a terror attack in 2004, Spain pulled out of cooperation with the Americans not only in the Islamic world, but within NATO as well.
- Speaking of NATO, the allies' financial contribution to their national defense was weak during the Cold War. After the Cold War, their defense spending dropped so much that the French cannot move troops without renting tourist ferries,

while the bulk of Germany's submarines, surface ships, tanks, and jets cannot even deploy.
• Canada is just so damn snarky. All. The. Goddamn. Time.

Many condemn Donald Trump for destroying the global Order. Let's be real here. If there is one thing that Americans on both the Left and Right agree on, it is that the United States should pursue a more modest role in foreign affairs. The push for an American retrenchment did not begin with Trump, nor will it end with him. Besides, if a single American election can upend the Order—in an era when there is no nuclear-massed superpower foe—it was never as stable and durable as anyone thought.

A more accurate assessment is that despite Donald Trump's trademark brashness, American policy trajectory hasn't changed much. In the seventh year of George W Bush's presidency, the United States initiated a broad global drawdown of its troop levels. That disengagement continued both under Barack Obama *and* Donald Trump. At the time of this writing, the Americans now have fewer troops stationed abroad than at any time since the Great Depression.

The decline of the Order isn't the fault of a single country, much less a single personality in the United States or elsewhere. The Americans have lost interest in being the global policeman, security guarantor, referee, financier, and market of first and last resort. The people who remember the world wars personally are almost gone, and the oldest of those born after the age of the Soviet nuclear threat have already voted in three presidential elections. It is difficult to justify an Order when the people doing the justifying weren't even alive when the circumstances that necessitated its creation occurred.

It is not only that the Order—the concept that the Americans will trade economic dynamism for security primacy—is simply no longer working; *it hasn't been working for three decades.* Without a strategic foe to focus minds throughout the alliance, nothing but inertia is holding it together, so the global system—the first truly global system—has been breaking apart bit by bit.

It is difficult to draw a line under things that will *not* change.

The regions that have benefited most from the Order's security and structure are the regions that either produce or consume most of the world's internationally traded crude oil: the former Soviet Union and the Persian Gulf are the producers, while Europe and Northeast Asia are the consumers. The post-Order world faces the perfect storm: not simply a global energy shortage, but a series of interlocking instabilities occurring in the places the modern world depends upon most.

A close second for topics of concern are foodstuffs. Throughout history, food supply has repeatedly proven to be the most significant limiting factor. If a government couldn't feed its people reliably, well, let's just say that famine is the ultimate continuity-ending event. The Order's safety and openness enabled such massive agricultural investment and expansion that famine was banished from not only the imperial centers, but most of the world. The global population *tripled*. But in 2020 four-fifths of the world's food production is dependent upon inputs such as fuel and fertilizer, inputs that for most farmers originate on a different continent. Without the Order, much of the global gains in health, nutrition, and calorie intake melt away.

Not to be outdone, most of the world's raw materials—whether iron ore or bauxite or lithium or copper—are produced on one continent, processed on another, and consumed yet somewhere else. Even minor interruptions to global shipping will collapse the availably of the base materials upon which modern life has been built.

And that's just the wonky high-level stuff. By forcing order upon a disorderly world, the American alliance system inadvertently—if happily—improved nearly every aspect of human existence. It wasn't so much a rising tide that raised all boats as a fundamental transformation of the human condition.

This Order forced peace upon Europe. This Order dismantled the empires, freeing colonies the world over. This Order enabled the formation of the European Economic Community during the Cold War, and the European Union after. This Order's extension into the post–Cold War world is what enabled the rise of Brazil and India and China. This Order ended imperial predation in Africa and China. This Order enabled Brazil and Kazakhstan to grow row crops en masse. This Order allowed South Korea and Slovakia to industrialize. This Order enabled oil to flow from Saudi Arabia and Australia and Libya to France and Argentina and Singapore. This Order transformed Nazi Germany and Imperial Japan into democratic pacifists. This Order provides hope for a world beyond coal. This Order makes London and Hong Kong and Singapore financial centers. This Order makes container ships possible. This Order provides global markets for South African ore and Thai electronics and Ecuadorian bananas.

The benefits of global stability and continuity have manifested as deep-rooted improvements in the lives of the entire global population:

- This Order has increased education levels to their highest ever.
- This Order has so increased economic opportunities that the proportion of the global population living in absolute poverty has plunged from 60 percent to less than 9 percent.
- This Order has improved security and health and education enough to reduce the chances of children dying before age five from over 24 percent to less than 4 percent, while reducing illiteracy from 58 percent of the population to less than 15 percent.
- This Order has so improved global health that child immunizations have shifted from being the province of the rich to the global norm, with over 99 percent of children getting immunized at least once, eliminating smallpox, nearly eliminating polio, and even putting malaria on the run.

- This Order has massively expanded human connectivity, and not simply due to the development of the Internet—the number of people with phone connections increased by an order of magnitude.
- The Order has so fostered democracy that the governing system has spread from less than one-tenth of the world's countries to over half.
- This Order has put human rights on the human agenda.

But think of what it takes to foster all these outcomes. Education requires wealth and political stability. Economic growth in subpar geographies requires access to foreign finance, technology, and markets. Progress against child mortality requires imported medicines and the economic capacity to purchase them, as well as funding for sanitation infrastructure. The Internet—the foundation of the digital economy—works in non–First World countries only if transoceanic cables can operate without interruption (no one puts a server farm in Uzbekistan or Congo). Democracy necessitates a degree of economic development that enables people to think about what happens a year from now. Population growth and high life expectancy require a world where nutritious food and clean water and health care are available every single day, not to mention a world flush with educational and economic opportunities.

There are any number of ways to bottom-line just how critical the Order has been to the human condition. Here's the one that hits me like a two-by-four to the forehead. For more than half the world's population—in countries as disconnected as Korea and China and India and Iran and Saudi Arabia and Egypt and Algeria and Mali and Peru—life expectancy has increased by three *decades or more* since 1950. For just about everyone else, the figure is at least a decade. The technological advancements, security environment, and spread of health care have tremendously improved the lives of everyone on the planet, with the heaviest gains in parts of the world that have traditionally been poor and war-torn.

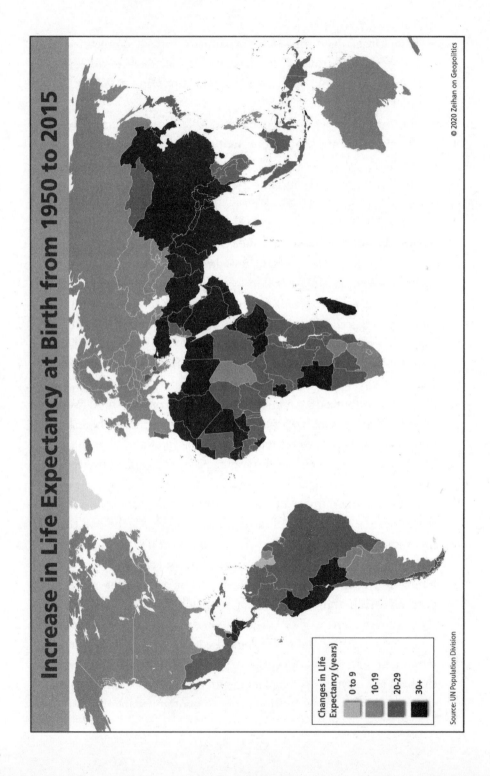

Increase in Life Expectancy at Birth from 1950 to 2015

Changes in Life Expectancy (years)
0 to 9
10-19
20-29
30+

Source: UN Population Division

© 2020 Zeihan on Geopolitics

All these gains and more are all *inadvertent* outcomes of the Order, not deliberate goals. Without the Order, all these and other gains are not simply in danger; the activities and infrastructure and economic wherewithal that enabled these gains to be made in the first place will fall away.

No one will want to give these gains up, whether in education or energy access or life expectancy or manufacturing markets. These things are worth fighting for. And so people *will* fight for them for the simple reason that scarcity prompts action. Those who lack direct access to oil, food, or industrial inputs—but who possess projection-based military capacity—will use that capacity to gain access to the others. In Europe, this will take a neo-imperial form as the former imperial powers will have little choice but to venture back into the lands of their old empires to secure resources. In East Asia, a sprawling, multi-sided conflict for control of sea lanes is more likely. The Persian Gulf is likely to be on the receiving end of both efforts even before local strategic competitions contribute their own fuel to the building fire.

It all leads to more or less the same place: the beginning of the fourth age, a global Disorder in a world without American overwatch. *That* is what the rest of this book is all about.

CHAPTER 2

HOW TO RULE THE WORLD, PART I

THE AMERICAN MODEL

Building and maintaining a truly global system isn't something any old empire can even attempt. To this point in history, only two countries have pulled off the feat: America and Britain.

Most casual—hell, most professional—observers of international affairs fear a third is on the rise: China. Evidence supporting this concern is all around us:

- China's economy grew from the seventh largest economy in the world in 1996 to the second in just fifteen years. If trajectories hold, China is set to overtake the United States in economic size around 2030.
- Add together expansive investment in infrastructure, low wages, and rapid gains in productivity, and it is only a slight exaggeration to say that all supply chains lead to China in one way or another. China has over a hundred metro regions with over a million people, which form a regional supply network that has fueled global growth for more than two decades,

in addition to making China by far the world's number-one exporter.

- The country's headline development program, Made in China 2025, is meant to put the already unnatural economic growth on steroids. Made in China 2025 intends to bring China into direct competition with the Americans' growth engines of the past few decades: high technology and advanced engineering.
- China's naval expansion has increased the number of active-duty vessels above three hundred as America's self-imposed spending diet has reduced its fleet below that level.
- The Chinese military now fields a ballistic missile it calls the Guam Killer, capable of striking the Americans' primary Western Pacific base.
- Chinese hacking and espionage activities have been wildly successful in both acquiring tech *and* in making the various American cyber commands look like Baby Boomers trying to figure out what emojis are.
- American investment often comes with legal and humanitarian strings attached, enabling the less picky Chinese to invest their way into influence in second- and third-tier—and a not small number of first-tier—countries the world over.
- The raw materials producers of the world, from South Africa and Australia to Russia and Saudi Arabia, now consider the Chinese market *the* most important market in the world. Whether it's China's rapid growth in oil imports (making up 15.5 percent of global imports in 2018) or its singular control of the soybean import market (59 percent of total imports), Chinese consumption seems to be the only consumption that matters.
- China produces nearly half of two little pillars of civilization called steel and aluminum and has near-total control over the production of rare-earth metals, which can be found in everything from advanced fighter planes to renewable utilities.
- The Chinese government is unafraid to wield the power of the

state to let its geopolitical competitors know of its displeasure. In recent years it has limited the outflow of tourists (to South Korea) and rare-earth metals (to Japan), as well as limiting the inflow of soy (from the United States) and steel (from the European Union). China, it would seem, holds all the cards.

Such sentiment isn't simply wrong, but hilariously so. Global Domination isn't an entry-level course. To be blunt, the Chinese are doing it wrong.

The American system of global management can be summed up as: entice everyone to be on your team. Sufficient allure to draw everyone to your global system is not easily attained.

The American Order is in essence based on bribes. Think of the American management strategy as having four carrots of success.

CARROT 1: ENSURE PHYSICAL SECURITY FOR ALL

The Order's primary benefit was absolute security for every country. The United States would protect all Order members from the military advances of Order and non-Order members alike. But the first piece of this process is something most who advocate for a post-American world gloss over: gaining trust.

A would-be hegemon must prove it is willing to bleed to protect the interests of *every* member of its new Order. The reason the Americans became entangled in wars in Korea and Vietnam was less to stop the march of Communism than to convince America's allies that the country would never step away from a fight to protect their interests. The two Asian wars were primarily fought to keep Japan and France in America's coalition—Japan being concerned that there would be no American footprint on the Asian mainland without South Korea, and France being Vietnam's former colonial master. It is a bit of a hoot to imagine the Chinese manning Checkpoint Charlie to protect the Germans from the Russians or getting involved in the Persian Gulf to keep the Iraqis out of Kuwait.

Anyone who wants to guarantee the security of others must first convince those others that such protection includes protection from the new would-be hegemon itself. As distrusted as the Americans may be around the world, they are still the only significant power that most want defending them. The reason is simple: the Americans are not their neighbors. Except for Mexico and Canada, the Americans border none of their wards. Such distance means any historical bad blood between the Americans and others is fairly thin; few countries have ever faced an American occupation.

Such is most certainly *not* the case for the Chinese. In contrast to the Americans, who have only been around since 1776—and in a strategically significant sense, since about 1895—the Chinese enjoy showcasing their millennia of history. Yes, such cultural longevity is an accomplishment worth celebrating, but also opens a trap: the Chinese have had a loooooooong time to burn bridges and ring themselves with enmities. The Americans have had meaningful long-term issues only with Canada and Mexico, countries with which it—by Chinese standards—enjoys cordial relations.

A quick look around China's neighborhood:

Whenever the northern Chinese have ever consolidated sufficiently to invade someone, the first direction they've marched is toward the Korean Peninsula. But the Koreans are no longer pushovers. In the twenty-first century, the peninsula is bristling with soldiers and nuclear weapons. There's no military walkover here.

Taiwan's hostility to mainland China long predates the Chinese Civil War of the 1920s to 1950s. The island of Formosa has long served as either a redoubt for this or that Chinese dynasty or a rebel or warlord force, or as a power center in its own right. Today it is both.

The Kazakhs—knowing the Chinese think of them as uncouth, horse-riding barbarians—moved their capital from the near-border city of Almaty to the former Soviet gulag of Astana while also getting buddy-buddy with the Russians to achieve a degree of insulation and protection from Beijing. That's worth underlining: the Kazakhs would rather be beholden to their former Soviet masters than integrated with China.

Japan has been the traditional regional bugaboo of the Chinese for the past handful of centuries, and Japan's at-times brutal raiding and/or occupation of much of the Chinese coast and northern interior regions leaves the Chinese itching for revenge. That's hardly the sort of thinking that earns one friends in Tokyo. India still smarts over territory lost to China under Mao, an experience that prompted New Delhi to develop its nuclear forces . . . and point them at China.

China's relationship with Russia is even more damning. The Russians' eighteenth- and nineteenth-century Asiatic expansions brought the sharp end of Russian power deep into China's business. Russian forces occupied what are now the modern Chinese provinces of Xinjiang, Inner Mongolia, Heilongjiang, Jilin, and Liaoning—the last serving as Beijing's access to the open ocean. The ill feelings are hardly ancient history: in the 1950s Stalin helped instigate the Korean War on China's doorstep knowing full well that if the Americans came, the Chinese would feel compelled to act. The Americans came, the Chinese intervened, and then Stalin charged his "socialist brethren" for the weapons the Chinese needed to fight it.

In the 1960s the Soviets took a straw poll to see if anyone would mind them nuking Beijing. The Americans were alone in saying no (paving the way for the Nixon-Mao summit of 1972). After the Soviet Union's end in 1991, the Kremlin bluntly warned that, should a rising China think it could seize Siberia from a falling

Russia, that Russian resistance wouldn't take the form of tanks or artillery or partisans, but instead nuclear-tipped intercontinental ballistic missiles (ICBMs).

The best example of the difficulty the Chinese face in establishing trust is the country that provided the Americans with their most memory-searing war: Vietnam. Agent Orange. Napalm. The Christmas bombing of Hanoi. America's war in Vietnam was messy and angry and lasted for two decades. In contrast, the Han Chinese fought the Vietnamese for two *millennia*. In 2020 the Vietnamese are eager to welcome American businesspeople and carriers because they don't think the war with the United States lasted long enough to qualify Americans as epic foes. In contrast, the Vietnamese view of China borders on the pathological.

Of course, it isn't as if the Chinese are the only significant country that has trouble earning regional—much less global—trust. Iran defending Saudi Arabia? (Or Saudi Arabia defending Iran?) German (or Russian) troops on Polish soil? Japan in charge of Chinese security? It isn't just that the Chinese can't displace the Americans as global security guarantor; it's that no one can.

CARROT 2: ENSURE MARITIME SECURITY FOR ALL

The Americans offered twin promises: they could and would deploy forces anywhere in the world on short notice to defend the Order, and they could and would guarantee absolute protection of all allies' merchant shipping. These promises required the Americans to patrol all seas at all times. Anyone wanting to displace the Americans as the hegemon of a globalized world would have to do the same.

No one can.

The core problem is capability. America's global position throughout the Order is a direct outgrowth of the country's naval dominance at the end of World War II. In late 1945 the American Navy was not simply on a wartime high in terms of force strength, experience, and deployment reach; the world's other navies were pretty much gone.

It wasn't that the Americans could sail anywhere they wanted unmolested—although they could and did; it was that in the war's aftermath all the world's good naval anchorages were momentarily unoccupied. The Americans spent the next thirty years entrenching themselves in ideal ports the world over while refashioning their naval forces to take maximum advantage of the new basing footprints. It took until the mid-1970s for the Americans to finally float their first modern supercarrier, the USS *Nimitz*—the first ship with the speed, range, power, and flexibility required to make the most of the basing sites the Americans had already controlled for a generation.

The United States was now not only the sole, undisputed global naval power; it now floated individual battle groups that were

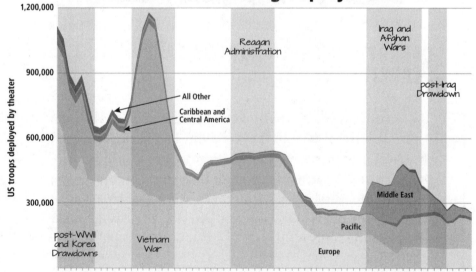

America's Diminishing Deployments

© 2020 Zeihan on Geopolitics

Source: US Department of Defense

more powerful than not only the entire Soviet navy, but arguably more powerful than *all* the world's other blue-water vessels combined.* During the next thirty-five years, the Americans gifted the *Nimitz* with *nine* sister ships (and have recently begun deployment of an even more powerful supercarrier class). This multidecade buildup would have been impossible in the world before 1939, when a half dozen other significant naval powers floated battle groups and still held most of those anchorages.

Any Chinese expansion that would replace (much less overturn) the American position doesn't simply begin after a long American head start, but must overcome an American naval global footprint that would take over a century to establish through force of arms—a footprint that is not replicable without complete victory in a world-spanning war that somehow manages to leave the Chinese mainland untouched.

This is meant less as a slam against the Chinese as it is a recognition of global naval realities. After all, aside from the Americans, *no one* floats even a single fully functional supercarrier, much less a supercarrier battle group, much less a global naval force.

The next-largest non-American carriers aren't even really operational. Russia's *Kuznetsov* appears to like catching on fire; the UK's *Queen Elizabeth* hasn't yet exited sea trials; India's *Vikramaditya* is a floating mass of technical problems; and the engines on France's *Charles de Gaulle* aren't powerful enough to bring it up to aircraft-launching speed except under near-perfect conditions. In fact, the next *nine* largest fully operational carriers are *also* all operated by the US Navy.

Much is made of China's *Liaoning*. It shouldn't be. On paper the *Liaoning* has perhaps one-seventh the combat capacity of a *Nimitz* . . . and the Chinese didn't build it. The Soviets did. In the 1980s. Sort of. This first-of-type vessel was never

---

* The Americans likely hold that distinction in 2020, although there's a healthy debate in defense circles about whether the soon-to-be-operational British and Japanese carriers might prove capable of tilting the balance.

completed. After the Soviet breakup, the Ukrainians stripped it for parts. The ship then languished, rusting at dock for a decade until the Chinese purchased it. Today it is used purely as a training platform. China's naval engineering *is* improving—the *001A*, China's domestically manufactured clone of the *Liaoning*, is undergoing sea trials at the time of this writing, but there are still decades of work to go before China can float something that can compete with a *Nimitz*. Carrier engineering isn't rocket science. It's a *lot* harder.

Beyond carriers, China is indeed floating an impressive number of missile frigates and destroyers that combine reach and lethality, but what the Chinese fleet lacks is operational *range*. Of China's three hundred–plus naval vessels, only one-ninth of them are major surface combatants that can operate over a thousand miles from shore (a little less than the distance from Shanghai to Tokyo) under ideal circumstances—ideal, as in no one shooting at them. Much less operate reliably at high speed. Much less operate as a tactically coherent group. Much less operate without air support. Most of these ships would be less sitting ducks than nicely roasted, ready-to-eat meals.

With its naval order of battle limited by its small number of long-range ships, China is horrifically dependent upon the ports of others. China's only reliable ally—stretching the meaning of *reliable* and *ally*—is North Korea. As China borders North Korea, North Korean ports—not positioned or designed to support large-scale naval power projection—don't exactly help the picture. China has therefore pushed a "string of pearls" policy to create a line of friendly ports between the Chinese coast and areas of interest. Malaysia, Myanmar, Sri Lanka, Pakistan, and Kenya, for example, serve as Chinese pearls on the all-important energy-shipping route from the Persian Gulf.

A big piece of China's One Belt, One Road (OBOR) diplomatic effort is to rope such countries into China's sphere of influence. The path the Chinese have followed is rather sophomoric: Beijing insists that the funds it has provided to these countries were not grants, but instead loans that need to be paid back. At the time of this writing, relations with all five countries have soured expressly *because* of OBOR to the point that China cannot trust any of them to serve as ports in any sort of military storm. (And that's before one considers the general hostility of their neighbors— most notably India—to China mucking about in their neighborhood.)

That leaves the Chinese with only one fully fledged foreign base: Djibouti. It is a facility over five thousand miles from China's shores, designed to serve as an operational node to help multinational anti-pirate operations. It isn't so much that the Chinese navy's only overseas base exists to hunt half-starved black dudes in speedboats. The point is that China had to go so far from home to find a country poor enough with few enough preconceived notions about the Chinese that it would be willing to host a full Chinese base in the first place. Not to mention that ironically— hilariously—China's Djibouti base is possible only because the Americans perceive it as helping uphold the lagging Order, and because Djibouti is so popular for anti-piracy bases that the Chinese hardly enjoy a strategic monopoly there: France, the United States, Italy, Saudi Arabia, and the United Arab Emirates all operate permanent facilities in Djibouti.

CARROT 3:
OFFER UNFETTERED MARKET ACCESS

There's more to a global Order than guns. There's also butter. The Americans used their naval power to create a global market, but just as important, they allowed *all* the Order's various members to access the American market with few limitations. At its core, the Order was a bribe.

The postwar economic relationship of the Americans and their allies wasn't anything like the sort of top-down control that had occurred in other postwar periods. Instead, it was as if the United States had *lost* World War II to *every* country, and all those countries imposed import requirements upon the defeated Americans. Such openness was extended decade by decade to any country that sought to join the Order: first the defeated Axis powers, then the neutrals, then the former colonial possessions, then Communist China, and eventually even the former Soviet satellites.

For their post-American "Order" to be suitably attractive to induce willing cooperation, the Chinese would need to replicate such open access. That's flat-out impossible.

In the United States, companies are primarily concerned with their bottom lines. They look after their profitability by developing, building, and selling products to consumers, whether those consumers be American citizens, foreigners, governments, or other businesses. Therefore American firms treat borrowing as just another piece of the process—whether that borrowing is part of product development, capital improvements, technological research, or sales financing. Any loan brings a cost, with part of that cost typically including a requirement to pay the loan back.

In the Chinese system . . . not so much. While China has become more market *oriented* since the death of Mao, it is still far from a market economy. The Chinese government is willing to put a hand (and all their friends' hands) on the scale, *particularly* to support state-sector companies or Party goals—triply so in times of crisis.

These interventions were meant to transition the Chinese Communist economy from a rural mess on the brink of collapse to a modern state all while—and here's the key—keeping the Party in power and avoiding the massive social upheaval that has accompanied just about all such transitions throughout history. It is, in the truest sense of the term, a political economy. It is a devilishly complex balancing act that involves a slew of entrenched interests, an ever-growing shitstorm of risk, corruption, and competing visions. Suffice it to say, the president's palms tend to be sweaty and greasy in equal measure.

One of the Party's favorite tools to accomplish its dual ends of rapid modernization and control is *dirt-cheap* credit:

- First and foremost, when the credit taps are on (and they usually are), the money flow enables favored companies to ignore pesky things like productivity and market forces and input costs whether that input is oil or copper or soy or cement or labor. Just take out another loan to cover costs and expand, expand, expand. Even poorly connected private companies see benefits from cheaper credit, though these companies don't have anything like the bowling alley gutter guards of their more favored cousins.
- Second, since these firms' credit access means they view the issue of cost as largely irrelevant, they can compete in *any* market on price. The Chinese can—and do—buy high and sell low because efficiency is secondary to the political needs of the Party (such as maintaining high employment levels).
- Third, there's only so much you can spend on fancy equipment and warehouses and shiny corporate headquarters. The money inevitably spills out and allows Chinese firms to acquire not simply tech and wood and workers and steel, but also other companies. The result since 2010 has been a massive Chinese corporate binge on foreign firms, whether it be to secure resources, eliminate competitors, or acquire technologies.

If this fire-hosing of cheap credit into companies to force breakneck growth seems rather pedestrian, rest assured there's more. *Private* financing has now run amok; it now includes a mix of shady practices expressly designed to avoid regulatory scrutiny. Some—like shadow lending—are akin to illegal hedge funds. Others—like peer-to-peer (P2P) lending—work a bit like crowdsourcing, but to fund personal *credit cards*.

All of this allowed China to go from an agrarian society on the brink of collapse to the world's second-largest economy in thirty years. But the Chinese development model has its downsides: risky corporate behavior brought on by a lack of consequences, bloating of expenses, and, of course, a mountain of debt that will never be repaid.

Chinese GDP has expanded by a factor of 4.5 since 2000, but Chinese *credit* has expanded by a factor of *24*. Total debt in China has ballooned to more than triple the size of the entire economy. According to Citigroup, some 80 percent of freshly issued private credit in 2018 *globally* is in China, while the Conference Board estimates productivity growth (how much you get from what you put in) has *declined* since 2012. The *Economist* now estimates three-quarters of the value of new loans does nothing more than pay the interest of loans issued previously. China is spending more and more to get less and less, and it has already resulted in the greatest debt run-up in history.

The rest of the world has seen—repeatedly—where this sort of expansion-at-all-costs development model leads: investment-led bubbles that collapse into depressions. In Japan, it resulted in three lost decades of deflation and near-zero growth. In Greece and Italy, the bubbles generated what are (so far) the largest sovereign debt and banking crises in history. In the United States, runaway lending created Enron and the subprime lending crisis.

Getting past something like this, assuming it is possible at all, is expensive. Indonesia used to follow similar models, albeit with a lower commitment to setting world records for size and much less . . . accounting creativity. The Indonesian economy crashed in the 1997–98 Asian financial crisis, with the claw-back costing 13 percent of GDP, a three-year recession, and the head of the country's president-for-life.

There are plenty of reasons to expect the Chinese situation to not end nearly as amicably. Most obviously, China has been doing this longer, more deeply, and with greater disregard for things like profitability and transparency than anyone else. In relative terms, China's debt mountain is easily an order of magnitude larger than Indonesia's was. Charlene Chu of Autonomous Research, probably the smartest person on the planet looking at this issue, puts the figure for Chinese loans that have gone completely bad at $8.5 trillion, to say nothing of general dysfunction throughout the broader system. For the point of First World comparison, the total value of subprime loans that went into foreclosure during America's 2007–9 financial crisis came out to approximately $600 billion. Moreover, the easy majority of Chinese debt is short-term, suggesting a faster, harder crash than anything the Indonesians suffered through in the late 1990s, much less the Americans in the late 2000s.

The Party has attempted at times to pull back the credit trough in order to rebalance, but there's a reason why many China-watchers describe the Chinese economy as a plane that lost power, gliding in for a hard landing. The political and economic consequences of a successful rebalancing would be severe, so any reform attempt major enough to trigger pain is immediately followed by a doubling down on the old ways.

How will it end? I suggest not obsessing about the details. This is a lot like watching a game of drunken giant jenga. There may be lots of *ooh*ing and *aaah*ing about this or that block or player, but you *know* it's all going to come down—in a fury of force and noise.

What makes China truly unique is not (only) the size and scale and concentration of the lend-lend-lend model, but its pervasiveness throughout all economic sectors and how un-entangle-able the web of Chinese credit has become.

A couple of examples:

First, agriculture. Traditional Chinese agriculture isn't farming (in which workers tend to fields), but instead gardening (in which workers care for individual plants by hand, like literally, each individual stalk of corn). It has to be. China has less than one-third the arable land per capita compared with the global average. When China finally started modernizing after Mao, the leadership relocated many of those gardeners to the cities to work in factories, while also applying scads of technology and capital to modernize the countryside and mechanize agriculture.

The industrialization of Chinese agriculture failed to generate the massive output increases that occurred elsewhere, because gardening tends to be more productive per acre than farming *so long as the labor is free*. Yet in terms of capital, the rural modernization was one of the most expensive things China has ever done. When China's financial system cracks, Beijing will face a stark choice: watch its modern food production collapse, or empty the cities and force industrial workers back into peasant gardening.

Second, housing. It is common for groups of Chinese housewives to pool their savings to purchase a house or condo as an investment, with most pooling generating multiple purchases over time. When they come up short, it is furthermore common for them to individually tap various forms of credit, whether it be a direct personal loan, maxing out a credit card, or tapping a P2P line of credit so they can collectively afford the new property. As these properties are intended largely as investments, few are either owner-occupied or rented out. This artificial demand generates follow-on demand in construction, steel, concrete, and other related industries, without creating anything that anyone actually uses.

By late 2016, the majority of residences sold in China were no longer first-home purchases, but instead investment properties. By early 2018, *third*-home purchases nearly outnumbered first-home purchases. In a time of recession or financial rectification, the entire Ponzi scheme of financing that led to the residence purchases collapses. The only way to prevent a cascade of financial carnage would be to sell a property or two to balance out the debts. That will work, so long as the market can absorb the housing stock. That is, in a word, unlikely. Best guesstimate? One-fourth of all urban housing in the country are unoccupied investment properties.

Simply put, when the China bubble breaks, it will take *every* sector down at once—including the one that grows the food.

In the meantime, this weird combination of bottomless finance and a gleeful disregard for efficiency generates massive *over*production, further disqualifying China as a global economic steward:

- This overproduction first surpassed domestic demand in the 1990s. Beijing subsequently directed the excess to massive infrastructure projects: airports and roads and rail lines and industrial plants and cities. Some were needed. Many were wasteful. China today sports several "ghost cities," designed to be home to a million residents, but which—aside from a handful of state bureaucrats holding down the carpet in government office buildings—are essentially empty.
- While China produces wildly insufficient volumes of the material inputs locally for products such as gasoline or steel or electronics or automobiles, mass subsidization enabled China to become the global low-cost producer and leading exporter of *all* of them.
- China's mass application of subsidies for state-run firms means mills and refineries run whether there is demand or not. Such subsidization has helped Chinese firms dominate electronics and automobile assembly. By the 2000s the overproduction so outpaced Chinese demand that it was dumped on the global system, wrecking markets near and far.
- In the 2010s the overproduction became so extreme that it surpassed *global* demand. Part of the rationale behind projects like One Belt, One Road is to dispose of this excess supply by building infrastructure to and in places that would never justify investment in the first place.[*]

---

[*] China's first completed rail OBOR project linked the country to war-torn, refugee-heavy, impoverished Afghanistan. It is unclear what the winner gets.

Cheap financing paired with global access enabled the Chinese to undercut almost everyone. A significant amount of the world's industrial hollowing-out can be laid at the feet of China's hyper-subsidization model, most notably the struggles of American manufacturing in the 1990s, 2000s, and 2010s. Much deeper hollowings-out have occurred in places that lack the infrastructure, educational, technological, and government advantages of the American system. Mexico, Indonesia, Brazil, India, South Africa, Romania, and the former Soviet Union—really any country that has attempted to modernize in China's wake—have been hurt particularly badly.

China cannot offer its internal market to anyone because it needs *everyone else's* markets to make its own system work—and not simply on the economic front. Part and parcel of Beijing's (master) plan isn't simply corporate expansion, but maximum employment. Give citizens jobs, so Beijing's thinking goes, and the people won't protest things like crackdowns on press freedoms or massive corruption or reeducation camps. Bottomless loans ensure enough economic churn to keep the masses' hands busy, the economy moving, and the Party in power. It isn't so much that the mere idea of China attempting something *other* than the domination of all sides of raw materials production and shipping and all sides of the creation and selling of consumer goods is amusing; it quite simply is that China cannot do anything but. Without its capital-flooded finance model and the outlet the Order provides, China's social fabric would burn.

It isn't as if anyone else could do better. There are reasons to malign the scooter-riding, SUV-driving, Spanx-and-Crocs-wearing, pill-popping, Harry Potter–reading Americans—but all those purchases make the United States by far the world's largest consumer market. That rampant, irrepressible consumerism is the economic half of what makes the global Order work. The Japanese consumer market is less than one-fifth of the American market in terms of final consumption. Russia's is one-*twentieth*.

The combined European economy is in the same league as America's, but after thirty years of trailing American economic growth, the EU has devolved into an Order-dependent export project.

In a world without America, mercantilism—the idea that exports should be maximized while imports minimized—will be the order of the day. Remove the Americans as the bottomless sink for global supply, and the desire for countries to be part of anyone else's global network suddenly wilts. A Chinese "Order" would be predicated upon the Chinese gobbling up as many of the world's resources as possible so that they can shove as many products down the world's throat as possible.

That's not what a bribe is supposed to feel like.

CARROT 4: FLOAT A GLOBAL CURRENCY

On top of ensuring security, enabling global shipping, and creating a global marketplace, the United States also provides an irreplaceable service by providing the sole global currency. And much like maintaining a global navy, not just anyone can float a global currency. There are some extremely strict requirements:

1. The volume of currency must be massive—large enough not just to lubricate trillions (of dollars) of economic activity that takes place far from the sponsor's shores, but so huge that ordinary transactions and business fluctuations do not affect the currency's day-to-day value. Otherwise the instability would scare away users.
2. Moreover, the sponsor's external trade must be so small relative to the size of its home economy that day-to-day changes in the currency's value don't over-upset the domestic economy.
3. The sponsor must be so nonchalant about the currency's value that it doesn't often intervene in currency markets to push the

currency up or down. A whiff of excess manipulation would put foreign assets at risk, reduce confidence, and prompt a search for alternatives.

4. The would-be hegemon must be willing to let currency flow in and out of its domestic market at the whim of others. If you can't get hold of the currency in reliably sufficient volumes when you need it, there is no point in using it in the first place.

The Chinese yuan may check the "volume" box, but it does so for the wrong reasons. The yuan has been one of the world's most manipulated currencies, with the Chinese on average printing five times as much currency a month as the Americans despite the fact that the yuan really circulates only within the Chinese system.

Around 2010 the Chinese began a piecemeal currency opening, the goal being the internationalization of the yuan. Much ballyhoo was made about the Chinese being on the verge of global domination. The reality proved different. As soon as Beijing allowed thin financial connections between China and the outside world, so many Chinese attempted to move their life savings beyond the reach of the Chinese Communist Party that Beijing quickly reversed the internationalization effort. After all, if the Chinese citizenry can get their money out of the country, it will no longer be available to fuel the Chinese lending system that makes a unified China possible. China's currency remains walled off; nearly all Chinese exports—to say nothing of imports—are denominated in US dollars rather than China's own currency.

The issue in Europe is not the currency per se, but instead Europe's banks. While the Europeans have a currency union, banks are still managed and regulated at the national level. Because those banks remain a source of intra-European competition, it is common for European governments to use banks for state funding . . . and then intervene on behalf of the banks so none go belly up when governments run into trouble. Such a bank-backed-debt stratagem makes nearly all European banks insolvent by American standards. Making matters worse, the Europeans developed a habit of confiscating insured bank deposits to pay for bank bailouts. Understandably, extra-European confidence in the euro is not high, and such continual mismanagement dents even the Europeans' view of their own currency. The euro is not only not used for much trade outside of the eurozone, but even within the eurozone itself the US dollar comprises some 40 percent of exchanges.

As to the other major currencies, the Japanese do something similar to the Chinese—even having a similar experience with capital flight back in the late-1980s. It got so bad that Japanese banks closed *all* international offices in the 1990s, knowing that an honest attempt to meet global banking norms would crash their entire system. That leaves the British pound—but if there is one thing the rest of the world agrees on, it is that the British should never be in charge of anything *ever* again.

Even if there were a competitor to the US dollar, the movie *Highlander* had it right: there can be only one. If there were more than one global currency, the currencies' independent movements would make a product—be it a computer or a bushel of corn—cost more in one location than another. Efforts to ship products from one zone to another to profit from the price splits would be massively destabilizing to producers and consumers alike, generating artificial shortages and surpluses—and that is before one considers things like government intervention in markets.

The scale of the US dollar's penetration into the global system is truly Herculean. Some 70 percent of global currencies by volume are linked in some way to the US dollar, in many cases being more dependent upon the US Federal Reserve for their macroeconomic stability than their own central banks. The dollar is the connective tissue between over 90 percent of global trade exchanges (nobody wants Polish zloty or Vietnamese dong or

Argentine pesos, so the dollar plays middleman in lubricating all global trade save that which occurs exclusively within the eurozone). And the dollar is the sole currency for *all* commodity trade—even in trades that don't touch the Western Hemisphere.[*]

MOVING BEYOND THE ORDER

The idea that the Chinese—or anyone—can meet these criteria doesn't gel. And by anyone, I mean *anyone*—including the United States. Part of what made the institution of the global Order possible back in the late 1940s is that the world was wrecked. The American economy was roughly the size of everyone else's put together, so subsidizing a broad-based alliance structure was possible. Since then the American economy has grown immensely, but—by design—everyone else's grew *more*. In 2020 the US economy is "only" about one-fourth of the global total. The sort of open-market access and indirect subsidization that was a light lift in the 1950s, '60s, '70s, and even '80s is now at the creaking edge of the Americans' capabilities.

The world has evolved past the point that *anyone* can provide the volume of carrots needed to hold the Order—any Order—together.

---

[*] Technically there are a few commodity trades out there that are not USD-denominated, typically run by countries that don't care for the United States. These make up much less than 1 percent of commodity trade and typically *still* use the USD as an intermediary because nobody has or wants Venezuelan bolivars or Iranian rials or Russian rubles—not even the Venezuelans and Iranians and Russians.

CHAPTER 3

HOW TO RULE
THE WORLD, PART II

THE BRITISH MODEL

The Chinese are incapable of providing sufficient inducements to bribe everyone into following them to global domination, but the carrot isn't the only path to global power. In less cooperative times, there has also been the stick. The Americans' former colonial master is by far the most effective stick-wielder in history. Perhaps the Chinese could follow in British footsteps?

The British model is far less complicated than the American system. There is no global set of rules. No paying swathes of countries to be on your side. No trade among nations to facilitate. No chronic need to militarily protect other countries. No guaranteed independence for weak states. There is only flat-out conquering of the world.

So the question is . . . how? The rules about mastering the arts of transport and security, economics and war still apply, but the British approach was wildly different. Think of the British approach as the three sticks of success.

STICK 1: AN UNASSAILABLE STRATEGIC POSITION

The lands of England are . . . OK. The Thames basin is a good—not great, just good—agricultural zone. The British diet is heavy on the baked, doughy, boiled, and fried in part because the sort of things you can grow en masse in Great Britain's short, cool summers that have a chance of lasting through the long, wet winters cannot be easily prepared in other ways.

The Thames itself is navigable, but isn't all that long of a river, limiting the capacity for capital generation. The island of Great Britain *does* have about twenty minor rivers punctuating its coast, deepening the economic potential of the Thames region, but even with modern engineering installing a bevy of locks, none of them punch inland more than ninety miles. Nature gives the United Kingdom the heft of a modest middle power. Not bad, but nothing special.

That is, until you consider that Great Britain is an island. That changes everything.

The temperamental English Channel, a mere twenty-five miles across at the narrowest point, is one of those funny quirks of geography that has literally shaped the history of the world. It was a barrier equal to the Alps in blocking armies but lacking limits for the seafaring English. Unlike everyone else in Europe, the English never needed to worry about an army getting bored and leisurely passing through, and so were freed from the burden of needing to field one of their own. Between cost savings and strategic insulation, the English were able to maintain a millennium-long continuity unmatched in world history, which allowed them to experiment with little things like capitalism and the sort of political devolution of power that would in time give rise to democracy.

The Channel also grants any government that can control Great Britain the ability to stick their noses into anyone else's continuity. Army movements are easy to predict and monitor: land forces must follow roads, cross bridges, avoid swamps, and so on. Not so much with navies. They can move with speed and stealth and show up at times and places most inconvenient, depositing marines or cannon for lightning strikes and then packing up and sailing away. Even at the British height, mainland Europe easily outnumbered the population of Great Britain by over seven-to-one. Yet the Brits' ability to come and go and raid and trade and wreck and bolster as they choose elevated them to become the Continent's first power.*

Great Britain's adjoining bodies of water are stiffer barriers than a map-glance suggests, layering in another bit of strategic depth. Most European attempts to breach Britain's moat met with disaster before even engaging English naval forces. Historians continue to bicker over the numbers and sizes of ships and cannon of the English fleet that faced down the famed Spanish Armada, but one thing is clear: a North Sea storm deserves at least half the credit for wrecking the ill-fated Spanish vessels. France's Napoleon and Germany's Hitler were obsessed with the Channel, undoubtedly staring broodingly across on more than one occasion, because they understood the danger the British navy posed. But they were also familiar enough with such maritime conditions and so pressed by the more immediate dangers of neighboring armies that they never attempted amphibious assaults, even at the height of their power.

China has no such insulation. Across its northern border lies Russia, a country that admittedly has had a rough time since the Soviet collapse, but which still boasts military technology that is nothing short of world class and a nearly million-man military and internal security system to boot.† To China's southwest lies India, a country

---

* It's a lesson the French learned better than most. The French capital is at Paris rather than further north at Orleans, so the English couldn't raid it.
† And that doesn't include the nearly 900,000 former troops that Russia reserves the right to call up in a time of war.

with just as many people to throw into the meat grinder of infantry combat as China. Making matters worse for the Chinese, Indian population centers are only a couple hundred miles from the Himalayan passes while China's Han-populated zones are a couple *thousand* and on the wrong side of occupied, restive Tibet—a region that would fall over itself to welcome Indian assistance. To the south are regional heavyweights Thailand and Vietnam. Neither has designs on southern Chinese territories, but with a combined population of over 170 million and proud histories of resisting Chinese power, neither can they be dismissed out of hand.

To the west lies Kazakhstan, a country about the size of the United States east of the Mississippi plus Texas, Oklahoma, Arkansas, and Louisiana, but with a population smaller than New York state. Kazakhstan isn't a threat per se, but China cannot easily project power to or through it. Central and western China is mostly desert wasteland, and the southeastern two-thirds of Kazakhstan isn't much better. What it makes for is a broad, open frontier whose patrolling—much less fortification—comes only at extreme cost.

Ironically, the Korean Peninsula is the least problematic of China's borders, but in times of Chinese chaos and Korean strength, even the Koreans have taken a bite out of China. Considering that the Koreas have about one and a half million men under arms between them, the South Koreans could develop a nuclear weapon in a few days, and the North Koreans have one *now*, the outright subjugation of the peninsula is fully off the table.

China isn't simply a constrained land power; it is in a box. And that's just the start of its strategic problems.

STICK 2: A POTENT, FLEXIBLE NAVY

The English aren't dumb. They realized they got a little lucky with their "win" versus the Spanish Armada and so shortly thereafter began investing in a navy worthy of the name.

There are many reasons for English success on the waves. In part, it is because they learned to sail on some of the world's most storm-wracked seas. Surviving sophomore English sailors had more skill and grit than most seniors, and they trained the following generations appropriately.

In part, it is because of Great Britain's isolation. The English never had to worry much about an enemy marching in and destroying their ports and shipyards.

In part, it is about concentration of efforts. Only in historically rare circumstances has the British navy ever had to compete with the army for resources.*

In part, it is about Great Britain's geography. England's neighbors are constrained. The French, Dutch, Germans, Poles, and Russians must constantly watch their backs (and fronts) or risk foreign land invasion. Sweden and Denmark have the benefit of island and peninsular geographies, enabling them to project more naval power relative to their size than their peers in the Northern European Plain, but both are bottled up in the Baltic Sea, limiting their ability to compete with—much less confront—the English. That leaves Ireland and Norway, countries whose entire populations couldn't fill London. They are, at most, strategic irritants rather than competitors. The English have restriction-free access to the ocean blue with their oversize navy.

But while size matters, it's really about how you use it. Unique among world powers, British naval acumen is multifaceted. As an island country, any trade the English engage in *must* traverse salt water. England's early (and successful) experiments with capitalism meant much of England's trade was a private effort rather than a purely state-dominated one. This did more than generate

---

* Ironically, the most notable of such periods is the here and now of the late Order. The United Kingdom's leaders fully realize that London's time as global leader is long past, and collectively believe attachment to the American system—whatever its form— is their best bet for wealth, security, and relevance. That means marching to war with the Americans whenever the Americans feel the itch. And because the Americans don't need naval help, that means the Brits have to contribute ground troops.

better economic, capital, and tax growth. It injected necessity-driven naval skill into the internal workings of the country's business class.

The mix of skills is as rare as it is potent. English traders plied the world's oceans, developing deep understandings of the political and economic intricacies of peoples near and far—understandings that could be pressed into either diplomatic or military service as the needs of the day dictated. A command crew with first-hand practical knowledge of how river, rail, and road trade works from the personnel, logistical, and infrastructure points of view found it breezily simple to disrupt the internal trade of an enemy or protect the trade of an ally.

China has nothing like this.

China is an inveterate *land* power that has fought major *land* wars with each and every one of the powers it borders. It simply cannot afford the sort of resource focus that made the British navy possible. Technically, the Chinese navy is known as the People's Liberation Army Navy, which should tell you everything you need to know about the pecking order of the navy in terms of prestige, political power, and skill sets.

Not only do the Chinese have none of the broad and integrated maritime skills the British are famous for, but they also have precious little practical experience with fleet operations. Few Chinese commanders have ever fired a shot in anything other than a drill. Industrialized China has *never* fired its naval guns in anger.

Sinophiles will point out here that China *has* had fleets in the past. In fact, at one point, China arguably boasted the world's largest naval force. Such "treasure fleets" roamed throughout East and Southeast Asia and on at least one occasion made it as far as East Africa. Their success didn't stick. For one thing, the sort of decentralization required for establishing meaningful trade with the outside world is not the sort of thing the Chinese leadership has ever been keen on. More important, when the treasure fleets returned home with stories of the wonders of the world, the Chinese emperor got spooked that perhaps China wasn't the greatest country on Earth after all. During the next half century the naval program—indeed, *all* ocean-going vessels—were scrapped completely. Needless to say, fleet management skills were lost.

China's coastal waters aren't all that treacherous. Roughly paralleling the Chinese coast are a line of archipelagos that break up the storms, winds, and currents of the wider Pacific—including the famed First Island Chain, which includes the countries of Singapore, Malaysia, Indonesia, the Philippines, Taiwan, and Japan. Chinese skippers don't simply hug the Asian coast; they learn to sail in the equivalent of a giant bathtub.

The sort of naval war the Chinese are preparing for speaks volumes. Knowing they cannot compete with Japanese air defenses or American carriers and submarines in a head-to-head fight, the Chinese instead prepare for mass area denial: the idea is to use China's air force—the world's largest—and scads of cruise missiles to sink any fleet within the Chain. So the thinking goes, if opposing fleets cannot safely operate within the island chain, then the Yellow, East China, and South China Seas become a Chinese lake. A particularly stupid American admiral might even lose a carrier if he sails it too close to the Asian coast.

But that strategy fails to address four critical issues.

First is the fish-in-a-barrel problem. Between China's lack of naval experience, lack of operational range, and the constrained operating environment within the First Island Chain, Chinese ships are easy targets. Should the Americans find themselves in a shooting match with China, the American Navy would have an easy week's work sinking anything flying a Chinese flag outside the First Island Chain, followed by some serious dynamite fishing within it. All a Chinese lash-out would accomplish would be an end to China's ability to access external sources of raw materials for supply, and merchandise demand—an "achievement" that would bring the Chinese economy, the

Chinese Communist Party, and a unified China tumbling down. Easy peasy.

Second is the cost imbalance. Navies will *always* be far more expensive than armies and air forces of equivalent power. For example, an American *Nimitz* carrier costs about $10 billion— about the same amount as *three hundred and fifty* F-16s. In a straight-up fight between such forces, the carrier would be so defeated as to be erased from memory. China would need to break out of the Chain using relatively expensive *naval* forces, but anyone on the Chain can fight back using *land*-based forces.

Third, a vastly successful mass area–denial effort that sinks everyone else's local fleets makes China's situation *worse*. A shooting war off the Chinese coast would condemn the Yellow, East China, and South China Seas to being no-go zones for the very merchant shipping China needs to survive.

Finally, a naval victory within the Chain is insufficient. Meaningful trade requires a security environment in which unguarded, civilian cargo can sail to and fro at all times. A stick-driven Chinese empire would require the Chinese to sub-jugate large portions of the Chain to establish and maintain such access, a task that would require China's navy to be nearly as powerful as the Americans' *combined* carrier fleet. And even *that* doesn't solve the problem of projecting power out *beyond* the First Island Chain to the resources and markets the Chi-nese need, or to and through the waters of other naval powers who might take issue with shoot-your-way-out commerce.

STICK 3:
A MASSIVE TECHNOLOGICAL ADVANTAGE

British businessmen in the later decades of the empire grappled with a sprawling, interconnected set of problems. Britain itself wasn't exactly resource-rich, so the money that drove the initial imperial expansion came from the sales of commodities generated within the British colonial systems. When the Americans broke away, much of the money went with them. If the Brits couldn't earn income by distributing commodities, they needed to distribute value-added goods—but there were sharp limitations on what was possible.

Crafting goods was slow work that required skilled craftsmen. Training those craftsmen took years, if not decades, and anyway, Britain's population was tiny compared with Europe's, much less that of the rest of the world. The Brits needed a way to surmount the one-at-a-time manufacturing throughput process.

Using the profits from their seaborne trade, a host of British businessmen combined a series of nascent technologies, including coal, oil, the assembly line, interchangeable parts, and eventually steel and electricity to develop new production systems. Britain's Industrial Revolution began in textiles but quickly spread to tools and furniture and railroads. Within a half century, the industrialized Brits were producing vast volumes of cheap, high-quality goods that could outcompete those slow-moving, skilled pre-industrial craftsmen in almost any product set.

Between the empire and its statelet-like corporate entities, the British had direct access to one-quarter of the world's population, the largest captive market in history. But the really interesting stuff happened when one extended considerations beyond the purely economic. Shoving cheaper, superior goods into a *hostile* market—something made possible by the empire's savvy, worldly-wise, globally dominant navy—caused no end of entertaining problems.

In the United States, it deepened the wedge between the industrializing North, which saw British imports as a threat, and the agricultural South, which was willing to trade cotton for cheaper gadgets. The disconnect nearly split the country in half during the War of 1812 and the Civil War. Somewhat similarly, in Prussia, British goods shattered the economic backbone of the local guild system, contributing to the inter-German wars and mass emigrations of the mid-nineteenth century. In France and Russia, the flood of British goods gave the riotous peasants a taste of the good life, nudging both countries in the direction of mass revolt. Spain's industry collapsed, plunging the once world-spanning empire into a funk it didn't recover from until the late twentieth century.

And it isn't as if British industrialization stopped with civilian goods. The new techs proved easily adaptable to military power: muskets gave way to rifles, sailing ships to steamships, cannon to artillery. The British navy was already the best in the world, but with the industrial overhaul, it became unassailable. Combine the new products and the new military applications with the Brits' already global reach, and the British could outcompete their Europeans peers and utterly dominate any non-European people they came across. For the better part of a century, the Brits topped the world because—quite literally—they'd bring guns to knife fights.

China has indeed made impressive strides by any measure during the past generation in infrastructure, education, industrialization, and modernization. Between 1980 and 2020, the Chinese have quintupled their share of global economic output. But global technological leaders they are not.

China's position in the global import market is made possible not by technological edges, but by subsidized production and risk-free transport, all made possible by the American Order. What technologies the Chinese command often have a theft component to them. Deep, pervasive hacking of foreign systems is a key component to Chinese success. It follows three basic forms: establishment of joint ventures abroad where the tech is stolen (Germany, Brazil), establishment of joint ventures within China in exchange for market access where tech is stolen (the United States, Japan), and the remote hacking of systems to steal technical advances (pretty much everywhere with an Internet connection). Such institutionalized theft is the basis for China's Made in China 2025 development program. When the Trump administration and European Commission started taking a harder line with the Chinese on forced technology transfer and cybertheft in mid-2018, by late 2018 the Chinese had already quietly dropped the program from internal propaganda, knowing that they lacked the educational capacity and skilled labor forces required to develop the requisite research and manufacturing base without foreign sources. China *has* made a great deal of progress in generating its own technology, but let's not get ahead of ourselves. China's domestic patent system only went into effect in 1984.

The mass application of stolen tech throughout the immense spread of the Chinese population fuels Chinese advances in production and market size and economic bulk, but do not confuse that with a technological edge that gives them a leg up in the British style of global strategic competition.

There *is* an exception worth mentioning (and debunking). The Chinese have poured an inordinate amount of resources into developing artificial intelligence (AI) systems, raising concerns in board and war rooms the world over. That makes it critical to understand what AI actually is.

Narrow AI, sometimes called applied AI, is exactly what it sounds like. An artificial system laboriously trained for one specific task so that a computer system can take a thin bit of the load from an otherwise human-labor-intensive system. Applied AI *can* do things better than humans, but only within the confines of strict rules, and so long as everything goes according to plan. It's great for assembly lines and spreadsheets, but is most definitely not ready for most real-world situations. China, with its teeming masses of coders, has the manhours to spend on a tech that in essence reduces the need for manhours. But put an AI trained to play chess in front of a game of checkers and *pbbbbbt*.

What we don't have—what no one has—is general AI. This is where AI can be "introduced" to a novel task or even see a common item but in a novel situation—like a stop sign but on the wall of a restaurant and covered in graffiti—and can figure it out. It's the magic combination of background knowledge and context and intuition and reasoning that we call basic human intelligence.

Most American tech firms are less excited about applied AI because it isn't all that high-value-added, and are instead focusing on the dream of general AI because it will change the world. When it happens. Which isn't now. Or in 2030. Or probably even in 2050. *That* tech breakthrough isn't so much around the corner as on a different continent.

Which means that, while applied AI *can* prove impressive in industries as varied as agriculture, finance, transport, manufacturing, security, and military affairs, this is not the sort of technological breakthrough—such as deepwater navigation or industrialization—that fundamentally alters how humans interact with their environment. Put another way, even if China *does* master applied AI before the rest of the world and somehow can maintain that edge, it doesn't help China break out of its strategic box.

THE REALITY OF CHINA

The problem isn't that China cannot build and maintain a huge, outward-looking navy (although it probably cannot sustain the effort).

The real problem is that China cannot build and maintain a large, outward-looking navy *and* a huge defensive navy *and* a huge air force *and* a huge internal security force *and* a huge army *and* a huge intelligence system *and* a huge special forces system *and* global deployment capability *at the same time*. For China to be a global power, it would need *all* of these.

Compare that to what the Americans needed in the Order.

A large outward-looking navy was obviously a prerequisite for American global power projection, but American strategic insulation meant the other items were necessary only to maintain the allies.

Which means for the Americans, it is only going to get easier. As the world falls into Disorder and American strategic commitments wither, the United States' strategic toolkit can be smaller. Defending the American homeland is pretty straightforward—float a sizable navy and back it up with some domestically stationed air force assets. The sort of overseas engagements the Americans are likely to embroil themselves in are likely to fall into the hands of its intelligence apparatus and its special forces teams. While not free, such tools of power are far cheaper than deploying and supporting divisions of ground troops in a different hemisphere.

China's strategic regional geography means it cannot downsize in that way—under *any* circumstances. The question is not whether China can be the next global hegemon. It cannot.

The *real* question is whether China can even hold itself together as a country.

CHAPTER 4

HOW TO BE A SUCCESSFUL COUNTRY

Historically speaking, most countries don't last long. Carving out a little space—whether through a revolt against one's master, clawing one's way out of a collapsing empire, taking advantage of a climatic hiccup, or benefiting from exceedingly smart leadership—is the easy part. But establishing and protecting a continuity? Far less so.

The Order changed the rules. Its peace, prosperity, security, and growth prevented the wheels of history from grinding down the upstarts. Near the end of the Imperial Age in 1920, there were about fifty countries. As of 2020, the count is over two hundred. Remove the Order and what has enabled many of these countries to form, survive—even thrive—will fade away.

Dozens of assets contribute to national survival and power, but these are the big four:

1. Viable home territories, with usable lands and defensible borders
2. A reliable food supply
3. A sustainable population structure

4. Access to a stable mix of energy inputs to participate in modern life

Master these factors, and your country is all but doomed for greatness. Fail, and your country will excel at the other thing.*

1: TERRITORIAL VIABILITY

Throughout history, the shapes of a country's lands, waterways, borders, and coasts have been the single greatest factor in determining whether the country would succeed or fail. The specifics *have* changed with the technological age. For example, once humans figured out how to sail beyond the horizon, island nations tended to get a power-up. But geography's immutable nature hasn't left a lot of wiggle room. Some guidelines:

INTERNAL WATER TRANSPORT. Because it is far easier and cheaper to float items from A to B than to drag or cart them, water transport is roughly one-twelfth the cost of road transport. Countries with naturally navigable rivers tend to have an easier time than those without.

In part, it's about money. Cheaper transport enables a local density of economic interchange at rock-bottom costs that otherwise would necessitate a global merchant fleet.

In part, it's about cultural and political unification. If it's easy to move around your system, then either that constant interaction makes all your people tolerant of one another and so enables a cultural melding, or it empowers one group to dominate and in time assimilate the others.

In part, it's about engineering. The difficulty of building and maintaining roads was so onerous that we didn't transport things

---

* Spoiler alert: it doesn't look great for the Chinese.

overland even over short distances—even *food*—until the Industrial Age brought us asphalt and industrial cement production and the steam and internal combustion engines.

In part, it's about continuity. If you can't quickly and easily distribute foodstuffs or generate the capital locally to defend yourself, odds are civilization-burning famines and invasions are within your experience, and therefore your country crashed hard.

PLAINS. Plains are the key building block for successful countries. Easy to build on and easy to move across, low-development-cost land frees capital for other things, like roads and education, which in turn lead to more advanced societies. But it's about a lot more than money. With no internal barriers to movement and communication, countries with open interiors find it far easier to achieve political unification.

There can be a darker side too. If it's easy to move goods and people and ideas, it's also easy to move troops. Cultural merging, or assimilation, is not the only way to "unify" a territory. Hunting down dissident and minority groups is more straightforward. Again, the American Midwest is a premier example. The Native Americans never had a chance. Similar unification-via-obliteration unfolded in most of the world's countries with flatlands, most notably in the North China Plain, the Northern European Plain, and the vast open spaces of North Africa and the Middle East.

TEMPERATE CLIMATE ZONES. Climate is key. Deserts lack the water required for broad-based civilizations, and the level of organization required to control said water often turns desert cities into isolated tyrannies. The tropics are great for vacations but are also great for bugs of both the insectoid and microbial sort, enervating human health and agriculture alike.

Seasonality also injects some useful quirks into cultures by providing regular pressures. Planning for the cold of the winter pushes for developments in materials science and construction.

Planning for the brief summers necessitates the organization of labor and capital. Planning for the starving season—spring is when most cultures face food pressures because it will be some time before the harvest comes in—encourages advancements in logistics and mathematics. It's hardly that cultures in nontemperate environments are lazy or less intelligent, but rather that the pressures they face don't encourage the sorts of technical advancement associated with stronger and more durable states.

Rivers plus plains plus a temperate climate help guarantee a sharp upward technological and economic trajectory.

COASTLINES. These are fickle things. Most of the world's coasts are swampy or desert or frozen or cliff or rocky or shallow or obscenely tidally variable or otherwise seemingly designed to be death traps.* But *sometimes* there's a magic mix of a flattish coastline with access to a flat interior with a deep coastal shelf and a large curved bay and maybe a river mouth that makes the coast accessible from both land and sea. When that happens, the opportunities for economic activity are explosive.

The key issue is *distance*. The farther away the origin and destination points of trade, the more likely the goods in question are going to vary. If western Iowa and eastern Nebraska trade, there's only so much enthusiasm that can be generated from exchanging different varieties of corn. But if San Francisco and Bordeaux trade, life gets considerably more interesting. The global spice trade was the quintessential example of how exoticism commands top dollar.

Rivers plus plains plus a temperate climate plus an accessible coastline all but dictate that a region will become a significant economic and military player, able to leverage the benefits of its local geography onto a larger stage.

* The Portuguese didn't name the Namibian Riviera the Gates of Hell lightly. A combination of subsea geography, reefs, and dangerous currents made it a graveyard for galleons and whales alike.

FRONTIERS. External geography is just as important as internal. For borders, the gold standard is to have the *opposite* of what makes a good internal zone. Flat, open borders may encourage trade, but they also allow attack. If one of your navigable rivers flows through another country on its way to the sea, your access to the wider world is hostage to the whims of another. Hills and swamps limit contact, but determined invaders can still push through them, and both are just habitable enough to house groups that might resist central rule. Mountains are better. But the best by far are wide, wild oceans.

In essence, what you're after is an outer shell firm enough to protect your own continuity, plus a useful interior zone large enough to support internal economies of scale. That magic combo is a precious, rare thing—and one country's tops them all.

When it comes to useful flatlands, it's impossible to beat the United States. The Piedmont coastal plain of the East Coast, the Columbia Valley, and California's Central Valley also all score well up in the global rankings. That all these regions teem with navigable river frontage is a bonus.

But America's Midwest is a place apart: The Greater Mississippi system includes over thirteen thousand miles of naturally navigable, interconnected waterways—more than the combined total of *all* the world's non-American internal river systems—and it almost perfectly overlaps the largest contiguous piece of arable, flat, temperate-zone land under a single political authority in the world. The overlap enables American farmers to export their high-bulk products to global markets at lower costs than most countries can shuffle grains about within their own countries. And the ability of Americans to move easily around their own internal systems led to cultural unification with a speed and ease largely unknown in human history.

In contrast, the outer crust of the American system is world-class thick. Hundreds of miles of desert and mountain separate

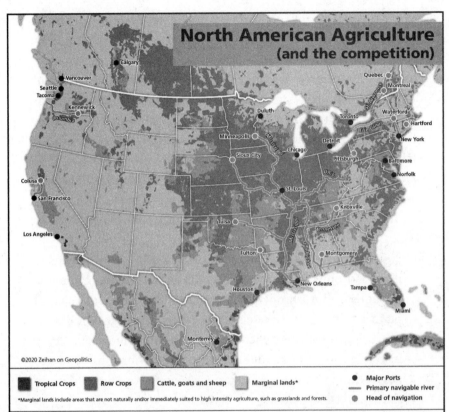

North American Agriculture (and the competition)

©2020 Zeihan on Geopolitics

■ Tropical Crops   ■ Row Crops   ■ Cattle, goats and sheep   ■ Marginal lands*

● Major Ports
— Primary navigable river
◉ Head of navigation

*Marginal lands include areas that are not naturally and/or immediately suited to high intensity agriculture, such as grasslands and forests.

Name	Total Land (million hectares)	Total Arable Land (million hectares)	2020 Population (millions)	Arable land per person (hectares)	Navigable Waterways# (miles)
Argentina	273.7	39.2	45.2	0.87	1,900
Russia	1,637.7	123.1	145.9	0.84	0^
United States	914.7	152.3	331.0	0.46	20,650+
Brazil	835.8	81.0	212.6	0.38	0
France	54.8	18.4	65.3	0.28	1,350
Turkey	77.0	20.4	84.3	0.24	300*
Iran	162.9	14.7	84.0	0.17	50
Germany	34.9	11.8	83.8	0.14	2,400
Saudi Arabia	215.0	3.5	34.8	0.10	0
China	938.8	118.9	1,439.3	0.08	1,950
Japan	36.5	4.2	126.5	0.03	600*

Source: UN Population Division and Food and Agriculture Organization

Includes only waterways navigable for at least nine months of the year at a depth of at least nine feet. Includes shared waterways for their shared length, such as portions of the Oder and Rhine. Does not include otherwise navigable rivers, which flow through unpopulated regions such as the Yukon, Lena or Amazon.

+ Includes the Intracoastal Waterway and Great Lakes system, which add approximately 3,000 miles of navigability each.

^ Russia's Volga system was heavily engineered to force partial navigability under Stalin. Its total useful length including tributaries is approximately 2,500 miles, but Russian winters make its use highly seasonal and subject to extreme variability.

*Japan and Turkey lack navigable rivers, but their inland seas (the Seto and Marmara, respectively) are large, sheltered bays that function similarly to rivers.

the Americans from populated Mexico, while scores of miles of lakes, forests, and mountains keep most of Canada at arm's length. To the west and east, *thousands* of miles of nearly trackless ocean separates the United States from both the Asian and European landmasses. What tracks there are in those oceans comprise either American territories (Hawaii, Guam, Midway, the Aleutians, Puerto Rico) or are controlled by allies (the United Kingdom, the Azores, Bermuda, the Bahamas). Flat-out, the Americans have both the richest territories in the world as well as the most securable.

There's not a lot of this sort of magic mix elsewhere:

- Northern Europe has lots of flat, rivered areas, which make it second only to the United States in wealth and economic dynamism. But Europe's rivers transect the continent's coastal plain, splitting it into sections rather than interlinking its subregions. Instead of encouraging cultural homogeneity, they instigate competition among Europe's many nations. And because Europe's north is so flat, everyone's core territories are someone else's frontier.
- Africa's topography is wildly hostile to development, both in terms of climate (desert or tropical) and terrain (jungle and mountain). Even worse, most of the continent's population lives atop a stack of plateaus, making trade among the African nations *more* difficult than trade with the outside world. That same rugged terrain makes it next to impossible for the local countries to unify their political identities. The outcome is a sad mix of external exploitation and internal conflict.
- The omnipresent archipelagoes of East and Southeast Asia make development horrendously difficult. Only in a few cases is one island within each group powerful enough to float the necessary naval strength to even attempt consolidation in the first place. Achieving economies of scale is fiendishly difficult.

- Much of South America is so tropical that it forces local populations to move uphill to escape disease belts. Living on a mountaintop has its perks, but it comes at the cost of hideously expensive infrastructure, while reducing economic and agricultural opportunities and fostering a national culture that tends to fracture easily.
- The Middle East is arid to desert while Russian lands are, at best, on the cold side of temperate. Consequently, both regions suffer from a hard edge that often bleeds into continuity-endangering starvation and violence.

For its part, China has far more in common with the parts of the world that don't work very well than it does with the United States.

The Chinese core territories of the North China Plain are decidedly mediocre. While they sit at about the same latitude as the American Midwest, they are hard-up against the Mongolian Desert, making them prone to drought, while most of their rainfall comes from monsoonal systems off the East China Sea, making them prone to floods. The only way to maintain reliable agricultural output is to apply bottomless supplies of labor to manage water supplies. Due to the choice between state-organized backbreaking labor or starvation, Chinese history tends toward the less egalitarian side of things.

Paradoxically, Chinese history also tends to be on the less unified side of things. Again, the issue is geography:

With the sole exception of the unnavigable Yellow River (Huang He), there are precious few barriers to movement in the North China Plain. Living in a land where a rider on a fast horse can reach any location *from* any location within a couple of weeks has consequences. On the upside, *cultural* unification is easy: the Han emerged as the region's dominant ethnicity over two thousand years ago. On the downside, *political* unification is nearly impossible: the plain's huge size and the lack of redoubts within the Han core make it painfully easy for any force to seize control of this or that piece of territory for a time, but equally painfully easy for them to be swept away.

Anyone can "sweep." A local warlord. A competing dynasty. Outsiders can play too, and they do so often. Anyone who can reach northern China has had a fairly easy time of dominating whatever chunks of it they find interesting: Mongols, Russians, Japanese, Koreans, Americans, British, French—even Australians. But they have just as much trouble *holding* that territory as any local authority. The result is a nearly complete lack of political and economic continuity in the Han core, with the Han themselves being just as disruptive as outside forces.

And that's just the part of China that's easy to integrate. Just to the south of the North China Plain, a series of broken, densely forested peaks hive off the Yangtze Valley. Bedeviled by floods and drought, the Yellow River is a silt-laden mess, but the Yangtze is one of humanity's great rivers, boasting nearly two *thousand* miles of navigability. The Yangtze has been the economic centerpiece of every successful Chinese iteration.

That's a problem. It's an issue of continuity, but also of unity. The Yangtze's physical remove from the North China Plain enables it to thrive in wealthy isolation while northern China is busy devouring its own tail.

The most important spot lies at the river's mouth: Shanghai—a city that is China's equivalent of the financial center of New York, the manufacturing center of Detroit, the entrepôt of St. Louis, the energy center of Houston, and the import/export node of New Orleans all in one (and a population easily larger than all of them combined). Which somewhat simplifies northern China's relationship with the often autonomy-minded Shanghailanders: it's all about the money.

In every phase of Chinese history, Shanghai has *always* been busy with the business of business. Shanghai trades not only with the rest of the Yangtze communities, China's southern coastal cities, and the northern Han core, but also with the Japanese, the Koreans, the Taiwanese, the Vietnamese, the Portuguese, the Brits, the French, the Americans—everyone and *anyone.*

Every bit of this makes Beijing perennially suspicious of Shanghai, and whenever the north manages to unify, Shanghai tends to be the first target of any wider imperial expansion. The city doesn't fall easy. Demographically, urban Shanghai alone boasts a population nearly as large as Texas. Strategically, the region's plethora of foreign connections come in handy when the arrows start flying. Tactically, Shanghai sits on the Yangtze's *south* bank, and because the Lower Yangtze is over a mile wide by the time it reaches Shanghai, attacking it from the north requires at least a modicum of a navy.

But Shanghai's best bargaining chip is its wealth. No matter how concerned the northern Han are about national unity, they don't want to kill the cow. Even in 2020 after four *decades* of a stilted financial system to help the rest of China catch up, the Greater Shanghai region *still* generates a quarter of China's GDP from barely more than a tenth of the country's population.

Upriver at the Yangtze's head of navigation is the bowl valley of Sichuan. The bowl's floor is warm and fertile enough to easily feed its entire population, while its oil and natural gas fields make it one of only three provinces in contemporary China that can fuel their own needs. Sichuan's access to the Yangtze enables it to trade with downriver and oceanic partners, making it nearly as wealthy as mighty Shanghai.

Northern China's beef with Sichuan is threefold. First, Sichuan is by far the most culturally distinct of China's Han-majority regions, sporting its own (awesome) cuisine and a dialect so dissimilar from Mandarin that anywhere but in centrally controlled, propaganda-heavy China it would be called a language in its own right. Second, Sichuan is *big*. So big that more people speak Sichuan as their first language than speak French or German.

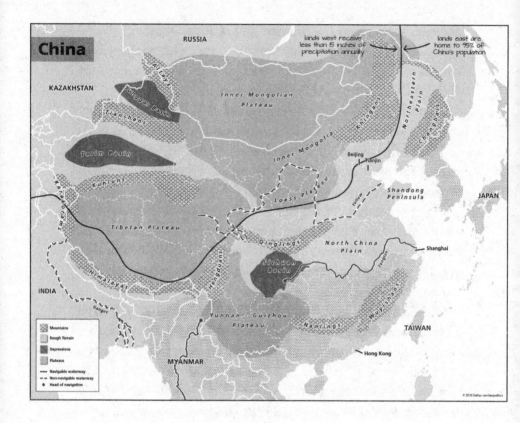

Third, the Sichuanese realize just how distinct and big and economically viable and remote they are from Beijing—a realization that often curses them with delusions of independence. Throughout most of Chinese history Sichuan has existed beyond imperial control, contributing to its long and fruitful reputation as a land of troublemakers. Sichuan has played host to this or that rebel leader or force right up until modern times. In the Chinese Civil War of the twentieth century, Sichuan was one of the last spots on the mainland to stand against Mao.

Move south of the Yangtze and the land changes again, edging into the subtropics and becoming incredibly rugged. Those simple facts cast the cities of southern China into the role of permanent collaborators and successionists—because they don't have a choice. Inland it is simply too rugged for either local infrastructure connections—the first road and rail systems that we would recognize as "modern" were not fully operational until the tail end of the twentieth century—or food production. Wealth and full bellies come from only one place: the sea. That means dealing with whoever the local naval power of the day happens to be. Hong Kong, the quintessential southern Chinese city, is the crowning example of how a separate economic life easily leads to a separate political destiny.

Between these "interior" zones, which make up China's crunchy mess of an inside, and its outer shell lie vast buffer territories.

To the southeast are tangles of forested and jungle mountains about twice the size of California, jam-packed with minority groups. Managing them generates a sort of low-grade headache that never goes away. The lands west of the northern Chinese core are more of a migraine. Centralish China contains enormous empty stretches. *Beyond* those vast swathes of nothing live ethnicities almost pathologically hostile to the Han—most notably the Tibetans on their namesake plateau, and the Uighurs of Xinjiang. These peoples command the gateway territories between the wider world and the Han's core territories—the Indian subcontinent and the former Soviet Union, respectively.

In times of Han strength, the Tibetans and Uighurs fall under Beijing's rule, and Beijing does all it can to impress upon them that flirtations with outsiders will not be tolerated. In times of Han weakness, the Tibetans and Uighurs are effectively independent, remember how the Han treated them, and tend to roll out the welcome mat for anyone willing to subvert the Hans' grip. Beijing itself is an expanded garrison town along what was the traditional invasion route through just such a buffer zone.

What we think of as "China" is in reality less a political entity and more a culture that has a damned hard time keeping itself together. Periods in which the territories of China are both politically unified *and* under centralized Han control are painfully thin, amounting to less than three centuries of the Han ethnicity's multi-millennia history. It doesn't take a lot of reading between the lines to get the gist. China's historical periods include labels like Ten Kingdoms or Five Dynasties or the bleedingly obvious Warlords Era. It's no wonder that the contemporary Chinese government—the Communist Party—expends so much effort on nationalist propaganda.*

The Han don't think twice about whatever flavor of repression or genocide they think might work on any given day: outlawing Tibetan Buddhism, burning down Tibetan cultural infrastructure, using tanks to run down protesting students in downtown Beijing, crushing any hint of political dissidence in Hong Kong, forcing Uighur families to house government informants in their homes, placing a million Uighur in concentration camps. It's all a day at the office.

2: AGRICULTURAL CAPACITY

In today's hyperconnected world of bursting supermarkets, on-demand organic avocados, and rideshare food delivery, the concept of food security often seems . . . quaint. It is not. Throughout history most conflicts have had a food-production component. Alexander the Great planned his assault routes based on the location of agricultural lands and harvest times so his army wouldn't have to transport its own food. Rome salted the fields of Carthage to cause famine among the enemy. The Persians (and Greeks, Romans, Ottomans, French, and British) conquered Egypt to use it as a breadbasket to feed their imperial expansions. The fertile

---

* Fun fact: China's entrance into the world of global film started only in the 2000s, and nearly all the films included a not-so-sly piece of propaganda implying that it is better to suffer under the unified rule of a murderous tyrant than to have parts of China ruled by different folks who might be less psychotic.

lands of France made it far less vulnerable to political disruption than its peers.*

Anyone sufficiently arrogant to think the poor will simply starve in silence has a particularly weak grasp of not only biology, but history. Far more cultures and governments and dynasties and countries and empires have collapsed throughout history from famine and failures in food distribution than have been wiped out by war or disease or revolution or terrorism.

When it comes to issues of national survival and expansion, not all crops are created equal. Bedrock nutrition products provide calories and protein *and* can be stored for at least a year. The chief row crops from which most people get the majority of their calories—wheat, rice, corn, and soy—top the list. Also included are pulses and legumes and tubers—everything from potatoes and yams to lentils and chickpeas.†

What do these critical foods have in common? They grow best on flatlands in temperate climates to keep output-per-acre high and cost-per-output low.

These simple requirements eliminate well over three-quarters of the world's land area from serious contention as breadbaskets. Much of the best chunks of arable land in the world in 1900 were—far from coincidentally—also the seats of the major imperial systems. Food reliability enabled government continuity, which enabled outreach to and domination of less stable systems.

---

* Yes, that's right, the land of the guillotine is one of Europe's *more* stable states.
† The various animal proteins in contrast are *not* required for successful countries. While I'm not suggesting for a moment that if Americans were denied access to hamburgers and pepperoni there wouldn't be . . . political consequences, when discussing what makes a country successful, I'm primarily concerned with whether a people can have a sufficiently varied diet of both calories and protein to avoid malnutrition and/or starvation. A pound of meat consumes an order of magnitude more land, water, and/or grain than a corresponding pound of vegetable matter. In addition, animal products by definition must travel a great distance from ranch to plate, and the chain of custody has to be clean and efficient to avoid spoilage. In contrast, corn can literally be stored in an unguarded pile. A high-meat diet is less a requirement *for* national success and more a characteristic of a country that *already* has achieved a degree of success.

The major exception to this imperial characteristic was the United States, in large part because it took the Americans the entire nineteenth century to grow into the lands made vacant by the departing French, the genocidal Spanish and English, and the smallpox and syphilis that all three brought from Europe. But then the Americans caught up quickly. The American Midwest is the largest chunk of high-quality, temperate-zone, arable agriculture on the planet, giving it solid baseline production potential. The jet stream carries moisture west to east across the North American continent fairly reliably, and on the odd year that it fails, the Greater Midwest still has a solid chance of getting moisture from tropical systems that spin up from the south off the Gulf of Mexico. Toss in a hard winter reset that kills off bugs and replenishes soil nutrients, and no agricultural basin anywhere else is as reliable.

The Americans also boast more variety within their country than most continents: summer wheat from the cooler upper Midwest, winter wheat from the warmer lower Midwest, corn and soy from the core Midwest, cattle from Texas, tree fruits from interior Washington, citrus from Florida, cotton from the Piedmont, and a bit of everything from California's Central Valley. In any given year, the Americans typically grow about 35 percent more calories than they consume. If the Americans didn't eat so much meat, the ratio of exports to consumption would be in excess of two-to-one.

Whether the year is 1850, 1945, or 2020, America's agricultural position is unparalleled. It is the rest of the world that has changed.

That's a problem. It isn't so much that things have changed under the Order, but *why* things have changed. The Order's imposition granted places the world over the ability to do one of two things: import food from abroad without first seizing control of someone else's food-producing territory, and/or shrug off foreign control and import the various inputs required to produce food themselves. These options are largely responsible for the fourfold expansion in food cultivation since 1946 and thus also the tripling of the global population.

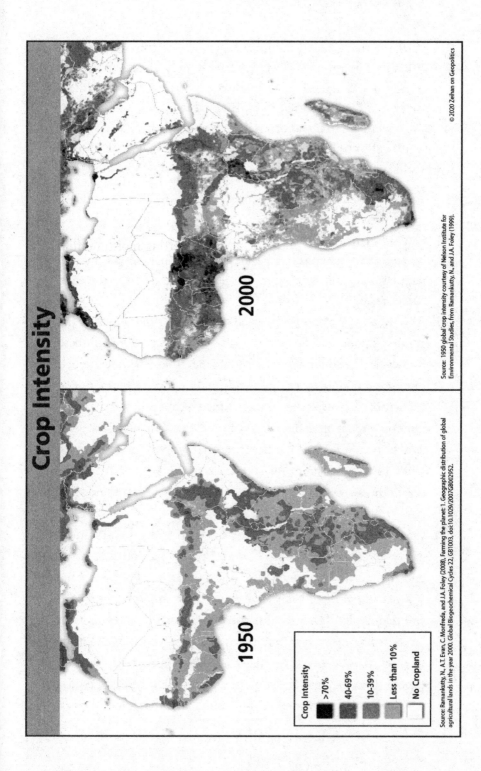

Crop Intensity

Crop Intensity
- >70%
- 40-69%
- 10-39%
- Less than 10%
- No Cropland

1950

2000

Source: Ramankutty, N., A.T. Evan, C. Monfreda, and J.A. Foley (2008). Farming the planet: 1. Geographic distribution of global agricultural lands in the year 2000. Global Biogeochemical Cycles 22, GB1003, doi:10.1029/2007GB002952.

Source: 1950 global crop intensity courtesy of Nelson Institute for Environmental Studies, from Ramankutty, N. and J.A. Foley (1999).

© 2020 Zeihan on Geopolitics

Without the Order's stabilization, much of this process will now happen in reverse. Fewer, shorter supply chains mean less demand for raw materials means less export income means less capacity to afford imported food. Break down the global supply-chain system for fuels and fertilizers, and much of the world's agricultural output becomes far less efficient. That's the clinical way of saying that without the Order, a billion people are going to starve.

China faces a quadruple bind:

1. China's margin of error *starts* razor thin, and not only because its lands are below subpar. One downside of China's massive population is that the country has less farmland per person than *Saudi Arabia*.

2. As China's population urbanized under the Order, much of the country's good(ish) farmland was paved over, pushing Chinese farmers farther inland into ever-more marginal territories, which require more and more inputs to produce the same amount of foodstuffs. Unsurprisingly, the Chinese economic sector that is most dependent upon the expansion-at-all-costs financial strategy is agriculture. When the Chinese financial system cracks, not only will China face a subprime-style crisis in every economic sector simultaneously, but it will also face famine—even if nothing goes wrong externally.

3. On the surface, it appears China has sufficient oil and natural gas production to maintain domestic production of its fertilizer and fuel needs for its agricultural sector. After all, while China is a net importer of both fossil fuels, unlike most of Europe it still retains significant local production capacity. Not so fast. In any sort of constrained import environment—such as problems in the explosion-heavy Middle East—the Chinese will have to choose what they will let go of. Electricity? Motor fuels? Fertilizers? There won't be enough to go around, and that forces choices.

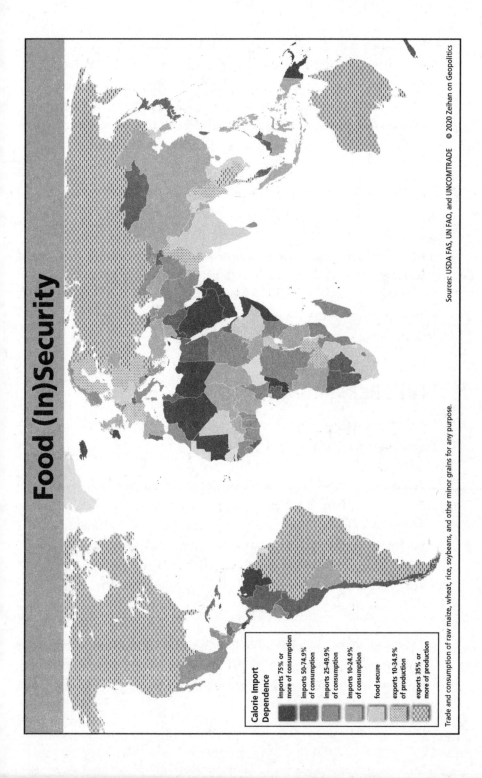

Food (In)Security

Calorie Import
Dependence

imports 75% or
more of consumption

imports 50-74.9%
of consumption

imports 25-49.9%
of consumption

imports 10-24.9%
of consumption

food secure

exports 10-34.9%
of production

exports 35% or
more of production

Trade and consumption of raw maize, wheat, rice, soybeans, and other minor grains for any purpose.

Sources: USDA FAS, UN FAO, and UNCOMTRADE   © 2020 Zeihan on Geopolitics

4. China now isn't simply the world's largest importer of rice, barley, dairy, beef, pork, fresh berries, and frozen fish by tonnage. It absorbs more globally traded sorghum, flax, and soy than the rest of the world combined. The ongoing import of those products requires both the American Order *and* the ability of the wider world to produce the products in the first place.

Unfortunately, it isn't as if the Chinese are alone in this dilemma. Roughly one-fifth of the world's agricultural produce is internationally traded, while four-fifths is dependent upon externally supplied petroleum-derived inputs, mainly fuel and pesticides. Remove the stability and shipping options and supply chains the Order makes possible, and many, many countries must figure out how to feed their populations with limited outside assistance. Most will fail.

3: DEMOGRAPHIC STRUCTURE

Let's start with a pair of facts:

- As of 1800, over 80 percent of Americans and Europeans lived on the farm. Now less than 2 percent of them do.
- The ratio of four children to three young adults to two mature adults to one revered elder has more or less held true since the dawn of human civilization: 4:3:2:1. But today many countries are closer to 1:2:2:1, signifying both rapid aging and shrinking of the population.

These shifts are a by-product of industrialization. The Industrial Revolution brought us concrete and steel and electricity and reliable food production and long-haul transport—the building blocks of *urbanization*.

That's changed the fundamental structure of how we live:

- On the farm, children are free labor, and everyone loves free things. In an urban apartment, children are luxury goods, and not everyone can afford more than one. If that.
- On the farm, the extended family works together, and a larger brood leads directly to greater economic and physical security. In the city, families scatter; large numbers are a boon only during holidays and on moving day.
- On the farm, we married young—typically during to just after high school—because agricultural work requires a strong back. In the city, we put off marriage until after university because a more technically inclined economy requires a more technically inclined workforce.
- On the farm, most technology of relevance relates to land, water, leaf, and hoof. In the city, technology is a world of its own—much of which is designed to help us live easier, longer, more leisurely lives.
- On the farm, we marry young, work young, and die young. In the city, we marry old, work old, and die old.*

Consider South Korea, the country that has seen the most dramatic security, economic, and demographic change in modern times. At the end of the Korean War, the South Koreans were one of the poorest, most rural populations the world. The subsequent American strategic overwatch combined with global economic access transformed their country to one of the world's most sophisticated technocracies. People surged to the cities, higher education became the norm, and the family structure

---

* My extended family is a case in point. On my mom's side of the family my farming grandparents had nine kids. Of the next generation, those who remained in agriculture had larger families than those who did not, but none had more than five children. My parents—townies—had only three. I live half on the road and only have two—cats.

collapsed from multiple generations toiling in rice paddies to tiny families cramming themselves into even tinier apartments. In 1955 a kimchi pot was the height of the South Korean technological experience; as of 2020, the Koreans sport complicated remotes for *toilets*. South Korea's birthrate dropped from among the world's highest to its lowest, radically overhauling the country's population structure.

And everything else. It all has to do with how people of different ages live.

Young workers are the big consumers in society: education, cars, homes, and feeding and clothing children. Mature workers are the big producers: with decades of work experience under their belts, they produce most of the goods, services, and taxes that make a system work. More young than mature workers? The country can consume most of what it produces. But more mature than young workers? Then the balance must be exported.

Korea's population structure inversion has deep implications for every aspect of the country's functioning.

At first, having no children is an unmitigated positive. The money that was once spent on childcare, diapers, Jell-O, and grade schools can be spent on cars, condos, pizza, and higher education. Consumption, efficiency, and output all boom. As the childless twenty-somethings pick up lots of technical work experience and age into fifty-somethings, productivity skyrockets. So long as they can export their output, they enjoy rapidly improving standards of living. The government finds itself swimming in tax revenues—funds that are typically applied to mid-career training and the raising of a truly technocratic infrastructure. South Korea enjoys the world's best Internet and cellular systems for good reason.

But the South Koreans are on borrowed time. In the 2020s, Korea's mature workers will retire en masse, and there is no replacement generation working its way up through the ranks. Upon retiring, mature workers shift from being massive suppli-

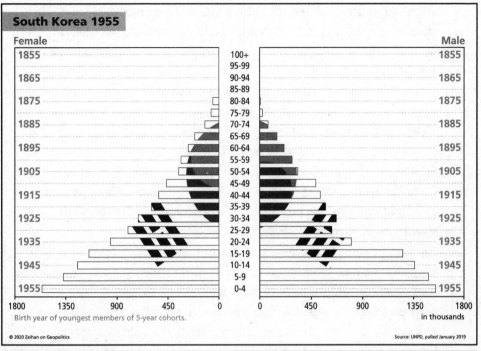

South Korea 1955

Female / Male

Birth year of youngest members of 5-year cohorts.

in thousands

© 2020 Zeihan on Geopolitics

Source: UNPD, pulled January 2019

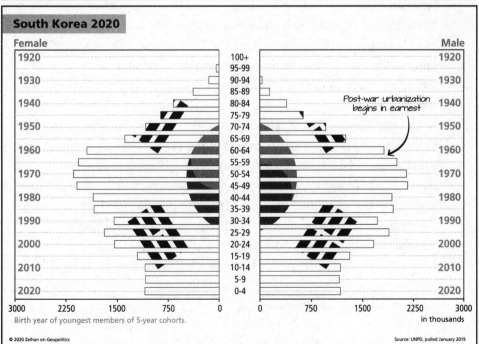

South Korea 2020

Female / Male

Post-war urbanization begins in earnest

Birth year of youngest members of 5-year cohorts.

in thousands

© 2020 Zeihan on Geopolitics

Source: UNPD, pulled January 2019

ers of capital to massive consumers of state spending via pensions and health care. Politically, geriatric populations tend to be . . . grumpy, adapting poorly to changing economic and social norms.*

The Koreans are hardly alone. The demographics are similar in nearly all the advanced countries, especially Russia, Ukraine, Belgium, Germany, Italy, and Japan (the farthest down this path, by far). In those places and more, demographic twilight before 2030 is both inevitable and imminent. But the aging of the "younger" countries has occurred even more quickly: Brazil, Iran, and Thailand are not far behind. All these and more are about to run out of mature workers, but they ran out of young workers forty and twenty years ago, and children a decade or two before that. No mature workers means no capital. No young workers means no consumption. No children means no future.

Even that dark outlook assumes that these countries last long enough to age into mass retirement. Many will not. Countries with an excess of mature workers relative to young consumers cannot possibly absorb all the products they produce. So long as the Order exists, no problem. End the Order, however, and their entire economic system comes crashing down. All are becoming *more* dependent upon the international system just as it is falling away.

On the flip side, there are but a handful of countries that have succeeded in keeping their birthrates high enough so this demographic crush is at least not imminent. The first cluster is a short list of First World states that managed to keep their birthrates higher until the 1980s. They still face the downward slide into decrepitude like much of the world, but instead of dealing with the issue during the next decade, their reckoning will not occur

---

* The grumpification of the United States' Baby Boomer generation is in part responsible for the election of Donald Trump.

until after 2035. These are the United Kingdom, the Netherlands, Sweden, and Switzerland.

The second cluster is countries whose Baby Boomer generation *had* kids, and so may face aging, but it will be a graceful process that can fairly easily be halted or even reversed with the right mix of circumstance and policy. France, New Zealand, and the United States top the rich world's rankings in this regard, while Argentina, Indonesia, Vietnam, and Mexico have—so far—bucked the trend in the developing world.

That leaves China, which has a demographic that is the worst of all possible worlds.

China—like nearly everyone else—experienced a baby boom after the Second World War. It's understandable; the Japanese went from having won the mainland war against China one day to being forced to retreat from all their gains by the victorious Americans the next. The Chinese got their homes back. People celebrated. Children were the natural outcome.

Twenty years later, the Chinese population had increased from just over a half billion to over 800 million. While Mao wasn't willing to allow economic reforms, he *was* willing to apply the power of the state to limit population growth. The result was a Two-Child Policy haphazardly enforced by the same gentle hands that brought about two of the most chilling episodes that you may have never heard of: the socioeconomic redesigns of the Cultural Revolution and the Great Leap Forward, the latter resulting in the greatest man-made famine in history.

After Mao's death in 1976, a group of technocrats did an audit on Mao's policies. Only two survived. The first was the Two-Child Policy, which was intensified into the more (in)famous One-Child Policy. The second was the diplomatic warming with the Americans. Nixon's trip was doubled down upon, and Deng eventually met with Jimmy Carter to formalize the normalization

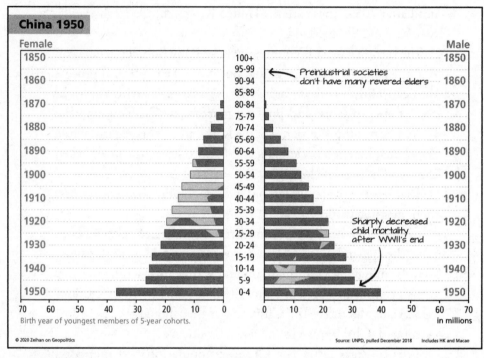

China 1950

Female   Male

Preindustrial societies don't have many revered elders

Sharply decreased child mortality after WWII's end

Birth year of youngest members of 5-year cohorts.

in millions

© 2020 Zeihan on Geopolitics   Source: UNPD, pulled December 2018   Includes HK and Macao

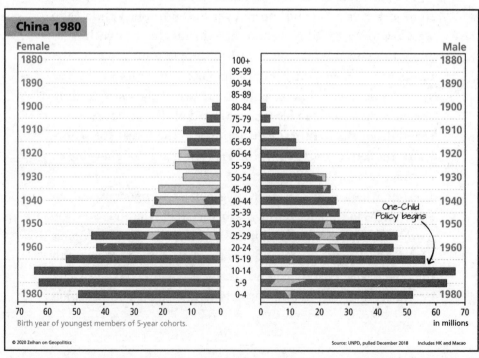

China 1980

Female   Male

One-Child Policy begins

Birth year of youngest members of 5-year cohorts.

in millions

© 2020 Zeihan on Geopolitics   Source: UNPD, pulled December 2018   Includes HK and Macao

of relations. China joined the Order. Which meant China started to industrialize and participate in global manufacturing. Which meant China started to . . . urbanize.

At the same time that the Chinese government was nearly panic-stricken about out-of-control poverty, unrelated trends were pushing the Chinese off the farm and into urban apartments. Such powerful forces shifted China from having one of the world's highest birthrates to one of its lowest.

In a way, this has helped China shine. Between 2010 and 2020, the Chinese demographic sported a bulge of citizens in their mid-twenties—people at the peak of their consumption experience. Everyone came to think of China as a boom state, but this was just a one-off. In the 2020s this cohort will be in their thirties and forties. A consumption reduction is baked in, as is increased dependency upon the American and global markets.

It's worse than it sounds.

All new births in China are now to the thin generation born under One-Child, whose median age increases from 37 in 2015 to 45 in 2040. In comparison, the United States ages pretty gracefully: from 37.6 to 40.6. As soon as 2030, China will have four pensioners for every two taxpayers for every one child. By 2050, one-third of the Chinese population will be over 60. Four decades of depressed birthrates have swapped the children who'd be the next generation of taxpayers and wage earners for a generation of pensioners who will never again pay into the system. China's 2010s boom was really just the beginning of the end.

Even this bleak forecast assumes no additional social disruption, of which more is already on deck. The Chinese preference for male offspring combined with One-Child resulted in widespread selective-sex abortion. A generation on, there is now—on average—a 10 percent imbalance between men and women born after 1979, the largest spread being in the urbanized coastal provinces where living space is at a premium.

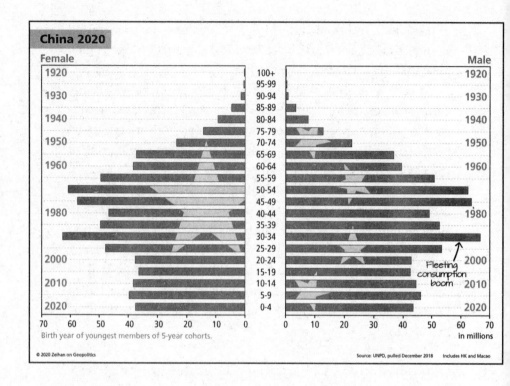

China 2020

Birth year of youngest members of 5-year cohorts.

© 2020 Zeihan on Geopolitics   Source: UNPD, pulled December 2018   Includes HK and Macao

Many make light of what it means for the Chinese population to have an extra 41 million men* under age forty who will never marry. The two most common concerns are how unmoored males might threaten social stability (legitimate) and how the Chinese government might be willing to throw a few million extra men into a military meat grinder just to get rid of them (also legitimate).

Let's shift that spotlight a bit and see how the imbalance intensifies the ongoing demographic crisis. It isn't pretty.

The Chinese preference for women in the light manufacturing industry resulted in mass female-only dorms in the coastal cities, while the men were left on the farm in the interior. After

---

* That's larger than the entire population of California.

thirty-plus years of the male birth preferencing, the socioeconomic roles of an entire generation have split along gender lines. Women are now more likely to live in cramped, coastal, urban quarters where they hold upwardly mobile white-collar or light manufacturing jobs their entire careers, while men are more likely either to live *in different provinces* in the poor interior or to work as undocumented migrants in China's economic underbelly. Interactions among the two increasingly economically, socially, and physically separated groups are now so rare as to threaten family formation. Chinese birthrates are now so low, they resemble those of countries that urbanized a *century* before them.

Immigration is often mooted as one solution for any demographic decline, but such a path only works in very specific circumstances.

First, the host culture must not culturally link the terms "citizen" and "ethnicity." In this the Anglo settler societies are best suited—they *began* their existence as various blends of English, Scots, Irish, Dutch, and French, with Germans, Poles, Greeks, and Italians following in successive waves as local definitions of what "white" meant expanded. Their preexisting experience with and tolerance for different foods, religions, and dress has eased the absorption of large numbers of Asians into Australia, Hispanics into the United States, and pretty much everyone into Canada. Spain has also managed this surprisingly well, but it did so because the immigrants it imported were in effect wayward children: the descendants of Spanish settlers who have "returned home" from the once far-flung Spanish Empire.

Second, the demographic decline being targeted cannot be rapid. Otherwise the simple math of in-migration will overwhelm the society.

An example: In 2015 Germany absorbed 890,000 refugees, the majority from Syria. The reverberations of this one-off event have shaped German internal politics ever since. If the goal was a demographic rebalancing, it was an utter failure.*

The issue is scale. *If* all of those refugees were young families with an average age of under 25, and *if* those immigrants maintained birthrates triple that of German ethnics, and *if* the Germans could import a similar number of immigrants for another *fifty* years, *then* Germany would return to its 1960s demographic health. But there won't be many ethnic Germans left. Immigration can help sand down some rough edges, but it is no panacea.

China's immigrant "needs" are an order of magnitude larger than Germany's, and of the countries near China, only one—India—has the size and age structure to even theoretically meet China's needs. Most would call the transfer of over 150 million young people to another country not immigration, but another word beginning with *i*.

4: ENERGY ACCESS

In the bad ol' days, sustainable civilizations were possible only in places where local food supplies were abundant and waterborne transport was possible. Everyone else was left to a meager, muscle-driven existence. The new industrial technologies of the steam engine and coal and fertilizers and steel drastically expanded agricultural output while also enabling food to be shipped across great distances. Such advances didn't so much eliminate the tyranny of geography as establish a new tyranny; the fuels that make long-haul transport feasible are not evenly distributed about the globe.

Maybe you've noticed: the process of getting oil from where it is produced to where it is consumed often gets a little . . .

* In more ways than one. Best guess is 85 percent of the inbound refugees were male. No boost to family formation there.

angsty. Only about 10 percent of the world's population is lucky enough to live within a thousand miles of the wells and mines that provide them with their energy. Now that the majority of the world's easily accessible oil and natural gas production comes from only two places—Siberia and the Persian Gulf—the angst is redoubled.

There are not a large number of countries where this works out well. Without energy there is no industrial lifestyle; everything from cars to cabbages to cell phones to condoms requires the petroleum economy. Europe imports some 90 percent of its oil and natural gas from, in decreasing importance, Russia, the Persian Gulf, and North Africa. East Asia imports a similar proportion of its needs, upwards of three-quarters of it coming from the Persian Gulf. Southeast Asia and South America—luckily—are in rough balance between imports and exports, although within those regions the petroleum is not distributed evenly. For the most part, the oil is not where the people are, and it is only the Order that has enabled the oil to reach the people with a minimum of fuss. It is no surprise that most of the major wars in the nineteenth and early twentieth centuries had an oil component to them.

There *are* a few ways to worm out of this problem.

The first has to do with the ugly sister of the fossil fuel family. Coal is not only the most easily produced of fossil fuels, but it is also the most common. Notably huge economically accessible coal reserves remain throughout the former Soviet Union, Western Europe, the Indian subcontinent, Australia, and the Americas.

Coal-rich countries can do more than burn the stuff to keep the lights on.

First, as the South Africans, North Koreans, and Germans showed us under Apartheid, Juche, and Nazism, if the price of oil is too high, you can turn coal into an acceptable substitute for nearly any standard petroleum product—including gasoline. It

Oil and Natural Gas Production Concentrations

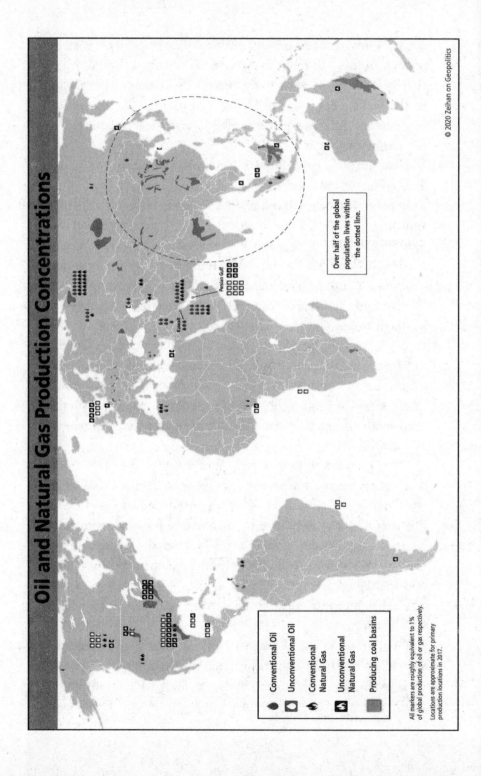

Persian Gulf

Kuwait

Over half of the global population lives within the dotted line.

Conventional Oil

Unconventional Oil

Conventional Natural Gas

Unconventional Natural Gas

Producing coal basins

All markers are roughly equivalent to 1% of global production of oil or gas respectively.

Locations are approximate for primary production locations in 2017.

© 2020 Zeihan on Geopolitics

isn't pretty, for all the normal environmental concerns about coal remain, and at about four dollars a gallon for just the processing cost to generate gasoline, it isn't cheap, but that doesn't mean it cannot work.

Second, not all oil and natural gas sources are "conventional." Instead of existing in large, easy-to-tap deposits, some are trapped in trillions of tiny specks within rock strata. Extracting it is difficult, time-consuming, technologically exacting, expensive work, yet the Americans have figured out how to do it more easily, quickly, and cheaply than the Russians and Saudis can produce conventional oil energy. American oil production has more than tripled, and as this book is being published in early 2020, the Americans have become a net exporter of crude oil, courtesy of their shale boom. They were already the world's largest exporter of almost every other conceivable energy product.

Third, technology now exists to wring energy out of something other than fossil fuels. Unfortunately, just as geography affects where you can put a city or grow crops, it also determines where you can generate wind or solar power.

The problems are not the technical aspects of installation, but instead a mix of climate and geography and transmission and, above all, density. Industrial and digital technologies concentrate people into urban environments. Services—whether private or governmental—work best in similar circumstances. Consider education: a single facility with a thousand students is far more efficient than a hundred schools with only ten each. Greentech likewise needs economies of scale: huge utility-level facilities in calm, sunny deserts or windswept plains—zones not well-known for hosting large populations.

It's worse than it sounds: using greentech requires continuity in the urban location, continuity in the generating location, and continuity across the territory that connects them. The circle can only be squared in very few portions of the world.

Understanding where greentech can work requires first factoring out all the world's territory that is not solar-ideal nor wind-ideal, or is more than a thousand miles from a population center.* Then it requires thinning the list further to eliminate places where greentech power generation is inappropriate. For example, you cannot site solar panels atop farm fields because crops need the light, nor wind turbines above about ten thousand feet in elevation because the air is too thin to spin the blades.

That still leaves several significant spots where wind and/or solar look fairly promising. The United States comes in a hard first, largely because it is the First World country closest to the equator and because the Great Plains are about as windswept as you can get. Argentina, Australia, and South Africa also look pretty good on both counts. Mexico can do solar, while the United Kingdom and Denmark can do wind.

But that's the end of the good news. For the bulk of the other 85 percent of the Earth's land area, greentech isn't (yet) productive enough to be all that useful at displacing fossil fuels at scale. The *really* bad news is in South, Southeast, and East Asia—home to half the human population. A combination of local factors largely gut alternative generation capacity, in no place more so than China. Persistent humidity-driven haze and rugged terrain and/or densely farmed landscapes eliminate solar power from China's coastal zones despite their proximity to the equator. Low air pressure due to high elevation, combined with inclement weather, eliminates wind power from Tibet. And what potential the Chinese have requires high-voltage wires stretching two *thousand* miles to reach the populated coast.

All those greentech investments in China you've heard about? All that talk about China being the world's green leader? Most of those installed solar panels and wind turbines are there only due to the same overinvested, highly leveraged, expansion-at-all-

---

* In the United States, five hundred miles of transmission is typically considered the absolute ceiling. At that point cost of the transmission alone is roughly equal to the all-in cost of locally generated coal power.

Global Wind + Solar Potential
within 1,000 miles of a major city

Over half of the
global population
lives within the circle.

Global Wind Potential
■ Ideal: 6.6+
Units: meters per second at 80 meters

Global Solar Potential
■ Ideal: >2,500
Units: kWh per year per square meter

■ Ideal solar and wind
□ Unsuitable

● Population ≥1m

Sources: NREL, DOE, IRENA, and EPA

© 2020 Zeihan on Geopolitics

costs development model that has made the entire Chinese economy a grotesque approximation of Enron in nation-state form.
Greentech is no solution to China's energy problems—and problems they most certainly are.

Between the shale revolution increasing American energy
output and the ongoing efficiency gains reducing American
energy demand, the Americans are no longer the world's top
importer of crude oil. They haven't been since 2014. That title
now falls to China, which imports more crude than Germany,
France, Italy, the United Kingdom, Spain, Poland, the Netherlands, and Turkey combined. There was once a faint expectation
that the Chinese might copy the Americans' experience in the
shale revolution, until they discovered that the only one of their
shale deposits that might be economically viable is in Sichuan—
and the last thing Beijing wanted to do was give the sometimes
secessionist-minded Sichuan a leg up.*

China's external options are, in a word, subpar. It all comes
down to relative dependencies.

Oil by itself isn't all that useful. It must be refined into a series
of daughter products, which are in turn processed into things like
gasoline and tires and nylon and fire extinguishers. Those processes require an army of gargantuan refineries and petrochemical
plants, and while these facilities can achieve some wondrous
feats of chemical engineering, what they cannot do is *move*. Such
facilities are placed where the oil is—primarily near a port or
along a pipeline route.

China's primary pipe import source is Russia, a country that,
since the Cold War's end, has repeatedly interrupted shipments
to consumers to achieve (geo)political goals. If the Russians ultimately choose not to play the energy card against China, China

---

* It isn't just the Chinese who have proven unable to replicate the American shale
experience. At present the magic mix of preexisting infrastructure, private property
rights, ample finance, deposits' proximity to population centers, and a large skilled
labor pool does not exist anywhere else.

would be the *only* Russian customer not to feel the sting of that particular whip.

Then there's an annoying little geographic detail. Russia and China are neighbors, but being next to each other isn't the same thing as being close. The vast bulk of Russia's oil and natural gas deposits lie in northwest Siberia—largely virgin territory that's a startlingly inconvenient five thousand miles from the densely populated Chinese coast. Most of the route at present lacks even a goat path, but let's assume for the moment that a railroad or pipeline exists linking Russia's Siberian fields to eastern China. The operating cost *alone* for transporting the crude would run at least thirty dollars a barrel. Even *that* assumes total security. Any pipe linking northwest Siberia to eastern China would be impossible to defend. A single bomb or missile could easily puncture the line, putting the entire endeavor on the wrong side of pointless. Needless to say, no one has fast-tracked large-scale Russian-Chinese pipeline construction.

If you've been wondering why navies keep coming up over and over, in part it's because over four-fifths of the world's internationally traded crude oil is waterborne. Under the Order, should this or that supplier get in a snit, it isn't all that complicated for an importer to source imports from a different party. Countries that import their crude by water have most of their energy-processing facilities in ports, so switching supplier partners doesn't require starting over.

But that's where the good news ends. The security of maritime supply is still subject to geography, so a country's ability to import crude via waterborne methods is wholly dependent upon its ability to maintain maritime security throughout *all* the various geographies between its crude source and its ports.

Americans are only aware of this to a degree. Whenever something would boil over in the Persian Gulf region between 1973 and 2010, oil prices would tend to jump up and so the Persian Gulf is a region Americans closely associate with energy politics. But Americans rarely consider the rest of the world's energy map.

Even at the height of their imports, between 2005 and 2007, the Americans received twice the volume of crude from Canada, Mexico, and Venezuela than it did from all Persian Gulf countries combined. Such imports faced exactly zero maritime threats. Compare that to the small sliver of US imports to the East Coast from the Persian Gulf.

The route starts by braving the drama of the Persian Gulf with all its normal rivalries: Kuwait-Iraq, Iraq-Iran, Iran-Saudi, Saudi-Qatar, and so on. The Persian Gulf squabbles manifest most intensely at the Gulf's opening, the Strait of Hormuz, through which almost all the oil must flow and where Iran has often threatened to sink all shipping.*

Upon leaving Hormuz, cargos take a sharp turn to the south and hug the coast of Oman before entering the Indian Ocean proper and sailing southwest for the east African Coast. At that point, they are more or less free and clear. Most of the route to the United States' East and Gulf Coasts transits deep ocean (the Indian and Atlantic), passes by countries with no desire to disrupt shipments (Madagascar, South Africa, and the Dominican Republic), or is so close to the United States that American military power can almost lazily prevent any untoward actions (Cuba). Not necessarily a lot of fun, but post-Hormuz sailing is not exactly problematic.

Such isn't the case at all for oil going to China. The Americans maintained the global Order, and part and parcel of that was making sure oil could flow from the Persian Gulf to the wider world to power global trade. China's success is the quintessential example of how that works. China—as the world's largest importer

---

* It is impossible to know which cargos of crude are going where, and oftentimes crude cargos are sold—sometimes multiple times—while still on the water. So if the goal is to cause havoc for any particular oil importer, Iran has no choice but to cause havoc for them all.

of *every* energy product—has a vested interest in keeping Persian Gulf oil flowing. What it lacks is the capacity to guarantee what are the world's (read: its own) most vulnerable import routes.

For China-bound cargos, clearing Hormuz means the fun is only getting started.

Rather than head southwest to Africa, China-bound cargos must first continue along the Iranian sphere of influence for another five hundred miles. Then they reach the relative safety of . . . Pakistan. The problem isn't Pakistan per se—formally, China and Pakistan are partners—but rather India. India and China maintain an unhealthy rivalry over who is the true Asian superpower, while India and Pakistan have fought four wars since the end of the colonial era in the late 1940s, and they point a few dozen nukes at each other. Oil shipments to China are dependent upon (a) the Indians and Pakistanis not shooting at each other, (b) India not feeling the need to nick a few cargos from the river of ships sailing by Indian shores, and (c) Indians never feeling the need to deliberately shut down China's energy supplies. Such concerns last for about two thousand miles, until the ships pass the Andaman and Nicobar Islands (an Indian military zone) in the far eastern reaches of the Bay of Bengal.

There a new problem arises. Instead of the relatively wide-open seas around India, in Southeast Asia the islands get bigger while the seaways narrow. In short order, ships sail by Indonesia, Malaysia, the Philippines, and Vietnam—four countries the Chinese have attempted to coerce and intimidate into accepting Chinese control of the South China Sea. If a bunch of Somalis with speedboats can capture a supertanker on the open seas, imagine what countries with actual navies and missiles and jets could do when large, slow-moving China-bound tankers bearing oh-so-delicious crude oil pass within sight of their lands.

Ships have now made it to extreme southern China, but that is not where most Chinese live. If the tanker's destination is Fujian or places north, the ships must pass by Taiwan—a country China

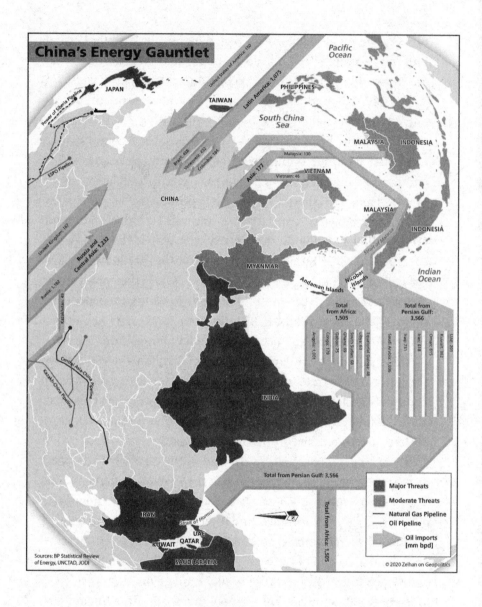

considers a wayward province and habitually, almost reflexively, threatens with invasion. Only then can tankers reach Shanghai and northern China—at which point they are sailing within the national security envelope of Japan.

IT'S THE END OF THE WORLD AS WE KNOW IT . . .

Most of the world's territories don't meet this list of requirements. Too rugged. Too flat. Too cold. Too hot. Too dry. Too wet. Too variable. Too vulnerable. Therefore, throughout most of human history, most of the world's territories were not home to "countries" in the way we define the term today. Blips of civilization in specific spots held enough economic potential and security wherewithal to generate intensely local continuities. Such spots were little more than city-states sporting thin trade links—and thinner political and security links—among one another.

Now don't get too depressed. While a country with high marks in all four categories—geography, agriculture, demography, and energy—is certainly going to have a breezier time of things than those that fall short, these are less prerequisites and more strong recommendations. Russia's geography has always been vulnerable, yet it has existed in one form or another since shortly after the Mongol retreat because most of its neighbors are in a worse position. India's agricultural capacity is so poor that much of the population lives at the subsistence level, yet between its bulk and its relative isolation, it is undeniably a durable country of significance. Singapore is the size of a postage stamp and has only like twelve people, yet its position as the linchpin on the world's greatest trade route grants it global relevance. Context is king.

But China fails on *all* counts. Allow me to detail the full unfurling fucking disaster.

The core problem is that it was not the Chinese who created the environment that made all of contemporary China possible. That was the Americans.

By putting the imperial era firmly in the grave, the Americans eliminated the powers who had been preying upon China for centuries. In granting freedom of the seas to all, the Americans gave the Chinese access to external finance, technology, resources, and markets. By giving the Chinese control over their own coast, the Americans enabled Mao to force Shanghai, Hong Kong, and all the coastal cities in between into domestically driven supply-chain systems rather than leaving them free to link to the outside world. In gaining access to the American market, a capital-starved, technology-starved, calorie-starved country was able to get all the inputs required to develop, *as well as* have a place to send its manufactured goods. In gaining access to the global market, the Chinese could import bottomless supplies of any resource they needed, oil being the most critical of all.

Strategically, the First Island Chain is the real Great Wall, only it boxes China in rather than protecting it. Demographically, China is in a state of not-so-slow-motion collapse. China's import-sourced, export-oriented economic model does not work without outside support and absolute security of the seas, yet China does not control end markets, resource sourcing, or even approaches to its ports. China cannot feed itself without imported material, either foodstuffs directly or the inputs required for China to grow and raise its own. China's internal politics and financial system are borderline explosive even in the best of times.

Much of the "China story" of the past generation is simply the economies of scale uniting China's long-fractured political geography into a single economic space: three millennia of pent-up demand combined with a long-overdue technological update applied to the ultimate greenfield environment. The ultimate economies of scale combined with the ultimate—artificial—continuity.

However, lose that political unity, lose those economies of scale, lose that continuity, and it all goes into screaming reverse. Most worrying to Beijing is the furious understanding that there are *parts* of China that *can* be successful statelets: the Pearl River Delta, Shanghai, Sichuan—maybe even Tibet or Xinjiang.

If the almost magical confluence of factors that enabled China's rise shifts out of alignment, China will suffer a cataclysmic flameout every bit as impressive as its rise to power. And since those factors were always and still remain beyond China's control, the question isn't *if*, but *when*.

Under the Order, the Chinese have kicked some serious ass, experiencing some of the fastest economic and military growth in history. In 1996, 42 percent of Chinese lived in extreme poverty. Two decades later, the figure was 0.7 percent. That's shockingly impressive. But before the Order, the Chinese were riven by internal differences, as they had always been—set upon by outsiders. Poor. Starving.

The future of China is not the utopian myth of Chinese domination. The future of China isn't even the somewhat blasé expectation of inevitable regional dominance. China is powerless to defend or maintain or replace the Order upon which its economic existence and political cohesion is predicated. The future of China is that of a people literally fighting to the death to continue to exist as a unified country at all. And even *that* somewhat dark possibility assumes that all else is equal in their broader environment.

It is not.

CHINA'S REPORT CARD

BORDERS: Vast emptiness to the west, jungles to the south, nuclear powers to the north and southwest, and superior maritime powers to the east. China doesn't so much secure its borders as manage them the best it can.

RESOURCES: China didn't get really serious about industrializing until the 1970s, so its local resources were all tapped more or less at once. That served China well . . . until now. China is on the verge of running out—of everything.

DEMOGRAPHY: Oy! Breakneck urbanization combined with Maoist population controls gutted the Chinese birthrate for decades. The one bright spot is that China's demographics are not the worst in the world. Yet.

MILITARY MIGHT: China is BIG and its military is modernizing quickly, but that doesn't mean its military is well suited to the challenges of today. Or tomorrow.

ECONOMY: The Chinese system is both highly leveraged and highly dependent upon international trends it cannot shape or preserve. Every system that has followed China's path has crashed. So too will China.

OUTLOOK: Only Russia has worse relations with its neighbors. When the Order ends, everything that has made China successful will end with it and *no* one will reach out with a helping hand.

IN A WORD: Overhyped.

CHAPTER 5

JAPAN

LATE BLOOMER

Societies and cultures are products of their time and place. A community's relationship to its physical environment and the challenges or blessings that come with it play a significant role in shaping "national character." Populations that inhabit mountainous regions as diverse as Yemen, Chechnya, Nepal, Anatolia, and Appalachia exhibit a similar propensity toward clannish communal groupings and a certain pugnaciousness engendered by their isolated, resource-limited environment.

And then there's Japan, the ultimate land of contradictions.

Rugged and volcanic in origin, none of the Japanese archipelago is what you'd call a breadbasket. Ranging from subarctic Hokkaido in the north to subtropical Okinawa in the south, the territory of the nearly seven thousand islands that make up Japan is about the size of Montana, but of Japan's 127 million people, 125 million of them live on scraps of land that are collectively smaller than West Virginia, only one-sixth of the total. The medieval Japanese rarely knew security. Chunks of flat, arable, or otherwise usable land are few and far between. The smaller the plot of agricultural land, the greater the likelihood that some weather or political event results in a poor harvest and the people have to struggle with less.

Yet the same geography that fractured Japan politically protected it strategically.

To the east lies the vast expanse of the Pacific Ocean, and to the west are a handful of countries that for much of their history were too busy sorting things out on the mainland to conquer—or even pester—Japan.

Consequently, Japanese preindustrial history was a never-ending litany of infighting, with local warlords—the daimyos—willing to bleed the population and one another in the pursuit of their own interests. The Japanese emperor was little more than a figurehead with a dusting of religious connotations. Real authority rested with his military commander: the shogun. Unfortunately for the shogun and the emperor, "imperial" power rarely reached much beyond the tips of their troops' weapons. Rather than think of the shogun as all-powerful, it is more accurate to consider him the most powerful daimyo.

Because of Japan's isolation and hillbilly hostility, the European empires largely ignored it. They preferred larger, flatter, richer, more attractive colonial targets—think India and China. Distance played a role too: it took a year for a sail-powered ship to make the trip from Europe to the Indies. Not a lot of folks wanted to spend another half year getting to even more distant, resource-poor Japan—especially when Japan's fractured geography and hypermilitaristic culture made the Japanese pretty effective at fighting back. The only other would-be colony that fought back harder was New Zealand, where the natives tended to *eat* European colonists well into the Industrial Age.

The traders being uninterested, it fell to Europe's missionaries to lead the charge. From the Japanese point of view, European colonialism had the feel of a door-to-door Jehovah's Witnesses marketing effort—so the Japanese simply turned the porch light off. For a full two centuries, the Japanese had a Girl Scout–cookie relationship with Europe: contact was limited to a single port—Nagasaki—and a single external partner—the Netherlands—that could keep its moralizing to itself. Japanese who left the country without express approval from the shogun forfeited their lives should they dare return. Outside Nagasaki, Europeans were killed on sight.

Politically, this enabled the Japanese to focus their efforts on unification, bit by bit at their own pace. Culturally, if the early wave of deepwater imperial powers affected Japan at all, it was simply to underline to the Japanese at all levels that outsiders were . . . odd. Inferior. Strategically, the inward-focused obsession made the Japanese a maritime people without a navy.

THE RISE OF THE RISING SUN

From these myriad weaknesses the Japanese have found strength and thus risen into the ranks of history's great powers. It all has to do with the Japanese's solution to its geographic shortcomings: with land-based options so poor, the Japanese had no choice but to go to sea, but they did so in a manner wholly different from the English.

Internal barriers within England are on the thin side, so the English identity consolidated hand in hand with the rise of the English state. There was never any doubt that London would be the seat of English power, and since London sat on the Thames, the English became a land, river, and sea power more or less in parallel. The English navy was a natural outgrowth from this unity of experience.

Japan couldn't have been more different.

Medieval Japan was a mountain-riven land with no clear core. Flattish portions of land were too far apart to unify easily, so strategic bickering was the norm. The subsequent Japanese "navy" reflected the disunity, and sea-facing daimyos each fielded their own forces of militarized junks. No individual daimyo could boast a large enough naval force to sustain a trade route (much less a mainland Asian colony) while still defending his territory back home, so Japanese interaction with the wider world was far more adversarial and far less disciplined than that of other naval powers. Not so much fleets, as mobs on water. Less imperialism, more piracy.

And yet Japan as a nation was forged in and by this naval chaos. Japan's topography made roads so difficult and expensive to construct and maintain as to be irrelevant . . . but once you could float a boat, towns near and far were just a sail away. Such naval connections—both hostile and friendly—weren't just Japan's primary infrastructure, but also its political unification process. In 1800, roughly a millennium after the Japanese cultural emergence, all of coastal Japan finally was at least nominally under a single government, the Tokugawa Shogunate.

But before the Japanese could explore what that meant, the world rudely intruded into their affairs.

Concerned that imperial European chocolate was being put into their Asian peanut butter, the US dispatched Commodore Matthew Perry to Japan in 1853 to force negotiations about American influence into the region's last European-free chunk of territory. Because Perry was the guy who operationalized the Americans' first steam-driven naval vessels, he didn't feel he needed to be very polite about it. His small task force demonstrated its ability to easily overpower everything the Shogunate could float, with sufficient firepower in reserve to shower destruction upon Tokyo.

Whether the Japanese liked it or not, modernity had arrived—at gunpoint, no less. Shortly after Perry's visit, the British, French, Dutch, and Russians visited, demanding similar concessions. In a mere fifteen years, the sudden introduction of the outside trade and the industrial technologies that went with them shredded

Japanese social, political, and economic norms, which had yet to fully absorb the consequences of Japan's own unification.

The result was a fundamentally different kind of empire. In part, the difference was because the rationale was different.

Going abroad gave the Europeans opportunities to get bigger and better and badder so they could continue to duke it out in the European sandbox. But the European centers had become empires because their home geographies were pretty good to start with. Japan's land wasn't in that league.

All Japan had was a natural progression of its national unification process. Japan had to have a navy to unite its country, and it had to open up to the world to avoid being plowed under by the empires.

In part, the difference was about regional competition . . . or the lack thereof.

Europe may have always had a first power, but at any given time from 1500 on there were a dozen or so significant players who balanced one another. Not so in East Asia. As recently as 1800, Japan was the *only* local power that had any semblance of unity, and after Japan's forced opening to the world, it was the *only* local power with steamships and firearms. Unity enabled the country to take full advantage of the new industrial technologies, and Japan instantly became the dominant regional power.

In part, the difference was about *imperial* competition . . . or the lack thereof.

It's a long way from Europe to Japan. Once the Japanese mastered the making of cannons, the Europeans simply could not compete effectively so far from home. Even the Americans left. Not long after Perry invited himself into Japan with all the subtlety of a mafia protection salesman, the Confederate Army was bombarding Fort Sumter and Yankee attentions turned elsewhere.

In part, the difference was about the speed of Japan's naval rise.

Floating a navy is difficult and expensive. Half the cost is the difficulty and expense relating to the training of the sailors. Pre-unification Japan's shattered political map under the daimyo system meant that collectively the Japanese maintained a far larger pool of ships and sailors than a unified nation would. When Japan flash-industrialized, it had the highest proportion of sailors of *any* country in the world to draw upon.

Less than twenty years after the Perry expedition, Japan had upgraded from junks to steam-powered destroyers. In 1894–95, Japan easily trounced the Chinese up and down the East Asian coast in the Sino-Japanese War. In 1904–5, Japan conquered all of Korea while also sinking the entirety of both Russian fleets in the Russo-Japanese War.

In part, the difference was about culture.

It is easy to believe in your cultural superiority when you defeat a people who are a half millennium behind you technologically (the Chinese). It is very difficult *not* to believe in your cultural superiority when you defeat a people who supposedly are your technological superiors (the Russians).

In part, the difference was about the speed of Japan's technological growth.

Industrialization is traumatic. Any large-scale shift in a society's technological suite upends old patterns—in employment, labor relations, social hierarchies, and governance. Few countries manage the process gracefully. Most have the odd riot. Some face revolutions. For a point of reference, the whole Karl Marx and world wars thing was part and parcel of the German industrialization experience.

Japan was different. Not only were the Japanese already on the path of wrecking their old social order when they started absorbing the industrial and deepwater technologies, but their societal organization proved capable of a rapid and (relatively) low-stress mastering of those technologies.

It all has to do with Japan's rugged geography. Preindustrial manufactures required highly skilled craftsmen—not just anyone can forge a samurai sword. In contrast, early industrial manufactures required no more than a bunch of bodies to run an assembly line. As the early industrial technologies reached Japan, things like mass-manufactured steel tools and chemical fertilizers started popping up throughout Japan's rice paddies.

For a while, the process was just like any other industrializing system. The new inputs pushed farm yields up, pushing down the number of people required to work the land. Former farmers migrated to Japan's industrializing cities to work in the new factories. But that's where the similarity between "normal" industrialization and the Japanese experience ended.

Japan doesn't have any big swathes of flatland. Japanese cities are crammed onto tiny footprints. They can neither easily expand nor easily integrate to achieve economies of scale. If they want to get bigger, they most go up, not out. High development costs forced Japan to move up the value-added scale as quickly as possible. It wasn't enough for the country to import and use the new technologies; its cities were too crammed to be competitive with the lower capital costs of other centers in the Industrial Age. Japan had to not only master the technologies, but also advance them.

Politically and culturally, the general population got swept up in the same modernizing, industrializing, nationalistic mind-set that had overtaken Japan's new, modernizing elite and their corporate expressions, the new *zaibatsu* ("money-cliques"). Strategically and militarily, Japan's newfound and rapidly advancing technical prowess combined with its appreciation for the geography of long-range naval warfare pushed Japanese engineers to construct the world's longest-range, hardest-hitting ships. Japan floated its first fully indigenous steel battleships in the mid-1890s and its first aircraft carrier in 1922.

But mostly, the Japanese Empire developed differently because the interplay of industrialization and imperialism put them on a treadmill.

No matter how a country industrializes, there's a list of non-negotiable inputs: labor for the factories, iron ore for steel smelting, and coal and oil to power the process. Of that list, Japan had only labor. Applying outside technology required that Japan venture out to secure industrial inputs. Modernizing and industrializing in an era without free trade demanded Japan become an empire.

From the day Perry arrived, Japan was condemned to transition from the world's most advanced medieval country to the most advanced, period.

Japan's first stop was the island of Formosa, a largeish island just to the south of the Japanese archipelago and home to contemporary Taiwan. Though it was nominally under Chinese rule, the Japanese had little difficulty dispatching its defending forces in 1895. Japanese imperial forces now controlled the northern half of the First Island Chain as well as a military platform nearly within sight of the Chinese coast. Unlike the occasional raiding and pirating by Japanese naval forces during medieval times, now the Japanese could make their visits to the Chinese mainland last.

Next up: the Korean Peninsula in 1905. Korea's rugged internal geography mirrored Japan's and produced an early Shogunate-like political structure as well. Industrialized Japan faced few issues subjugating the politically fractured, preindustrial Koreans. Attention turned to Manchuria in 1931, a Chinese region replete with fertile farmland, coal, and minerals—nearly everything Japan lacked. With these new resources and their preexisting military presence in Formosa, the Japanese could easily project power up and down the entire Chinese coast.

In World War II's early days, imperial armies surged from Manchuria to every part of the northern Chinese core, reaching all the way to the Yangtze itself. Often launching from Taiwan, marine landings secured control of Shanghai, Wenzhou, Fuzhou, Xiamen, the Pearl River Delta, and Hainan Island. All the former European treaty ports in coastal China concessions, except Macao and Hong Kong, were now Japanese imperial territories. Less than two months after the fall of Paris to German forces, the Japanese seized total control over French Indochina because, well, no one else was using it.

There was but one fly in the emperor's ointment. The Americans occupied a choice piece of territory smack in the middle of it all: the Philippines. From that position in the middle of the First Island Chain, Americans could theoretically threaten everything the Japanese had and wanted. It didn't help that prewar American policy was something Washington called Open Door. Officially, the policy was designed to limit European predation of China, at that point a thriving industry over a century old. Unofficially, the goal was to muscle the US of A in on the action. Unofficially and *very* quietly, the intent was to box Japan out of the region completely.* Japan is best known in the American mind for the attack on Pearl Harbor, but ultimately the Battle of Pearl Harbor occurred only because the Japanese needed the Americans ejected from their Philippine foothold in the East Asian Rim.

Japan needed to—and did—clear the board.

In under six months, Japan had conquered nearly all European holdings in Southeast Asia, most notably the territories that today comprise Indonesia, Singapore, Malaysia, Papua New Guinea, and Myanmar. Collectively these lands provided the Japanese with everything they could need, from sugar to metals to oil. The empire may have been a bit gangly, but if there was one thing the Japanese knew how to do, it was how to manage an archipelago.

Less than a century after Commodore Perry's threatening of a "backward" nation, Japanese forces in World War II stretched from the Aleutians to the edge of India. Its navy vied with the Americans for control of the Pacific Ocean. It all occurred against

---

* The diplomatic jokes of the era were that the Japanese policy of Asia for the Asians really meant Asia for the Japanese, while the American Open Door was really about a Closed Door for Japan.

the cultural backdrop that allowed for events as horrific as the Rape of Nanking, the impressment of Korean "comfort" women, and the Bataan Death March.

It was a pattern that did far more than give the Americans pause. Assessing Japan's rapid technological improvements, lightning military advances, apparent lack of moral center, and the logistical restraints of maritime warfare a Pacific Ocean away from home ports, the Americans chose not to do battle with Japan's armies at all. Rather than duke it out island by island, the Americans seized only sufficient islands so that their naval and air power could wreck the shipping routes upon which Imperial Japan depended. Then, with the Japanese economy and military complex on its knees, the Americans declined ground combat one last time, opting instead for nuclear obliteration.

JAPAN UNDER THE ORDER

America's defeat of Japan in 1945 wasn't simply the Japanese losing a war; it should have been the end of Japan altogether. In many ways, industrialization is a straitjacket. The suite of industrial technologies improves literacy and mobility and reach and wealth and health, but without the inputs of oil, natural gas, iron ore, phosphates, bauxite, lead, copper, and so on, the whole process collapses in upon itself.

None of these are found in any appreciable volumes within Japan itself. Japan acquired all these materials via its imperial expansion, made possible largely by the Japanese navy, which under the terms of the unconditional surrender was to cease all operations and return home. Permanently.

No navy, no empire. No empire, no resources. No resources, no industry. No industry, and Japan returns to preindustrial standards of living, complete with famine, disease, and centralized political breakdown. There is good reason Japan had to be nuked to be forced into surrender. The Japanese knew full well that military defeat meant the end of Japan as a country. There could be no middle ground between a Greater Japan that was industrialized and the fractured nonentity of the Shogunates.

But the Americans surprised them, in large part because the Americans *needed* their defeated Pacific foe. The Order's core rationale was for America's new allies to stand between the Soviet Union and the United States—and to do so willingly. The United States achieved this by imposing security globally, crafting an international economic system, and granting unilateral access to the American market. In one fell swoop, the Americans provided the Japanese with everything Japan had fought for—and ultimately lost—between 1870 and 1945. A position under the American nuclear umbrella was tossed in as a cringe-inducing bonus.

Japan wasn't so much dismantled and rebuilt as upgraded.

Japanese factories that had made weapons were reconfigured to make sewing machines and household goods. Optical device companies began making cameras instead of gunsights. Heavy industries switched from tanks and planes to automobiles. Aside from a pair of newcomers—Honda and Sony—the who's-who on the list of most powerful Japanese firms—Hitachi, Toshiba, Mitsui, Mitsubishi—were the same names that had dominated the Japanese system before the war.

The term "miracle" to describe Japan's postwar boom is a misnomer. It was as highly planned, tightly regulated, and deliberate as every step of Japan's evolution since 1852, and the hoped-for outcome was the fruition of Japan's considerable domestic ambitions backed by the full force of the American economic, political, and military system. In a single generation, Japan recovered from the destruction and despair of its World War II defeat to become the second-largest economy in the world.

That achievement was notable from any angle: the unexpected preservation of the Japanese way of life, the ongoing success of the Japanese technocratic experience, the anchoring of Japan in the American alliance structure, and the prevention of a large-scale Soviet expansion in the Pacific theater. But in becoming so big so fast, Japan may well have been the first country to make the Americans second-guess the Order's very existence.

One of Japanese leaders' favorite Order-era tools to maximize their economic strength was currency manipulation. The central bank would print lots of yen and use them to buy dollars on international markets, driving the yen down in value versus the dollar, making Japanese goods relatively cheaper, and thus encouraging Americans to purchase them. Japanese exports to the United States—and from them the Japanese employment, economic, and political systems—would enjoy a nice little kick.

The Reagan administration didn't like that sort of behavior and in September 1985 called the Japanese along with other serial currency manipulators—most notably the French and Germans—to New York's Plaza Hotel to deliver an ultimatum: either cease and desist or risk the Americans abandoning the Order wholesale. After all, Mikhail Gorbachev had been at the Soviet helm for six months, and it was clear some sort of American-Soviet rapprochement was underway. If the Cold War was getting friendly, so the logic in the White House went, then there was no reason to give the allies as much economic room to maneuver as the Order typically granted. No harm in strong-arming the currency manipulators a bit. It worked. The Japanese—and others—shifted course, and within six months the US dollar had dropped by half.

The issue was a simple one: Japan had done very well for itself under the Order, capturing all the benefits of empire without any of the costs. That was all well and good. That was the plan. What was not the plan was that, by the mid-1980s, the average Japanese citizen became better off in terms of income than the average American. Subsidizing an ally to make them strong against a common enemy makes sense. Subsidizing an ally to the point that the ally becomes wealthier than you makes somewhat less. The Japan question quickly boiled up into a headline American political issue, featuring loudly in the 1988 presidential election among both the Republican and Democratic candidates, as well as attracting the ire of a New York property mogul by the name of Donald Trump. It was the first true crack in the edifice that was the global Order.*

Anything that spooks the Americans is typically not all that good for the spooker. Anything the Americans are paying for will always be subject to harsh evaluation. Anything that spooks the Americans that they also pay for is on borrowed time.

Three decades after the Plaza, Japan's time has run out.

SUPERPOWER, REBOOTED

Contemporary Japan faces a thorny series of difficulties, most of which are disturbingly similar to Japan's late-nineteenth-century concerns. Luckily (for Japan), just as the Japanese were able to massage several of their preindustrial problems into strengths, the same logic holds true for the Japan of today.

The first issue is the looming iceberg of Japan's demographic implosion. The niggardly amount of flatland in Japan that has so shaped the country's political, agricultural, industrial, and technological history has similarly shaped Japan's demographic structure.

Once one filters out countries that aren't really countries (think Monaco) and takes into account the fact that over 80 percent of Japan's land is uninhabitable, Japan is the world's most densely populated and fifth-most-urbanized country. Cramming every-

---

* Incidentally, the hatchet man Reagan dispatched to Japan to ensure Tokyo did not renege on its commitments was Robert Lighthizer. Three decades later Lighthizer became Donald Trump's chief trade negotiator, tasked with taking American trade policy into the post-Order era.

one into tiny urban condos generates some amazing economies of scale and wonderfully efficient city services, but it makes it damnably difficult to raise children.

Japan's ruggedness prevents the formation of something commonplace in America: suburbs. If you want kids, you cannot move outside the city and commute in; you must squeeze them into your postage-stamp-size apartment. (The average Tokyo apartment comes out to less than 275 square feet per occupant.) In such circumstances, there are a lot of only children, a fair number of childless couples, and a far from insignificant number of folks who never marry because they don't want to share their space.*

The demographic degradation has been going on since the majority of Japanese relocated to the cities just before World War II, and passed the point of no return shortly after the turn of the millennium. Japan can now look forward to an ever-rising bill for pensions and health care, an ever-shrinking tax base, and a deepening shortage of workers in every field.

There are a few bright spots. Japan has the indubitable advantage of having gotten (very) rich before becoming old. As the country with the highest proportion of retirees in its population, Japan has the incentive for finding better and more cost-effective methods of caring for the elderly, but it also has the financial muscle and high-tech economy to do so. Japan isn't simply a land with higher sales of diapers for adults than diapers for infants; it is a land where elder-care facilities are partially automated.

Japan is approaching the worker shortage of the twenty-first century in the same way it approached its higher cost structures in the late nineteenth and early twentieth—by being more advanced. Japan is the most technologically advanced society humanity has ever produced, and it continues to push the limits of what humans consider possible. You name it—automotive

---

* The movement to no or late marriage is so intense that there is an entire Japanese industry of "love hotels" that allow women to spend their time with men at an hourly rate without sharing their addresses.

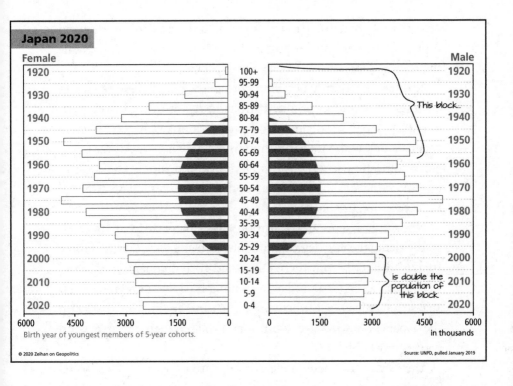

Japan 2020

Female / Male

1920	100+	1920
1930	95-99 / 90-94 / 85-89	This block... / 1930
1940	80-84 / 75-79	1940
1950	70-74 / 65-69	1950
1960	60-64 / 55-59	1960
1970	50-54 / 45-49	1970
1980	40-44 / 35-39	1980
1990	30-34 / 25-29	1990
2000	20-24	2000
2010	15-19 / 10-14 / 5-9	is double the population of this block / 2010
2020	0-4	2020

6000 4500 3000 1500 0 0 1500 3000 4500 6000

Birth year of youngest members of 5-year cohorts.

in thousands

© 2020 Zeihan on Geopolitics

Source: UNPD, pulled January 2019

engineering, automation, mass electrification, battery capacity, supercomputers—Japan is at the leading edge of all of them, not simply in development but also application. It isn't by accident, nor did it occur organically. The Japanese government literally has a national *robot* strategy.

None of which is meant to take away from the seriousness of the threat. Japan isn't just a rapidly aging nation. It is already the world's most aged nation—leading humanity's charge into demographic oblivion.

The demographic challenge has become tangled up with Japan's more traditional obstacle: materials access. Japan is just as resource-poor in 2020 as it was in 1852. In fact, in many ways, its exposure has gotten worse. As the list of dominant technologies expanded from rail to automotive to semiconductors, the list of required inputs has expanded from oil and coal and iron ore and copper to bauxite and uranium and silicon and lithium. Name the input—Japan doesn't have *any* of them.

Nor has the energy bugaboo gotten any better. Japan's terrain is largely volcanic in origin. Petroleum deposits are largely sedimentary. Japan's indigenous oil production is only ten *thousand* barrels a day while its consumption is four *million*. A year of Japan's domestic natural gas production would supply the country for only two weeks. Its uranium and coal reserves are so minimal to be functionally nonexistent.

Nor can Japan go green. Not only do Japan's steep urban landscapes make the very concept of rooftop solar a joke, but the Home Islands' distance from the equator and their frequent clouds and storms condemn Japan to having among the lowest and most variable solar intensities in the world. The same mountains that cram the Japanese into dense footprints also make appreciable wind power impossible.

The Order has made none of this matter. Japan has wallowed in unlimited access to every imaginable industrial input, making contemporary Japan far *more* dependent upon external access now than it was before World War II. With the American-led Order dissolving—taking its global continuity and economies of scale with it—Japan seems to face a complete structural collapse on a scale the Japanese feared would occur after their World War II defeat.

STRENGTH FROM WEAKNESS, AGAIN

In these interlocking problems, there lie some interlocking solutions the Japanese are already implementing.

First up, the Japanese are fairly nondenominational when it comes to where they get their electricity. They have to be. The enclaved nature of Japan's cities means there cannot be a meaningful

Industrial Commodities

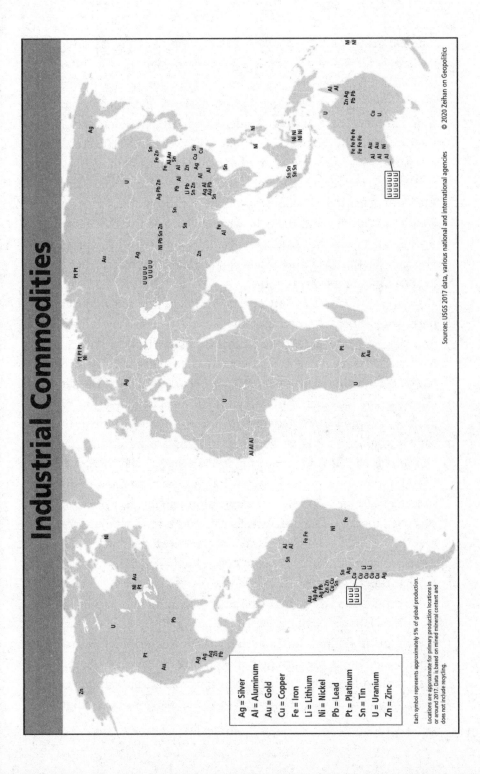

Ag = Silver
Al = Aluminum
Au = Gold
Cu = Copper
Fe = Iron
Li = Lithium
Ni = Nickel
Pb = Lead
Pt = Platinum
Sn = Tin
U = Uranium
Zn = Zinc

Each symbol represents approximately 5% of global production.

Locations are approximate for primary production locations in
or around 2017. Data is based on mined mineral content and
does not include recycling.

Sources: USGS 2017 data, various national and international agencies   © 2020 Zeihan on Geopolitics

national grid—only the Greater Tokyo region has any meaning-ful large-scale interconnections. Each urban center must maintain its own electricity system, and so each city has found itself forced to overbuild generation capacity and diversify it among several different fuel inputs so that, should one system fail due to lack of imported inputs, the others can take up the slack. Nuclear, coal, oil, natural gas. *Each* major city *independently* has them *all*.

The 2011 Tohoku earthquake and the subsequent Fukushima Daiichi nuclear power plant meltdown put the system's pros and cons on global display. The cons were obvious, as the Fukushima region—as one of the least densely populated parts of Japan—also had the least-redundant power system and so suffered black-outs and brownouts for months.

However, it was the only region to do so. The self-sufficient nature of each city's power systems prevented cascading failures; Tokyo largely recovered within a month. Since then the Japanese have steadily expanded interconnections to prevent something like the Fukushima brownouts from occurring again. Because every region has such vast amounts of surplus generation ca-pacity, the only way to generate even a regional blackout in the future would be a major war that puts foreign boots in Japan or shuts down all trade lanes for several weeks. The disruption would have to interrupt oil *and* natural gas *and* coal *and* uranium shipments. Taking out one or two wouldn't do much. Preserving this overlapping energy security is so important to the Japanese that they've been bringing their entire nuclear system back online even as other countries were so spooked by the Fukushima disas-ter that they're going nuclear-free.

Next up is the labor and materials problem. Demographic ag-ing means Japanese labor is expensive (and getting more so). Jap-anese industrial inputs are huge and varied (and getting more so). Japan's position at the far edge of Asia gives it some of the lon-gest, most vulnerable supply lines in the world. Importing ever-larger volumes of ever-more diverse materials from ever-longer distances for processing and manufacturing by an ever-shrinking and -aging workforce is a recipe for failure.

So Japan is changing its industrial model.

Most are familiar with terms like *out*sourcing (shifting production overseas but shipping the product back to the home market) or *re*sourcing (returning production home). Japan has become the master of *de*sourcing: shifting production to another country to serve that specific market (aka "build where you sell"). Doing so does far more than place Japanese products on the right side of currency, military, political, and tariff barriers. It pre-positions Japanese industry *within* the handful of countries with stable-to-growing demographics (and thus stable-to-growing markets). It gives the host country a vested interest in protecting industrial and energy input supply chains that indirectly benefit Japan. It generates scads of hard currency that can come back home to mitigate the loss of income tax from a shrinking worker base. And in the long run, it buys the goodwill of the host country, which Tokyo hopes to cash in on other issues.

The desourcing trend has *already* become so deeply enmeshed in the Japanese industrial system that Japan itself no longer produces a large percentage of its products for export. In the auto industry for instance, only a fifth of Japan's internal manufacturing is meant for markets outside of Japan. Japan is now one of the world's *least* trade-dependent countries. It keeps much of the high-brainpower work—especially design—at home. If a supply-chain system needs to be broken up, the higher-end and final-assembly work often goes to the United States—Texas, Kentucky, Alabama, and South Carolina being favorite spots. It has been a long time since 1980, when Japan was a global export leader.

Desourcing doesn't solve everything. Even with mounting technological advances that squeeze lower-skilled labor into a smaller and smaller piece of the process, Japan still cannot rely on high-cost Americans to do all the work. And just because the raw materials don't need to make it to Japan doesn't mean they don't need to sail *somewhere*. The next part of the solution is firming up relationships with countries that co-locate both industrial inputs and those initial processing steps that are low-skilled-labor intensive. For the most part, the countries that check those boxes in the industries and markets Japan cares about are in two regions.

The first is the Western Hemisphere, where a combination of American action and sheer distance is likely to keep the chaos of the Eastern Hemisphere at bay. Because there are no competitors—no powers whatsoever—between Japan and the Americas, Japan should be able to retain access.

The second is Southeast Asia. Except for Thailand and Singapore, all are resource-rich and boast young, growing populations. For their part, Thailand and Singapore are far more technologically advanced and are already heavily integrated into Japanese manufacturing systems. That Southeast Asia and the Western Hemisphere have the two greatest concentrations of the foodstuffs the Japanese prefer is a bonus.

Raw materials and processing in South America and Southeast Asia. End markets in Southeast Asia and throughout the Western Hemisphere, with an emphasis on the United States. It's a neat fix with only one itty-bitty problem.

DEALING WITH CHINA

There are plenty of reasons to think the Chinese system will implode spectacularly without the Japanese feeling the need to do a thing.

The Communist Party's authority is predicated on improving economic conditions, and for a mix of reasons China now faces interlocking crises in finance, consumption, housing, energy, materials import, merchandise export, pensions, and manufacturing. Each of the eight will hit the Chinese harder than the Americans got hit during the 2007–9 financial crisis, and all will hit more or less simultaneously. One of the downsides of single-party rule is, when recession hits, there's no one else to scream at.

Making matters worse, there are likely to be some serious tin-ear issues. Few in the Chinese bureaucracy have any experience—or even memory—of dealing with a real economic slowdown, much less an existential crisis. The Chinese have known nothing but increasing stability and wealth since the post-Mao consolidation of the late 1970s. A late-stage, lifelong bureaucrat in 2020 would have been no older than twenty-six the last time the Chinese knew civic breakdown, political chaos, and famine. There's no institutional memory of or skill in dealing with the political and cultural fallout that recessions bring, much less something more typical of Chinese history. Balls will be dropped. Minds will be lost.

Chinese history provides literally dozens of ways China can fall apart, most involving the Chinese system seizing up from top to bottom and then breaking into factions:

- Along economic lines: the north, center, south, and interior don't cohere well unless forced.
- Along class lines: the urban rich of the coast have far more in common with outside powers than one another, much less with the seething interior populations.
- Within the Communist elite: the culture of bottomless financial resources has generated a mass of corruption that if vomited forth would either break China into mutually antagonist pieces or devolve it into a kleptocracy—and it isn't clear which would be worse for the citizenry.

In these scenarios China doesn't so much stew in its own juices as boil in its own blood, and that is *before* the Communist Party has to make any decisions about how violent it might be in its efforts to preserve a unified China. The last thing on the Chinese mind will be venturing out into the wider, more dangerous world.

On the off-chance Beijing *can* keep it together in an environment of epic disruption and civil breakdown, the idea that the central government might consider a *Blammo!* approach to East Asia cannot be entirely discounted. Let us be clear. Such an effort will absolutely fail. China is utterly incapable of shooting its way to resource security or export markets or a diversified domestic economy. Just as important, the country on the receiving end would *not* be the United States. The Americans are out of reach, and even a mild American counteraction against Chinese interests would utterly wreck everything that makes contemporary China functional.

Instead, a failing, belligerent China would be Japan's to deal with. Gun-for-gun and ship-for-ship the Chinese should be able to overwhelm Japan, but a sane Chinese leader can read a map and knows full well that any conflict with Japan is not about equivalency; it's about range and position, both of which Japan has but China lacks.

In an environment in which global energy shipments become compromised because of destabilization elsewhere in Eurasia, there would not be enough industrial inputs—first and foremost oil—available for everyone. Southeast Asia consumes about as much as it produces, so it is out of the game. The Europeans retain a relevant mix of both naval reach and political links to their former colonies in Africa to secure those supplies—supplies that will no longer be available for China. That leaves the Persian Gulf—five thousand miles distant from Shanghai—as the only significant remaining source.

Even China's limited expeditionary capacity is not as good as it sounds. Nameplate operational range assumes maximum fuel efficiency and not sailing at full combat speed, which gobbles up fuel more quickly. There's also a problem of scope. Oil supertankers with the pedal to the metal (operating above safe design speed) still need *nineteen days* to make the trip from the Persian Gulf to Shanghai. China needs six two-million-barrel supertankers every *day* to keep its system running. But the Chinese have only about seventy surface combat vessels very roughly suited to the task that could even theoretically make the trip in the first place. Along the entire route, they will be operating in or near potentially hostile powers—such as Vietnam, the Philippines, Malaysia, Indonesia, and India—and doing so without a smidgeon of air support. Defending a string of—at a bare minimum— eighty-four slow, fat, supertankers sailing through moderately to extremely dangerous waters at any given time is simply impossible with the navy China has.

Japan, in contrast, has *four* aircraft carrier battle groups that *can* make it that far. The point is less that the Japanese would use carriers for convoy duty, and more that the Japanese could scrub Chinese naval power out of existence from Hormuz to Malacca with a minimum of fuss. In a shooting war, the only tankers that reach East Asia are the ones the Japanese let through. Even worse (for the Chinese), the Japanese only have one-third of China's oil import requirements, and Japan will have the option of sourcing fuels from the Western Hemisphere to boot.

China would find itself outreached and outmaneuvered and out of options, and ultimately out of fuel. And because the Chinese cannot compete with the Japanese in the Persian Gulf or Indian Ocean, the only option left would be to strike at Japan directly.

In any *real* shooting war, the Chinese can do a *lot* of damage. China's air force and missiles could probably sink everything floating within several hundred miles of its shores, which takes out pretty much everything within the northern three-quarters of the First Island Chain. Longer-range missiles could rain down on Japan to great effect: Japan is heavily urbanized, and Japan's cities often do quadruple duty as population centers, seaports, naval bases, and air force bases. Even with American assistance, most would suffer significant damage. Without American anti-ballistic-missile defense, it would be much worse. Civilian casualties would easily reach the hundreds of thousands.

For a mentally untethered Chinese bureaucrat with no sense of history or context or consequences and facing massive stress throughout the entire Chinese system, war might seem the perfect release of cathartic nationalism. But that's all it would be. China lacks the naval wherewithal to follow up such an assault with a First Island Chain breakthrough, much less an amphibious assault on the Home Islands. Aside from killing (a lot of) Japanese citizens, it would achieve *nothing*—well, nothing that would work out well for China.

First, China is a trading country that imports nearly all its energy and most of its raw materials. Sinking the ships near China's shores means no ships will sail near China's shores for a good long while. The Chinese will have then caused their own economic collapse, social breakdown, and famine.

Second, Japan is no nobody. Japan may have surrendered unconditionally at the end of World War II, but that doesn't mean it disarmed. The Americans needed the Japanese equipped and standing upright to help face down the Soviets. Consequently, military production in Japan never went away, and it is doing more than making machine guns. The stealth F-35 jet that will form the backbone of American airpower for the next two generations? Mitsubishi Heavy Industries runs licensed production of it in Japan.

Japan doesn't only build but also designs its own naval vessels and *has since the 1880s*. Japan's navy is easily the second-most powerful expeditionary force in the world. China's first designs date back only two decades, and most of its advanced designs are little more than clones of foreign ships. But most of all, the issue is range: the entire Japanese fleet is blue-water capable. That reach, combined with the First Island Chain disconnect, enables Japan to interrupt shipping to and from China, far from where China's air power could complicate their efforts.

And never forget those all-important aircraft carriers. Japan's larger pair—the *Izumo* and the *Kaga*—will soon be carrying the aforementioned F-35s and so will pack more punch than nearly any ship in history save the American supercarriers. These mobile air bases enable the Japanese to engage in offense or defense wherever they want and, for the most part, out of range of Chinese anti-ship defenses.*

Japan's air force has sufficient reach to strike the Chinese mainland in any war scenario, and would eagerly prioritize any targets that might grant the Chinese future military options. At this point the Chinese will have lost their entire navy, the dry docks that would enable them to float more ships in the future, and the energy pipelines from Russia, which provide China with the bulk of its imported energy that doesn't come in via ship. Those fat container ports that crowd the Chinese coast would certainly be reduced to TV- and shoe-strewn craters.

Third, China's situation vis-à-vis Japan is more than a bit like Japan's position vis-à-vis the United States at the dawn of World War II. The Japanese attack on Pearl Harbor in 1941 failed to sink the most powerful units of the American navy—its carriers— which were out to sea. Japan's navy is fully blue-water and doesn't spend a lot of time in port, a habit likely to intensify if geopolitical tensions are running high. Hitting Japan not only wouldn't remove Japan's navy from the board, but it would also give the Japanese full justification to treat all Chinese merchant shipping anywhere in the Pacific and Indian Oceans as prey. In less than a month, China's entire global position would dissolve into dust. That time frame assumes that the Chinese do not fall prey to a Japanese first strike, and that the Indians, Koreans, Vietnamese, Americans, and others remain neutral.

One way or another, this will all end excruciatingly badly for

---

* China does have one ace in the hole: the DF-26D ballistic missile. With a 3,500 mile range, it was designed to strike those far-out carriers. The complication is targeting: the DF-26D requires eyes on the target, typically by satellite. In any future war scenario with China, the opposing power will have to quickly remove China's eyes in the sky. Luckily for the Japanese, they have an indigenous space program.

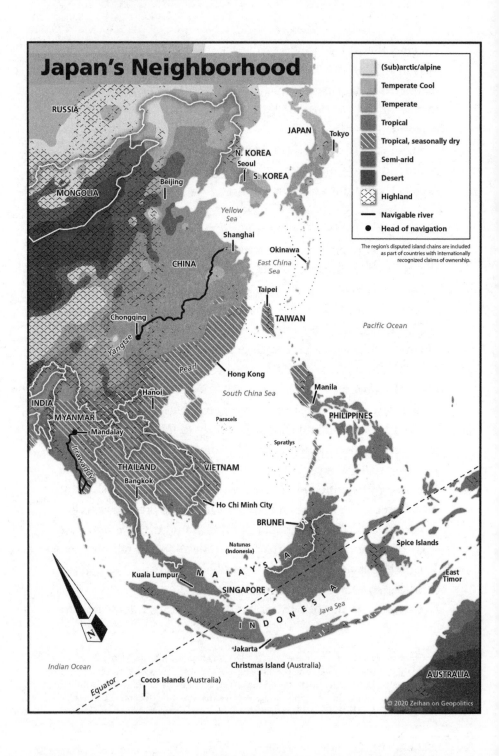

Japan's Neighborhood

(Sub)arctic/alpine
Temperate Cool
Temperate
Tropical
Tropical, seasonally dry
Semi-arid
Desert
Highland
Navigable river
Head of navigation

The region's disputed island chains are included
as part of countries with internationally
recognized claims of ownership.

RUSSIA

JAPAN
Tokyo

N. KOREA
Seoul
S. KOREA

Beijing

MONGOLIA

Yellow
Sea

Shanghai

Okinawa

CHINA

East China
Sea

Taipei

Chongqing

TAIWAN

Yangtze

Pacific Ocean

Pearl

Hong Kong

South China Sea

Manila

Hanoi

INDIA

MYANMAR

Paracels

PHILIPPINES

Mandalay

Spratlys

Irrawaddy

THAILAND

VIETNAM

Bangkok

Ho Chi Minh City

BRUNEI

Natunas
(Indonesia)

Spice Islands

MALAYSIA

East
Timor

Kuala Lumpur

SINGAPORE

INDONESIA

Java Sea

N

Jakarta

Indian Ocean

Christmas Island (Australia)

AUSTRALIA

Equator

Cocos Islands (Australia)

© 2020 Zeihan on Geopolitics

China, and even if the Chinese land a series of sucker punches on the Home Islands by launching the largest assaults on civilian targets since World War II, it is Japan—not China—that will be the last man standing.

And then there's point four: Japan would not be fighting alone.

ASIA AFTER CHINA

The countries most concerned about Chinese power are the countries best positioned to do something about it—and to do so by allying with the Japanese. Let's work from the outside in.

India *must* be a Japanese partner whether the Japanese like it or not. As powerful as the Japanese navy is, it is not omnipotent, and it certainly can't be omnipresent. The energy supply line upon which Japan depends is 7,000 miles long, some 2,500 miles of which are within India's military envelope. Japan cannot keep its lights on without a degree of Indian cooperation, and Japan lacks the military reach and power to intimidate India into silence. A friendlier approach is required. A bit of trade. A bit of investment. A bit of shared tech. The odd bribe. Undoubtedly, those oil convoys the Japanese are ferrying back home will need to stop off in Indian ports and share some of their cargo. Luckily for Tokyo, Japanese relations with India are as good as China's relations with India are bad, and that difference alone might prove enough to cause China's defeat in a war with Japan.

Next up are the littoral states of Southeast Asia. China has done pretty much everything possible to aggravate all of them. Economically, the Chinese have attempted to lock them all into dependency relationships via the One Belt, One Road system (especially the Philippines and Malaysia). Politically, the Chinese don't hesitate to inflame internal tensions, especially when there's a bit of historical umbrage in play (especially in Vietnam), or a Chinese population that can be riled up (especially in Malaysia and Indonesia). Strategically, the Chinese have attempted to seize the entirety of the South China Sea, expanding atolls and emplacing significant military assets throughout the area (which bothers pretty much everyone).

At a glance, it is easy to see why Beijing feels it can get away with being bossy. The Southeast Asian navies are piecemeal at best. But this isn't about confrontation. It's about access.

Indonesia and Malaysia are well beyond the reach of the bulk of China's navy, but together they control the all-important Strait of Malacca, gateway to Persian Gulf oil and the European consumer market. Closer in, Vietnam and the Philippines flank the west and east sides of the South China Sea, the first leg of the long journey from the Chinese mainland to those same destinations. China *must* have at least passive acquiescence from *all* of them to maintain its import and export shipping. All it would take to transform the South China Sea and Malacca into no-go zones for Chinese shipping would be a few dollops of military assistance from an eager Japan.

Taiwan is an even more obvious recruit. For China, the "wayward province" propaganda line is from the heart. Moving against Taiwan just might provide the Chinese people with a patriotic victory. All those shiny new ships the Chinese have that cannot penetrate the First Island Chain or attain global reach are more than enough to broach Taiwanese defenses. The more economic, cultural, financial, diplomatic, and military pressure China finds itself under, the more economic, cultural, financial, diplomatic, and military pressure China will put on Taiwan.

But the Japanese can read maps as well; Taiwan's physical position is critical. It serves as an unsinkable aircraft carrier that could end Chinese internal *coastal* shipping between northern and southern China, and as Taiwan is littered with anti-air defenses, nothing less than a full amphibious assault can take it out of the equation. Even worse (for the Chinese), it really wouldn't matter how an assault on Taiwan would end, because even an outright Chinese occupation of Taiwan doesn't solve China's problem of being far from its resource needs and end markets. It all still ends with Japanese regional primacy: a war-wrecked Taiwan would be formally folded into Japan's military sphere of influence, while an intact Taiwan would be formally folded into Japan's economic sphere of influence.

The shifts in circumstance will be most extreme for the two Koreas. Both Seoul and Pyongyang spent the past seven decades attempting to play Washington, Moscow, Beijing, and Tokyo (and each other) off one another in attempts to carve out a bit of geopolitical space. With Russia's decline (more on that in later chapters), the United States' disengagement, and China's choice of collapse or retreat, most options have vanished.

The smart play would be to seek de facto economic fusion with reemergent Japan. Japan will control the regional security alignments that are absolutely required if the South Koreans want to continue with their import-driven/export-led economic system, and the Japanese–Southeast Asian axis will prove just as central to ongoing Korean economic development as it will for Japan's own. In a China-less Asia, North Korea will have lost its primary sponsor and source of both raw materials and consumer goods. Economically, it is a clean, easy decision.

Politically, it is anything but. Korean history on both sides of the DMZ is replete with examples of defeat and humiliation at Japan's hands. The most pressing Asian issue of 2030 onward will be how the two Koreas relate—or fail to relate—with Tokyo. It is far from a minor issue. North Korea is *already* a nuclear power, South Korea could become one nearly as quickly as Japan, and both Koreas are armed to the teeth.

There *might* be room for some version of China in a Japanese Asia, regardless of whether the Chinese opt for a war of national destruction or a less-explosive national disintegration. The question is control. Japan will undoubtedly be willing to fold bits of China into its new system of resource supply and market access, but *only* if those bits accede to Japanese security primacy. Southern coastal portions of China will find that just peachy. Areas farther north will prefer the word "traitorous." The age-old internal Chinese wheel of imperial center versus rebellious periphery will spin once more. Most likely a host of southern Chinese coastal cities will again be folded into economic networks that have nothing to do with their countrymen.

A FEW WORDS OF WARNING . . .

For the handful of Americans with a sense of history, all these changes will trigger a few disconcerting waves of déjà vu. The establishment of a Japanese thalassocracy creepily echoes some of the Western Pacific economic and security patterns that led the Americans to station a few aircraft carrier battle groups at Pearl Harbor back in the 1930s.

The correct word *is* "echo." Japan's Co-Prosperity Sphere policy of the early twentieth century was about the Japanese flat-out imposing their will. While contemporary Japan's technological advantage over its new partners is towering, none of the nations the Japanese will interface with are preindustrial, so there will be no absolute rollovers. Nor can Japan's aged and aging demography support the sort of large expeditionary army required to bring the region to heel in a manner that still enables large-scale resource extraction. Simply put, the Japanese need local buy-in. It's easier and better to secure that with trade and money and tech than bullets and bombs and troops. Just because no one can replace or replicate the Americans' *global* Order doesn't mean that a potent regional power couldn't make a less holistic version *in its own neighborhood*.

The question will be whether the difference between the old "co-prosperity" and the modern version is stark enough to soothe

the Americans. An American retrenchment from the world isn't quite the same thing as Americans paying no attention to anything beyond their shores. If the Americans choose to venture out of their continent *at all*, Southeast Asia will undoubtedly be a stop. It is one of the few locations on the planet with demographic stability and industrial potential unconstrained by security complications. There is more than enough room in the region for both Japanese and American interests, so conflict is hardly hardwired in, but it is absolutely imperative to Japan that the Americans do not perceive Japan as a threat. After all, in times of strategic retrenchment, the Americans tend to have a potent military with a lot of free time. That tends to make Washington somewhat . . . twitchy.

Between here and there, Northeast Asia's convulsions will spill over into the rest of the world. By itself, contemporary Northeast Asia is responsible for half of global manufacturing, half of global finance, half of global maritime transport by miles shipped, and 90 percent of global shipbuilding. That's the *out*put side of things. On the *in*put side, the region comprises the largest center of demand for iron ore, bauxite, copper, wheat, corn, soy, rice, lithium, steel, aluminum, cement, and nearly every other raw and refined commodity imaginable—including natural gas and oil. No matter who emerges most intact from the region's inevitable convulsions, most players in most sectors will lose their biggest markets, their biggest suppliers, or both. Adjusting to the new reality will generate literally thousands of follow-on complications and competitions, which will reverberate for decades.

Finally, a few words of warning from the far side of the oil-supply run.

Few countries on Earth have as positive a relationship with all sides of the Persian Gulf as Japan does. Japan's status as one of the very few countries of the world that can bring naval power to bear in the Persian Gulf will induce the region's quarreling countries to take any requests from Tokyo *very* seriously. It is less gunboat diplomacy and more a client who leads a coalition who pays in cash and guards their own deliveries.

But the Persian Gulf is still the Persian Gulf. Regardless of how China's fall and Japan's rise manifest, the Japanese *still* must ensure sanctity of supply, both for themselves as well as for anyone they wish to be in their orbit. For the states of the Persian Gulf, their end markets will be wholly at the discretion of the only naval power that can reach the Gulf, ensure product delivery, and care enough to do so regularly.

That will no longer be the United States. It will be Japan. One way or another, the politics of the Persian Gulf are about to be a *Japanese* problem.

JAPAN'S REPORT CARD

BORDERS: Japan's island geography provides covetable standoff distance from the region's major land power, China. But the archipelagic nature of Japan has made intraregional connectivity costly and difficult.

RESOURCES: The most resource-poor of the world's major powers, Japan is also located at the very end of several major supply lines. The Japanese import nearly all their energy. No potent navy, no modernized Japan.

DEMOGRAPHY: It's not just the monarchy of Japan that's the world's oldest. Japan's population is also aging, with 28 percent over age sixty-five.

MILITARY MIGHT: On paper, Japan doesn't have a military. In reality, they are close training partners with the United States military and possess the most capable indigenous navy in Asia, arguably the second best in the world.

ECONOMY: Japan was once one of the most globally exposed industrial economies, but Japan has been in the rare position of having the capital to face its challenges head-on. The Japanese have invested heavily in automation to offset a shrinking labor pool, and have moved much of their supply chains offshore.

OUTLOOK: The Japanese have the capital, navy, technological know-how, and geographic insulation to step into the space left by a retreating United States better than any other regional power. They also don't have a choice.

IN A WORD: Jefe.

CHAPTER 6

RUSSIA

THE FAILED SUPERPOWER

Contemporary Russia is heir to an empire. At its height, the Soviet Union was the largest country of the time and the third-largest empire in history. Its technological achievements pushed the boundaries of human knowledge in calculus, physics, chemistry, biology, rocketry, and aerospace. Its military cowed dozens of countries, and only a wildly expensive, globe-spanning alliance of nearly *all* previous imperial powers could keep the Soviet Union from spilling forth across the world.

But the most relevant fact about the Soviet Union is this: it is gone.

Today's Russia is a sad, pale shadow of the Soviet colossus that came before. Its infrastructure crumbling. Its political structure ossified. Its people dying.

Within a generation or two, the government that runs the territories of today's Russia will be vastly different, not in that it will no longer be neofascist, but rather that it will not be run by the people we identify today as Russian. The Russian ethnicity itself is vanishing from the world.

Nor are the Russians alone. The Germans face a similar fate.

The world thinks of contemporary Europe as safe, a land united by common culture and common cause, but this condition has occurred *only* in the context of the Order. In *any* era before 1945, Europe's defining characteristic was intense, visceral, perennial competition—always political, cultural, and economic, and often military.

In some ways, the Russian and German departures make these next two chapters the most transformative of this book. Yes, Chinese power will dissolve and China itself will likely break up as well, but China has never been a global player—in fact, it has rarely even exercised power beyond its current borders. China's degradation is simply a reversion to the mean.

In contrast, the fears and fates and follies of the Russians and Germans have shaped regional and global trends for centuries. Over the long term, their disappearance will reshape the human condition in ways we can barely fathom.

But let's start with the basics. It is no minor miracle that Russia was *ever* successful.

A CURSED LAND

Russian rivers are so useless as to be largely cosmetic. Those that are navigable tend to flow through useless terrain on their way to emptying into remote, unpopulated seas. Most of them flow north, which sounds innocent until one factors in Russia's trademark harsh winters.

South-flowing rivers in the Northern Hemisphere freeze from their headwaters to their mouths. Snow and ice in the river's upper reaches do not flow into the river until spring, so during winter the river's level drops considerably.

In contrast, north-flowing rivers freeze from their mouths to their headwaters, meaning the upriver flows do not necessarily drop, but they flow *into a frozen river*. The river's force breaks up the ice but pushes it farther north into ever-colder areas where there is *more* ice. Eventually, the ice accumulates to the point that it forms an ice dam, which forces the semifrozen water up and over the banks. Spring brings no relief because the spring thaw swells the river's flow . . . which has nowhere to go but into the remaining downriver ice. If you're watching from satellite, this entire process is fascinating. If you're watching from the river banks, you're dead.* It's a geographic quirk that takes what in the preindustrial era would have been the most reliably productive farmland—a river's floodplain—and instead makes it a particularly horrific seasonal swamp.

Even worse, Russia utterly lacks the maritime frontage of the Northern European Plain, which greatly moderates European weather. One result is that most Russian lands are fairly arid in the summer and bitterly frigid in the winter.

And the size. *Oh, the size!* The Northern European Plain spills open from Poland into Russia, Belarus, and Ukraine. What had been a coastal strip three hundred miles wide becomes a vast, often featureless frontier of two thousand miles, from the Russian Far North in the Arctic to the Caucasus Mountains and the deserts of Central Asia in the far south. This territory is fundamentally impossible to develop by any means that would be recognized in Europe or North America. It is as if the dry, erratic climate of the Great Plains of North America— from southern Alberta to southern Texas—also stretched east-to-west the *entire* width of the United States. Growing food, building infrastructure, defending the periphery—none of it can be even attempted without masses of labor treated as something far less than human. Hordes.

These Eurasian Hordelands suffer under a multifecta of issues that wreck cultures: endless distances, expensive transport, min-

---

* There is a bit of a (un?)healthy competition among Russian air force pilots over who gets to go out and bomb the rivers to break up the ice dams.

imal barriers to invasion, short growing seasons, and horrific flooding followed by horrific droughts. Even when Russia is not at war and its farmers can at least try to tend their crops, it is fairly common for Russian wheat fields to either burn due to fires or be consumed by the truly epic swarms of locusts created by the flood-and-drought sequence.[*]

Bereft of capital-providing rivers, reliable weather, and high-productivity farmland, most preindustrial Russians scraped out near–subsistence level existences. Subsistence farmers require a lot of space per unit of food produced, so preindustrial Russia had the lowest population density of the European nations.[†]

This pattern is an opportunity and a problem.

First, the opportunity. Much of the Russian rise in the eighteenth and nineteenth centuries was a straightforward population-growth story. Winter in Russia often sucks up half the year, leaving relatively short periods to grow crops. Near-subsistence farming on Russia's short-farming-season land encouraged serfs to have lots of kids so sufficient labor would be available during the intensive growing and harvesting periods. Protecting broad swathes of territory necessitates a big army. Luckily the serfs had kids with little to do half the year. When the army wasn't in use, those kids were released to establish (marginal) farms of their own. And so the wheel turned, and Russia got bigger and bigger both in population and size.

Next, the problem. With Russians spread out thinly over such huge tracts of short-season land suffering from such onerous transport constraints, economies of scale could never happen. Which meant that capital-intensive advancements like industrialization could never occur organically.

---

[*] Such swarms are also a miracle of nature to behold. Again, by satellite.
[†] We're just looking at populated Russia here—a rough triangle between St. Petersburg, Omsk, and the Crimean Peninsula. Obviously, when one includes the vast wastes of Russia's Far North and Siberia, Russia is nearly empty compared with almost any country, but even in "populated" Russia the human footprint is light by global standards, much less European standards.

THE POLITICS OF HEAD-CRACKING

Such obstacles to industrialization didn't overbother Joseph Stalin, who forced the issue. Under his rule, Soviet authorities herded Soviet citizens into tiny substandard urban apartments adjacent to pop-up factories. Below-cost labor combined with slap-dash industrial plants in a country with at least double the population of its European peers? It might not look pretty, but the Soviet version of industrialization was still industrialization.

The Stalinist approach would haunt the Russians for generations.

First, it gouged the country's demographic structure. In the cities, the Soviet industrial experience's impact on birthrates and family size was more or less what you would expect. Take people off the farm and cram them into small spaces, and the birthrate drops. That, however, was *not* all she wrote.

Most identify industrialization with cement and steel and automobiles and high-rises and manufacturing—with the cities. That vision isn't so much inaccurate as incomplete. When a country industrializes, the technologies penetrate every nook and cranny. That includes the countryside. Combines and spreaders and shredders and balers and grain elevators and barbed wire are industrial-era technologies too.

Like any preindustrial society, the 1920s Soviet Union didn't begin with any of these features. The Soviet peasantry—all working on near-subsistence plots—couldn't afford them. Again, Stalin wasn't deterred in the slightest. He forced Soviet peasants off their tiny plots. Some were sent to the new factory cities, while others he concentrated into industrialized collectives with shiny new equipment. The change in living conditions had a similar impact on birthrates within the collectives as moving to the city. But that was only the beginning of the damage.

Stalin saw the output of the newer, more modern farms as no longer the bounty of the peasant class, but instead the property of the Soviet state. The Soviet government confiscated their agricultural output to support the factory workers living in antlike conditions. With no hope of prosperity, the farmers quit trying to grow surpluses at all, using their new equipment to grow just enough for their own families. Annoyed by such "traitorous" peasants, the Soviet government took every single grain of output. The result is known by different names in different parts of the former Soviet Union, but one word everyone can agree on is *famine*. Depending on whose numbers you use, somewhere between six million and thirteen million people died during the Soviet Union's collectivization crisis in the early 1930s. Most of those who perished lived in the countryside, where birthrates were highest.

Russia's demography never recovered, but as bad as it was, it was only one of a host of artificial disasters that plagued the Soviet population.

Just before collectivization, in World War I, and just after, in World War II, the Soviet Union suffered immeasurably, with the bulk of the deaths in both cases among young men. While not as essential as young women to maintaining a nation's population, they aren't exactly optional. In all, some twenty-six million Soviet citizens perished in the pair of wars.

At Cold War's end, the Soviet Union didn't simply end; it *failed*. Government collapse meant the suspension of nearly all government services. Instead of defending the nation, the military moonlighted as drug smugglers, fostering a national heroin addiction on top of widespread, epic alcoholism. The health-care system ceased, besetting the country with drug-resistant tuberculosis and HIV crises. All of the above reduced birthrates while increasing death rates.

Russians—now free of government limitations on where they could live—voted with their feet, abandoning the serviceless Arctic wastes and the countryside for the threadbare services of Russia's secondary cities, while many in those secondary cities

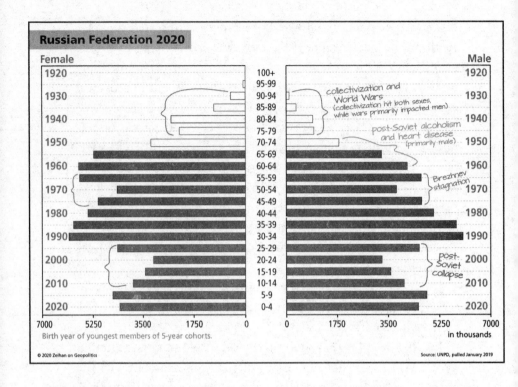

Russian Federation 2020

Female / Male population pyramid. Labels: collectivization and World Wars (collectivization hit both sexes, while wars primarily impacted men); post-Soviet alcoholism and heart disease (primarily male); Brezhnev stagnation; post-Soviet collapse.

Birth year of youngest members of 5-year cohorts.

in thousands

© 2020 Zeihan on Geopolitics

Source: UNPD, pulled January 2019

fled to the semiservices of Moscow, where they moved into the warmest apartments they could find. In the 2010 census, the Russian government discovered more than eleven *thousand* small towns, once home to over a combined one million people had been abandoned since 1990.

And just because Moscow provided the best government services in the country did not mean it was a good place to raise a family. Yelena Baturina, the richest Russian woman, is closely connected by blood and business to the city's largest organized-crime group, the Solntsevo Gang. Between 1992 and 2010, her husband, Yuri Luzhkov, just happened to be the mayor. The ultimate power couple, they ran Russia's capital like a protection racket, preventing most new construction so they could maximize rents. The result? Entire extended families crammed into single-room apartments, which, shall we say, limited the opportunities for the sort of alone time required to have a high birthrate, never mind the space needed to raise kids.

FAILURES OF LEADERSHIP

Russia has never been known for economic efficiency. Unfortunately, the country's economic trajectory is pointed firmly down.

State-centric economic planning has some natural advantages over laissez-faire capitalism, mostly when it comes to the achievement of state-defined goals. When imperial or Soviet Moscow said it wanted a rail line to here or a factory built there, the full force of the state was behind the action, and it simply happened. In a free-market system, the government has fewer resources relative to the broader system, private interests come into play, and the court system must balance the two in an environment in which law constrains rapid action. Centralized systems have none of those pesky problems, and so progress can proceed rapidly.

Clearly, a centralized system can be better . . . so long as no one ever makes a mistake. Disastrous decisions face just as few barriers to rapid application.

Stalin's collectivization of the countryside was hardly the only catastrophic miscalculation of the Soviet leadership. Stalin's breakneck industrialization efforts wasted massive amounts of resources on producing subpar products. Common stories include setting up factories to each produce parts of a product to be assembled at yet another location. That part is normal. What isn't normal is failing to have any communication about how many of each sort of widget are required so a complete gadget can come out of the process. Soviet-era tractors and automobiles rarely lasted long, assuming the finished products worked at all. If you've ever found yourself trying to put together a bookcase and you're missing a screw, you might know how that feels . . .

but with less fear that you will be the one who goes to the Gulag for it.

The housing programs of Stalin's successor, Nikita Khrushchev, may have put roofs over heads, but the resulting apartments were so tiny that they lowered birthrates nearly as much as World War II. In the 1970s the Soviet Union suffered under Leonid Brezhnev, a leader smart enough to realize that the Soviet system was failing and charismatic enough to face down all challengers to the throne, but not smart and charismatic enough to find a replacement for central planning. On his watch, the Soviet Union stagnated, depressing the Russian birthrate by half.

With the Soviet collapse in 1992, the Soviet bureaucracy—never known for being better than bureaucracies anywhere else—largely seized up. Most Russian industries responded by simply shutting down. Even industries like oil production, which by necessity were run by technocrats, lost coherence. Output halved.

More important for the long term, the Russian government's financial collapse meant that not only were the bureaucrats and military not getting paid, but neither were the intelligence services. Many operatives applied their skill sets to the private-sector equivalent of espionage, linking up with (or forming their own) organized-crime syndicates.

This is only the tactical tip of a strategic iceberg. A quirk of the Soviets' Russocentric, top-down, statist, information-controlled system was that the only people who had a full picture of what was going on were those in the top tier of the intelligence services. By the late 1970s, the leader of this group, Yuri Andropov, had privately come to the quiet conclusion that the Soviet Union had lost the Cold War. Ascending to national leadership in 1982, he and his disciples, Konstantin Chernenko and Mikhail Gorbachev, began an internal debate about how to manage defeat with honor. The Soviet collapse occurred without solving the problem, and the pipeline for new leadership recruits—never particularly robust—shut down completely.

The leader of the remaining crop of Soviet-era intelligence officers is Vladimir Putin, who has ruled the Russian Federation since late 1999. During those two decades, the bulk of his inner circle has been drawn from that same shrinking pool of talent, augmented by a sprinkling of oligarchs who looted the state during the 1990s chaos. Both are a limited resource, and Putin knows it.

Putin's success in stabilizing Russia is notable and should be congratulated. Most important, Putin's government has restored a faint semblance of a health-care system, which has nudged death rates down and birthrates up. But the case must not be overstated.

First, recent birthrate gains have little to do with Putin, and they're transitory anyway. In the late 1980s, the Soviet Union experimented with political and economic openings. That moment of hope generated a mini–baby boom. That boom's echo arrived in the mid-2010s—the grandchildren of perestroika and glasnost. For the first time since 1991, new births exceeded new deaths.

By 1988 it was apparent that perestroika and glasnost had failed to resuscitate the USSR. Collapse was imminent, and it had the expected impact upon the population. Annual new births dropped by half in just five years. It is this gutted post-Soviet generation that is now taking over as Russia's young parental generation, so the current "positive" demographics are the *last* bit of good news on the demographic front the Russian ethnicity will have. Ever.

Second, what limited gains in life expectancy Russia has garnered have really occurred only where government services have been reinstated: the larger urban areas. Borderline apocalyptic death rates continue to plague the countryside.

Third, contemporary Russian stability is a brittle thing, made possible largely by a surge in oil prices from under $20 through-

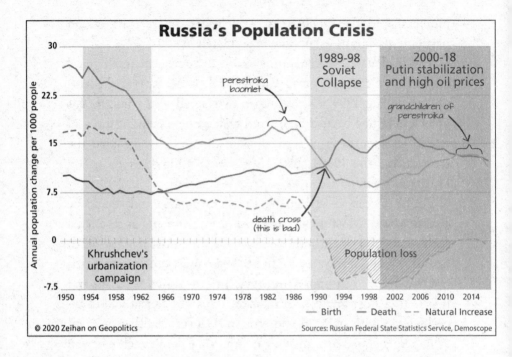

Russia's Population Crisis

1989-98
Soviet
Collapse

2000-18
Putin stabilization
and high oil prices

perestroika
boomlet

grandchildren of
perestroika

Annual population change per 1000 people

30

22.5

15

7.5

0

-7.5

death cross
(this is bad)

Khrushchev's
urbanization
campaign

Population loss

1950 1954 1958 1962 1966 1970 1974 1978 1982 1986 1990 1994 1998 2002 2006 2010 2014

— Birth   — Death   – – Natural Increase

© 2020 Zeihan on Geopolitics

Sources: Russian Federal State Statistics Service, Demoscope

out the bulk of the 1990s to over $70 a barrel in 2007–8, and by Putin's success in alternately co-opting and smashing the Russian oligarchs.

Fourth, keep the gains of Putin's Russia in perspective: they are improvements from abysmally low levels. Death rates for Russian males between fifteen and twenty-nine were *over six times* those of Iraqi men during the American occupation and subsequent civil war—which overlaps the period of "stability" Putin generated. Dead young people yesterday mean no young parents tomorrow.

In Putin's first years as premier, his government made many attempts to restart Russian industry with an eye toward this or that value-added sector. It has most assuredly *not* worked. The Russian educational system collapsed *before* the Soviet Union's end. In 2020 the *youngest* full crop with a "normal" university background is in their midfifties, and the average age of death among Russian men is probably under sixty.* Most of the Soviet Union's best and brightest—particularly in academia, engineering, energy, and aerospace, probably about 1 percent of the Russian population—emigrated to the United States or Europe during the 1990s crash.†

Even at its height, Russian industry was so inefficient and mismatched, it never generated the income required to power local consumption. And between outmigration and contemporary Russia's dead-in-the-water demography, Russia lacks the workforce required to staff new industries anyway, even assuming that Russia could somehow generate a sufficiently robust leadership cadre to forge and manage them.

Nor is outside help coming. By 2015, as punishment for various Russian aggressions against its neighbors, Russia had incurred a host of Western sanctions that, among other things, limited technology transfer. Therefore Putin more or less gave up on modernizing anything within Russia. From now until the end, the Russian economy will operate on little more than primary-resource extraction with some incidental value added in metals and energy. Aaaaaand that's pretty much it.

---

* I say probably because we don't know for sure. The last year the Russian Federal State Statistics Service issued data in a way that at least pretended to be focused on accuracy was 2000. In that year, the average life expectancy for a Russian male was 58.7 years. Since then the Russians have officially asserted significant improvements, the 2018 claim being 67 years. While the health of Russian men has undoubtedly improved in recent years, gaining a decade of life expectancy in only two decades of time would be a historical record in health improvement.
† We face the same problem with Russian emigration rates and disease rates that we see with life expectancy. All data flows from the State Statistics Service. The 1 percent figure is for Russians who formally surrendered their citizenship when they relocated. There are almost certainly that number again—also trending toward the younger and more highly skilled—who moved out of Russia without such official renouncement.

AND THEN THERE'S THE *REAL* PROBLEM

Russia's most severe, most unrelenting problem, however, is strategic.

It's an issue of size. Russia is comprised of vast tracts of low-to-moderate-productivity land stretched out across eleven time zones in an area with minimal internal navigable waterways. Moving people and goods and troops across such vast territories has always proven expensive and difficult. In preindustrial times, its size meant Russia was not only a technological and economic laggard, but a military laggard as well. Pretty much any organized state could outmaneuver the Russians with a smaller, more advanced force. Moscow's only hope of winning was to subject its foes to wave after overwhelming wave of conscripts, while praying for bad weather.

Industrialization changed the math somewhat. Russia still lagged (far) behind, but in 1904 the Russians completed their first transcontinental rail line, enabling them to shuttle troops from their western border with Imperial Germany to their eastern border with Japan in just a few weeks. Previously, such a logistical feat would have taken over a year.

Once again, it was Stalin who took Soviet industrialization to the next level, this time in military affairs. The superior organization and technology of German forces had overwhelmed Russian forces time and time again throughout history. It wasn't exactly a hard fight. In the opening days of World War I, in some battles the Russians had only enough firearms for about one-third of their conscripts. Russian battle strategy was for the unarmed two-thirds of the troops to rush into combat along with those lucky enough to have weapons, pick up the guns of those who fell, and keep fighting. Needless to say, it was a stupid plan that didn't work. The Russian defeat at German hands in World War I was so complete that it initially cost the Russians all their

Ukrainian and Belorussian territories. Only allied victory a year later restored these lands to now-Soviet Moscow.

But in World War II things had changed. Sure, German military tech and organization and logistics and tactical doctrine remained far superior, but Stalin had pushed military development hard enough that every infantryman had a rifle. It was no longer a competition between an industrialized and a non-industrialized force. Now it was a fight between an advanced industrial force and an early industrial force. One German infantryman versus five unarmed Russians lacking logistical support was one thing. One German infantryman versus five *armed* Russians with a logistical tail stretching by rail link back to militarized industrial towns well outside the reach of Nazi aircraft was quite another.

Russia won the Battle of Stalingrad not based on tactics or technology, but on numbers. The same proved true for follow-on battles at Kursk and the Mius and Belgorod and Kharkov and Smolensk and the Dnieper and Kiev and the Crimea and Narva and Debrecen. In each fight, the Russians brought wave after wave of disposable troops. Most of these Soviet wins took a second attempt. Some took *four*. In some battles, the casualty ratios were five to one against the Soviets, but they . . . just . . . kept . . . coming. Never forget the first rule of life in the Hordelands: fight with a horde. Soviet Russia could. Nazi Germany could not.

Thirty-four months after the Germans invaded the Soviet Union, the Red Army swept away the final German resistance and entered Berlin.

Stalin won the greatest war the Russians had ever fought, but he did *much* more than merely win; he solved the Russians' age-old challenge: the frontier.

The Russian interior is a great wide stretch of featureless, arid flats: the famed steppe. Even with all its hordes, the Russian Empire and subsequent Soviet Union lacked the manpower to guard it all, and the logistical capacity to engage in anything other than a mass, static defense. Russian leaders' strategic policy back to the very beginning was to expand, to absorb land after land to use as buffers, and eventually—they hoped—reach a geographic barrier the state could hunker down behind.

With industrialization, Stalin solved the logistical issue. The government could shuttle troops to-and-fro and so develop strategies other than a static forward defense. With the defeat of Germany, Soviet forces were anchored in the Danube Valley, the Northern European Plain, and the Baltics. That solved the access issue. With more logical borders and better infrastructure, the Russians became the most secure they would ever be.

It has pretty much all gone to pot.

Contemporary Russia's strategic crisis consists of five interlocking problems:

1. **ALIGNMENTS.** The post-Soviet breakup robbed Moscow of not only its satellite states in Central Europe, but also the constituent republics of the Soviet Union itself. Within two decades, the entire western swathe of the Soviet Empire—Estonia, Latvia, Lithuania, Poland, the Czech Republic, Slovakia, Hungary, Romania, and Bulgaria—had joined the European Union and NATO. Their military assets were now in the wrong column.

2. **BORDERS.** Due to its irregular boundaries, Russia is approximately one-third smaller in land area than and has half the population of the Soviet Union, but its external borders are actually slightly longer.

3. **APPROACHES.** Independent Russia's borders are also much more difficult to defend, particularly from the west. Russia has lost most of its Baltic frontage to Estonia, Latvia, and Lithuania, which are now NATO members, while the Carpathians are now seven hundred miles *beyond* the frontier.

4. **MANPOWER.** In addition to Russia's shrinking demography and loss of the former Soviet territories, rising disease and drug-addiction rates mean that the number of bodies available for Russia's defense is already down to less than one-fifth of what it was in 1989. By 2022 the Russian army will likely have shrunk to half of its 2016 size, making it incapable of defending the old Soviet borders, much less the longer, more vulnerable borders Russia has now.

5. **TECHNOLOGY.** Russia's labor shortages have already forced it to prioritize the aspects of its system it hopes to maintain. Moscow knows it cannot *both* develop new military technologies *and* produce them in sufficient qualities. Moscow had hoped China would help it square the circle: selling advanced weapons to China and then using the proceeds to fund the manufacture of its own. Instead, the Chinese did what they've done with everyone else: buy the floor model, reverse engineer it, engage in some light tech-theft, and start domestic production of whatever they can figure out. Russia *has* managed to develop a fifth-generation fighter jet—the scary-cool, stealthy Su-57—but Moscow has proven incapable of producing enough to field more than one lonely squadron. Similar bottlenecks plague the entire Russian arms industry.

Collectively, these issues are forcing the Russians to return to a system of static defense, a strategy that only works with a bottomless supply of manpower—which Russia no longer has.

Russia's future is bleak. Demographically the country faces inevitable collapse. While that collapse is in slow motion, it has already progressed far enough that, if Russia faces a large-scale invasion from any quarter, its only viable defense option would be its nuclear arsenal.

It gets worse.

CATCHING THE CAR

Much of Russia's post-Soviet foreign policy—most notably providing economic and military support to elements the Americans don't care for in Afghanistan, Serbia, Iran, North Korea, Syria, and Venezuela—is expressly designed to tie the Americans down in ancillary concerns and to challenge the very fabric of the Order.

Recall that the Order's primary purpose was to build and maintain an alliance of states to oppose the Soviets, states that otherwise might not have done so. In the late 1940s and the 1950s, the Order achieved that task admirably. But by the late 1960s, the European and Japanese economies had stabilized, and they *could* have gotten back to the normal ebb and flow of inter-imperial competition.

The Americans would not allow it. Part of the deal for the Order was that the Americans would manage and operate the alliance's security policy, including all anti-Soviet actions. By keeping the allies in line, the Order may have contained the Soviet Union, but it also focused pressure along fairly predictable paths. Because the Order constrained many of the Russians' traditional rivals, the *Soviet Union* had something the Russians of ages past never had: continuity. Ever since the founding of the Slavic national identities in the tenth century, the territories that would later compose the Soviet Union had weathered a constant storm of raids, counter-raids, and outright invasions from both within the Hordelands and from the lands beyond. In contrast, from 1946 right up until the Soviet end, the Russians didn't face a single assault, freeing them to focus on managing all those occupied territories.

The implications for Soviet continuity were intense. Politically, the Soviet Union was an often unwieldy multiethnic empire in which all the non-Russians were fully aware that they were occupied peoples. Had the Order not existed, the Soviets would have undoubtedly found foreign agents swarming over and exacerbating the rebellions in Hungary, the Czech Republic, and Poland in the 1950s, 1960s, and 1980s.

Instead, the Soviets enjoyed cold peace for a half century, enabling them to build out road and rail networks and industrialize all of Soviet society. Arguments that the Soviets did so incompletely and inefficiently and awkwardly all have merit,* but the fact remains: they *did* it. Without the Order, the Soviets would have been unlikely to have had a free enough hand to move much beyond Stalin's efforts in the 1930s and 1940s.

Of course, the more interesting bit is what is occurring now. For contemporary Russia, the Americans' abandonment of the Order is nothing short of catastrophic.

Due to demographic and educational constraints, the Russian economy is forever trapped in its current status as a commodities producer and exporter. *Exporter* being the key word.

Russia's exports of oil, natural gas, coal, wheat, steel and semi-finished iron, unwrought aluminum, diamonds, timber, copper, platinum, fertilizers, and frozen fish—which generate revenue in that order—mostly transit oceans or third parties. The only reason those exports can safely reach end consumers is the security provided by the Order. Remove the Americans, and the ability of the Russian economy to generate capital vanishes.

Just as damning, the Russian neighborhood isn't any friendlier now than before 1939. Here the Russians face a double bind. First and most obvious, all the powers the Russians have historically been concerned with—Japan, China, Uzbekistan, Iran, Turkey, France, Poland, Germany, the United Kingdom, Denmark, Sweden, and Finland—*are still there*. Most have long been nestled into the American security network and so have refrained from acting against Moscow directly. Remove the Americans from the equation, and these states will be forced to develop their own

---

* Oh so very, very much merit.

intelligence and military tools of state power and use them accordingly.

The Order may have been created to hem in and destroy the Soviet Empire, but now Russia is one of the states *most* dependent upon the dying Order for its ongoing survival.

And that's hardly Russia's only problem.

Russia's decline in reproductive capacity is not felt equally all over the country. While no part of Russia is what one would call balmy, the farther south one goes, the less cold it gets. Those warmer regions tend to have lower urbanization rates as well as stronger birthrates, but it must be noted that they are *not* traditionally ethnic Russian lands. These are territories peopled by the descendants of the Turkic tribes who migrated through on their way to the Sea of Marmara. Chechens, Ingush, Dagestanis, Tatars, Bashkirs, and more never saw their birthrates dip below replacement levels, yet they've benefited as much from the modest improvements in stability and health care in the 2000s and 2010s as the more urbanized ethnic Russians have.

The combination of the two is potentially deadly. Just as Russia faces the probability of a plethora of ethnic-based identities throughout its own territories agitating for autonomy or even independence, Russia also faces the near-certainty of a plethora of foreign powers reaching into Russian territories to assist such agitation. The 1990s Chechen rebellions—with 160,000 to 200,000 killed—are just one among many examples of how the combination of demographic shifts and ethnic identity can be a brutal, destabilizing mix.

The irony is nearly painful. The Russians have struggled since the late 1940s to tear down the Order so that they might defeat the rivals surrounding them in detail. It never occurred to the Kremlin that when that blessed day finally arrived, what passes in contemporary Russia for security and stability and wealth would *only be possible because of American involvement*.

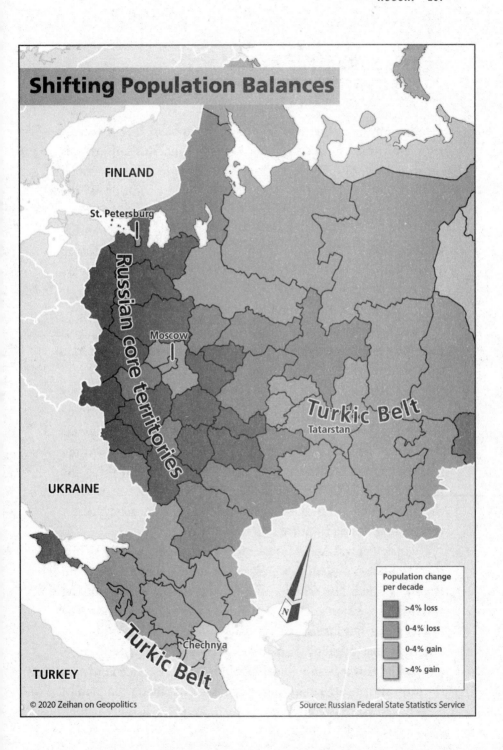

Shifting Population Balances

FINLAND

St. Petersburg

Russian core territories

Moscow

Turkic Belt

Tatarstan

UKRAINE

Turkic Belt

Chechnya

TURKEY

Population change
per decade

- >4% loss
- 0-4% loss
- 0-4% gain
- >4% gain

N

© 2020 Zeihan on Geopolitics

Source: Russian Federal State Statistics Service

ONCE AROUND THE BLOCK: CONTEMPORARY RUSSIA'S BORDERS

Fending off this future isn't an economic problem. The Russians, reasonably, instead obsess about security. The Russian army is shrinking by the day. Combine demographic growth among Russia's restive minorities with demographic shrinking of the ethnic Russian population and broadscale national economic degradation, and Russia will need more people for its internal security system. The only way to free up enough personnel is to secure more coherent borders than the indefensible disasters Russia has for borders today.

Not all borders are created equal.

To be blunt, there isn't a damn thing the Russians can do about their Siberian borders with China. What Russian populations exist in the Far East cling to the narrow corridor of the Trans-Siberian Railway with Chinese populations on the border's far side outnumbering the Russians at least ten-to-one. Even in Soviet times, a conventional standoff here was silly, so the Soviets threatened the Chinese with nukes. So it was in 1969, when the two powers had border clashes. So it is today. The *only* thing the Russians can do about approaches from this direction is what they're already doing: making it clear that the consequence for threatening Russia from that direction is annihilation.

Russia's Central Asian frontier is similarly indefensible. While the population disconnect between Central Asian states and Russia is much less scary—about three-to-one—many of the Russian citizens on the Russian side of the border are not ethnic Russians but instead part of Russia's Turkic Belt. With only one transport corridor in play—the Trans-Siberian Railway— any hostile power pushing north from Central Asia need only cut the railway in a single spot to relieve Russia of Siberia completely.

What Russia has going for it here, however, is distance. Much of central Kazakhstan and Uzbekistan are uninhabitable desert and near-desert. Beyond that zone, the politics of southern Central Asia are fractured among Uzbek, Kazakh, Turkmen, Kyrgyz, and Tajik populations. While the Russians trust *none* of these peoples, the Central Asians' mutual antagonisms are a blessing for Moscow.

There's only one power in the region that holds meaningful military capability: Uzbekistan. In an open brawl, the Uzbeks could fairly easily conquer most of the portions of neighboring Turkmenistan, Kyrgyzstan, and Tajikistan that are worth having, forming a Greater Uzbekistan with most of the region's Uzbek communities and water resources. It would instantly elevate Uzbekistan into a regional power of significance. That's not a future the Russians crave seeing. But the same geography that limits Russia's ability to project power south of the Central Asian barrens similarly limits any Greater Uzbekistan from projecting power north.

There's also the advantage of time. An Uzbek mass expansion and subsequent consolidation will take a couple of decades before it could even theoretically pose a direct military threat to Russia proper. Russia has bigger problems that are both closer in distance and time.

The last piece of Russia's southern borders, the Caucasus region, is by far the messiest. It isn't so much that the border itself is a problem—the Greater Caucasus mountain chain is pretty damn great as a border, comprising rugged, sharp peaks with few passes and narrow coastal strips on both sides. In addition, of the three of states south of it, Armenia and Georgia are one stiff breeze away from failed-state status even during the Order, while Azerbaijan just wants to be left alone. In and of themselves, none pose any real danger to Russia, and the countries that have threatened Russia from the other side of the mountains—Turkey and Iran—are likely to be locked up in issues closer to home for at least another generation.

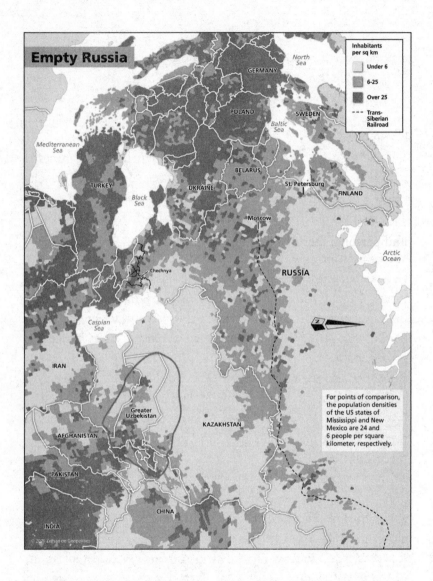

The region's *real* threats to Russia are the provinces on the mountains' *northern* slopes, on *Russia's* side of the international border. All are packed with almost universally hostile Turkic minorities. By far the loudest and largest of these groups are the infamous Chechens, who famously battled the Russians in a pair of wars during the 1990s and 2000s, and who less famously did so for a full two centuries before that. At the recent wars' heights, Chechen attacks throughout the North Caucasus became common, with high-casualty terror attacks as far north as Moscow also occurring.

The pacification of the Chechens throughout the 2000s is often cited as the greatest of Putin's achievements. The dirty secret is that it didn't occur as advertised. While the Russian army certainly poured a fortune in time, blood, and ammunition into the province, what really made the difference was Putin's direct and personal recruiting of one faction of Chechens to side with Russia in the war against the rest.

Chechnya today is not functionally part of the Russian Federation. Instead, it is the personal fiefdom of one of the most pathologically violent men on the planet, Ramzan Kadyrov. Ramzan is the son of Akhmad Kadyrov, the man Putin cut the original deal with. The younger Kadyrov fancies himself the leader of all Chechens, including the several hundred thousand who live elsewhere in Russia, most notably their greatest extra-Chechnya concentration in none other than Moscow itself. Kadyrov's influence via the Chechen mob extends so deeply into the Russian core territories that Putin has on occasion called upon Kadyrov to rid him of this or that meddlesome person.

The Kremlin has few illusions about this. Russia's top leadership is fully aware that, should Russian forces ever become heavily distracted—say by a war or uprising elsewhere—Kadyrov's forces will face few problems ejecting Russian power not simply from Chechnya, but from the entirety of the Northern Caucasus. This is far more dangerous for Russia than it sounds. While the populations of the Northern Caucasus regions live mostly in a ragged strip of plains just north of the mountains, a combination of rain shadow effect and aridity billowing in from the deserts of Central Asia turns the lands north of this strip into near-desert.

Further separating the Caucasus region from Moscow, an arm of the Black Sea—the Sea of Azov—slices nearly through the populated crescent of land arcing down from Moscow. A meaningful Northern Caucasus rebellion would banish Russian power from the region of the Don and Volga Rivers, unhinging Moscow's entire southern arc.

But much like Russian concerns with borders farther east, even this is not Moscow's primary concern. After all, a deal—such as it is—is already in place with Kadyrov. What *can* be done *has* been done. Instead, Russia's eyes are firmly on the West.

The Russians perceive their western horizon as a single, integrated battle space. In the north near St. Petersburg are the three tiny Baltic states of Estonia, Latvia, and Lithuania. Moving south is Belarus, sandwiched between Russia and the Polish Gap, which grants access to the Northern European Plain. Next is massive Ukraine and, finally, Moldova, a sliver of territory wedged just this side of the Bessarabian Gap separating the former Soviet Union from the Danubian Basin.

Russia has been invaded across *all* its frontiers, but these western approaches are what have brought the Russians their most searing memories. The Swedes raped and pillaged their way through Russian lands for centuries. The Ottomans raided Russian territories with impunity for centuries. And more recently, the Germans launched the twin wars that crashed the tsarist era and nearly overwhelmed Stalin himself.

Unlike the problems farther east, resolving Russia's western border issue isn't all that complicated. So goes the thinking in the Kremlin: occupying and absorbing *all* the countries to Russia's immediate west (except Finland) would rest Russian power against the triple barriers of the Baltic Sea, the Carpathian Mountains, and the Black Sea. Toss in the eastern half of Poland, and Russia's open frontage would shrink by three-quarters, and *that* is a line the Russian army could hold *while also* freeing personnel to help manage Russia's rising internal issues.*

Such a western-looking approach squares many of Russia's circles, but it comes at the cost of introducing the Russians to another problem.

A very familiar, very *dangerous* problem.

---

* Say what you will about the Russians, they have never had a problem thinking *big*.

RUSSIA'S REPORT CARD

BORDERS: Russia's borders are long and impossible to defend, prompting the Russians to endlessly expand outward until they hit significant geographic or military resistance.

RESOURCES: Russia is a huge producer of oil and natural gas, and its vast geographies sustain massive mining and even more massive grain production. Much of this activity is seasonal; most Russian territory vacillates between frozen and swampy.

DEMOGRAPHY: The horrific Soviet legacy and the post-Soviet birthrate collapse have fused with skyrocketing mortality fueled by alcoholism, heart disease, violence, tuberculosis, and HIV. Russia is suffering through a complete, multivector, unstoppable demographic collapse.

MILITARY MIGHT: Russia still invests heavily in defense, though much of the hardware is showing its age. Thirty-plus-year-old submarines and an aircraft carrier that habitually catches fire, but impressive tanks and aircraft and the world's largest nuclear arsenal—Russia's kit may be old, but it still packs a punch.

ECONOMY: Sanctions and an overreliance on commodity exports have made Russia struggle since the Soviet fall, but Russian geography never supported a successful, industrialized economy.

OUTLOOK: Russia is an aging, insecure, former power determined to make a last stand before it is incapable of doing so. American disengagement from the global scene couldn't have come at a better time, but the reactivation of Russia's traditional local foes couldn't have come at a worse one.

IN A WORD: Panicked.

CHAPTER 7

GERMANY

SUPERPOWER, BACKFIRED

Acumen. Efficiency. Precision. Speed. Effectiveness. Accuracy.

All these and more are things we think of as characteristically German.

Here's another one: dying.

Europe faces far worse problems than the withdrawal of the United States from its security and economic affairs, worse even than the disintegration of the European Union and eurozone. Europe faces the reintroduction of a Germany desperate to preserve its postwar gains in an environment where mere survival is beyond its ability.

THE CURSE OF THE PERFECT

Unlike Russia's chronic problems, which lead back to its terrible geography, Germany's issues stem from the opposite: from an economic point of view, Germany's geography is too *good*.

America's waterway network might make the United States the *largest* capital-generation system in the world, but Germany's is far more *concentrated*. Between the Rhine, Elbe, Oder, Danube, and a half dozen minor rivers, Germany boasts about two thousand miles of natural waterway for a chunk of territory smaller than Montana. Germany's rivers flow to locales even better than America's—the Rhine and Elbe access the North Sea, the Oder empties to the Baltic, and the Danube winds to the Black and from there the Mediterranean. Germany's river footprints grant it the best economic access of any territory in Europe—arguably of any territory in the world.

But what makes sense economically does not always make sense in other ways.

Squatting squarely amid the Elbe, Rhine, and Upper Danube are Germany's Central Uplands—a rugged, densely forested mountain knot that resists infrastructure even in the twenty-first century. Germans tend to live in an inverted V roughly following the Rhine and Elbe, arcing *around* the Central Uplands. The mix of great rivers and the Central Uplands gives Germany some awkward bones: German lands are traditionally both Europe's richest lands *and* their most separated. The lands at Germany's edges *are* its core territories. Historically, other, more consolidated European powers have had a breezy time of taking what they want from the Germans.

The early Germans dealt with this the only way they could: by organizing into something *better*—a stronger educational system, a better workforce, more reliable financing, more productive manufacturing systems.

Yet the geographic separation of the Central Uplands made it impossible for the preindustrial Germans to do this at the national level with *all* Germans. Civic governments became responsible for issues of finance and infrastructure and education and foreign policy normally reserved for national authorities. And because the German geography made everything a sink-or-swim issue, Germany's independent city-states tended to develop organizational capabilities at the *local* level far superior to those at

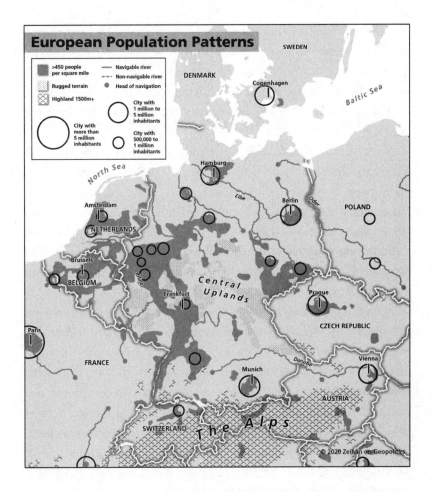

the *national* level of their much larger neighbors. Excellent civic government is part and parcel of what it means to be German. Governance at the national level? Not so much.

For most of German history, there was no *Germany*. Identities and competencies were a strictly local and regional affair. Nearly all German cities have a lengthy stretch in their past in which they were the core of a small German kingdom—all of which long predate the unification of the German people into the entity we think of today as Germany. Each was an independent political authority in the shifting morass of the thousand-year history of the Holy Roman Empire, often allying with one another or non-German powers in conflicts with the others. Only the most successful of these German statelets—Prussia, a heavily militarized state that originated in the northeastern German lands—ever got big enough to achieve any economies of scale.

Such advancement and independence radically shaped the Germans' development. The arrival of the industrial technologies in German lands in the 1840s and 1850s found fertile ground in dozens of advanced cities that could all apply all the technologies all at once. Among the first things this constellation of German urban regions did was run industrial road and rail links to one another. German industrialization made German unification not only possible, but inevitable.

It wasn't all beer and pretzels, though. Some German communities—used to functioning as independent powers—resisted. So did *all* the Germans' neighbors, who rather enjoyed having shattered, easily exploitable people next door. But with new roads and rails enabling direct inter-German commerce and cultural consolidation, they were denying reality. Between 1864 and 1871 the various German statelets found themselves on multiple sides of three major wars and countless negotiations among themselves and with neighboring powers. By the end of 1871, it was all over. A unified Germany was—finally—born.

The fusing of Germany's industrialization and unification processes shaped Europe's next several decades, as well as charted out Germany's challenges today.

WILL THE LAST GERMAN PLEASE TURN THE LIGHTS OUT?

Contemporary Germany's biggest problem is demographic. When Germany industrialized, the entire country did so at once—

which means it also *urbanized* all at once. Which means the German birthrate started *falling* throughout German lands all at once. And it never stopped. Birthrates might tick up on occasion during economic booms, but with most Germans living in cramped conditions, there's a big difference between deciding to try for a second kid and a third one.

About the only thing that encourages people to have more kids is for people to have more *space*. That's why Germany *did* have a baby boomlet during a period they understandably don't enjoy discussing: the Lebensraum era of the late 1930s, when Germany was busy annexing its neighbors and getting physically *bigger*.

Actions breed reactions.

Germany's World War II defeat wasn't simply economically, culturally, and militarily jarring; it physically reduced the size of postwar Germany by roughly a quarter. The postwar governments of Germany's neighbors forcibly ejected the Germans living in the formerly German lands. Postwar German authorities almost universally settled these expellees—who accounted for about 18 percent of the population of the combined West Germany and East Germany—in urban apartments with little space for living, much less procreating.

That was then. This is now.

Contemporary German social stability is predicated upon a generous, ironclad, cradle-to-grave system of government services. Germany has been able to afford it because of a triple quirk in its demographic structure:

1. Both East and West Germany enjoyed a brief baby boom between 1955 and 1965. That generation has been working and paying taxes for decades and in the 2010s is at the height of its earning and tax-paying years. The German treasury overfloweth.

2. Germany's Boomers didn't have many children, circumscribing expenses for services like primary education and childcare. The state instead directs the savings into value-added items, like higher education and infrastructure.
3. Some eight million Germans died in the world wars, gutting populations that *would* have been retirees in the 1970s, 1980s, 1990s, and 2000s.

High tax receipts plus few dependents equals excellent government finances, but this golden age of high tax receipts and low government payouts is ending.

The unavoidable fact remains that, with the exception of the Lebensraum and early-Order hiccups, German reproductive rates have been edging down for sixteen *decades* and are now far past terminal. Birth-based demographic growth in Germany has been impossible for a quarter century. It is about to get worse. Much worse.

The nearly childless German Baby Boomer generation enters mass retirement in the 2020s, cursing contemporary Germany with a collapse in productive capacity, a collapse in tax receipts, a collapse in consumption, and an explosion of state outlays for pensions and health care. As soon as 2030, the young generation entering the workforce will be less than half the size of the Boomer generation retiring. In any normal country, this would spell economic and social failure.

Such ruin assumes that economic trends in Germany are positive. They are not.

The quest for efficiency and quality so central to German identity comes at a cost. Like, literally it costs a lot of money to craft top-notch educational, industrial, and infrastructure systems. All those local and regional German governments try to capture private savings to fund community goals, leaving little in the way of financial resources for the development of

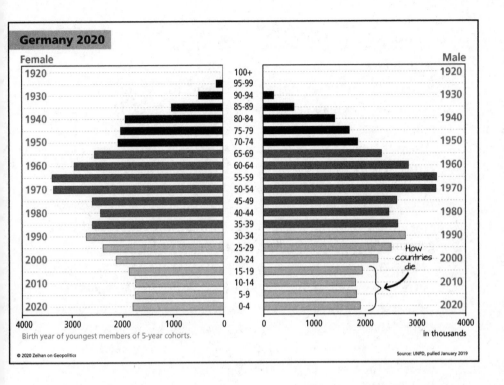

Germany 2020

Female / Male population pyramid. Birth year of youngest members of 5-year cohorts.

How countries die

© 2020 Zeihan on Geopolitics

Source: UNPD, pulled January 2019

a consumption-centric culture. Pair high production with low consumption, and the only option is to export the difference.*

Late-1800s Imperial Germany was able to absorb its output because then—like 1990s China—Germany was industrializing. But by 1900 over half the population had moved off the farm, making it the world's first majority-urbanized country. An achievement, yes, but there was nowhere else for its production to go . . . except abroad. The mix of overproduction and subsequent imperial trade barriers ensured a litany of recessions, price

---

* Part of why American-German relations cooled so during the presidency of Barack Obama was the American argument that governments around the world should expand spending to help the global economy claw its way out of the global financial crisis of 2007–9. The savings-obsessed Germans found such suggestions rude. The Americans saw the Germans' refusal to spend money combined with their enthusiastic export efforts to take advantage of American spending to be equally rude.

crashes, and general economic dislocation. European interstate competition—already political—became deeply economic as well. It was only a matter of time until it became strategic.

The first of those strategic competitions—World War I— ended with German defeat and kicked off the Great Depression, which struck nowhere worse than where it started: in Germany. Americana tells tales of Depression-era city folk returning to the farms to work the land during the Depression's deepest years. In the lightly populated United States, where most people were less than a generation divorced from the farm, that was an option. Not so in Germany, where the majority of Germans had been living in cities for decades. The only thing the Germans could do to keep costs down was to . . . not have children.

Germany's post–Cold War unification was—and should be remembered and celebrated as—a phenomenal national achievement, but that doesn't mean there were no side effects. In the spirit of unity, the new national government decreed that the value of the former West and East German currencies would be broadly equalized despite the East's far less advanced industrial plants. The valuation decision aimed to preserve the personal savings of the former East Germans—a political necessity in savings-driven Germany—but it has gutted the competitiveness of the East German economy ever since. Not only did this haunt the Germans with echoes of the sort of dislocation and unemployment they experienced during the Great Depression, but family formation, and from it, birthrates, in the former East crashed and never recovered.

There is never a good time to have a terminal demography, but contemporary European economic trends argue for now being among the worst. A rapidly shrinking working-age population paying a higher tax rate to support an ever-larger retiree population means Germany's already weak ability to absorb its own production weakens further. It is more important than ever for the Germans to have a place to send their wares. Germany's biggest exports by value are automotive, with 80 percent of the cars Germans manufacture at home being sold outside of Germany.

The European Union is no longer that place. While few countries have as poor a demographic structure as Germany, most of those in the running are also within the European Union. The rest of Europe has aged beyond being able to serve Germany as an export sink.

In addition, Europe isn't economically capable of filling that role even if its demographics were healthy. Whether technical or the more painful official kind, some portion of the Continent was in recession between 2006 and 2015. In part, that's Germany's fault. First, because the Germans superglued the monetary policies of places like Greece to their own without much consideration for dislocations this would create. Second, because the Germans provided the largest chunk of the financial assistance required to bail out the EU's failing members, but Berlin conditioned that support on budget cuts. Europe's weaker countries are Europe's weaker countries because they lack Germany's sublime economic geography. The state spending of these weaker states is part of what helped those weaker states close the gap with the Germans. German-imposed austerity gutted growth potential and, from it, Europe's capacity to absorb German products.

HAUS OF CARDS

Northern Europe with all its flats and seas is a rough-and-tumble region with everyone in everyone else's face all the time. Germans, right in the middle of it all, have little choice but to treat military issues the same way they treat everything else: by being *better*. Take anything as efficient as Germany and make it as big as Germany and as twitchy as Germany, and the nervous energy tends to spread. Germany's very strength invites panicked challenge. The rise and fall and rise and fall and rise and fall of Germany *is* European history.

The only way the Europeans have ever discovered to prevent this never-ending cycle of wars is to change the game. To bring in an external security guarantor who forces everyone to be on the same side. Who makes *all* major strategic decisions for everyone. Who enables access to raw materials without needing to resort to war. Who allows for the mass export of Germany's always destabilizing volumes of manufactured goods to somewhere beyond Europe.

That's the United States. That's the Order.

In Europe, the institution that operationalizes the American-led Order is the North Atlantic Treaty Organization. The NATO alliance shoehorned all European militaries—Germany included—into a single command and logistical structure under American strategic and operational authority. NATO's twin goals were to force an internal Western European peace while also securing Western Europe against Soviet aggression. As Hastings Ismay, NATO's first secretary general, so famously put it, the alliance's raison d'être was "to keep the Americans in, the Russians out, and the Germans down."

With the Cold War's end, Germany entered the best of all possible worlds. NATO expanded into the former Soviet Empire, surrounding Germany with *allies*. Germany's military ceased being relevant, largely because it ceased existing. Repeated past-the-bone budget gougings functionally eliminated *all* German air, naval, and armored deployment capacity. For all intents and purposes, Germany no longer *has* a military.

From 1990 until 2020, this has been fine. After all, Europe didn't face major security threats.

And so the spotlight shifted from NATO to the parallel European institution formed by the Europeans themselves: the European Union. Unlike NATO, which benefits from a very large-and-in-charge leader and a crystal-clear founding goal, the EU tends to muddle through. Is it an economic union? A financial union? A political union? A counterweight to the United States? A free-trade zone? As of early 2020, the EU has twenty-eight members (Brexit remains unresolved at this time), and if you ask the EU's collective leadership any of these questions, you will get at least a half dozen different answers—and that's before getting down to the nitty-gritty details of bank bailouts or debt levels or negotiating authority or political norms or voting parameters. Even relatively mild disputes require the heads of state hashing out the details during multiple all-night summits.

The *really* big issues—such as how to address a sovereign debt crisis or undocumented migration—occupy all leadership bandwidth for a decade or more. There's also the not-so-minor issue of who's in charge. Formally, the EU presidency rotates among all the members for six-month terms. When it's Malta's turn in the big chair, it's hard to take Europe seriously. Even in Europe.*

Under the Order, this slowly, softly, oddly ambling path to integration was fine. The Americans did all the heavy lifting on the security front via NATO, on the global front via their navy, on the energy front via their position in the Persian Gulf, on the raw materials front via their primacy in the Western Hemisphere and their alliance with Australia, and on the trade front via the World Trade Organization, as well as keeping all of Europe lined up on the same side. Europe could afford to have knock-down, drag-out multiday arguments with itself over cheese policy.†

For the Germans, all this dysfunction is part of contemporary Europe's beauty. NATO bars security competition among the European powers. No matter how messy or awkward or myopic the EU might become, keeping the field of competition purely economic enables Germany to be physically intact and

---

* Even in Malta.
† That's not a joke.

industrially dynamic. For Berlin, that's what matters. That's *all* that matters.

With wealth rising, stability assured, and the Germans quiescent, many Europeans have come to believe in a heady dream: that after centuries of living in the world's most blood-soaked lands, they had finally achieved a sort of historical escape velocity. It's a beautiful dream. I want to believe in it too.

But with the American abdication of responsibility, that dream dies. Europe is a region of military pygmies who face pressing military challenges. A region suffering demographic implosion, unable to absorb immigrants. A region peppered with xenophobes who cannot control their borders. A region rumbling with environmentalists who cannot generate green power. A region facing a banking crisis that lacks the means of even defining the problem, much less rectifying it. A region that has largely outsourced its relationships with its nearest neighbors of significance—Turkey and Russia—to the departing Americans. A region whose most coherent economic, financial, and military power—the United Kingdom—has left.

Europe now faces simultaneous, interlocking crises: currency, finance, banking, monetary policy, supply chains, inequality, migration, oil, natural gas, electricity, demographics, consumption, exports, imports, Libya, Syria, Turkey, Russia. (Perhaps even America?) It gets worse: Any response requires that all European states agree on how to prioritize and address each problem. Which means for the Germans, the worst of all worlds is about to arrive:

- Germany is *not* a sunny or windy country. Despite spending nearly 2 trillion euro on alternate energy infrastructure and setting up omnipresent regulatory structures to favor the greentech sector, which have doubled power prices from 2000 levels (which were already more than double average US prices), Germany receives less than 10 percent of its electricity needs from green power. Germany is *more* dependent upon fuel imports now than before it started its green surge.

- Half of Germany's exports and all its energy imports rely on access to countries *beyond* the EU. The United States is Germany's largest end market, while China ranks third.
- Nearly all of Germany's manufacturing supply chains rely upon access to countries *within* the EU.
- Germany's (and Europe's) demographic collapse deepens both the consumption and the production problems.
- With the Americans' departure, security competition is back on the table.

As in China, German economic growth has been nothing shy of fantastic. As in China, German economic growth has been possible only due to the changed local security environment and risk-free trade links to the wider world that the Order provided. Like China, Germany will suffer hugely from the Order's end.

Germany is now returning to the same rise-and-fall cycle that has so defined its history, but *unlike* China, the Germans have *celebrated* their demilitarization and in doing so willingly surrendered any tools that might enable them to look after their own needs.

The Germans are left with starkly limited choices:

- The North Sea has oil and natural gas, but other countries in the sea's basin—most notably, the United Kingdom—already absorb most of what is available via preexisting infrastructure. The sea's resources are unavailable to Germany. The Middle East has plenty to go around, but that region is on the wrong side of the European continent and Turkey. In a world of unsafe shipping lanes, Germany might as well be on another planet.
- Labor options are similarly limited. Within Europe, the Scandinavians, British, and French all operate at similar cost

structures, so there are few options for operational synergies.
- Many European countries keep most of their manufacturing supply chains in-house for reasons geographic (the United Kingdom, Sweden) or mercantile (France, Spain).
- Sometimes it is simply an issue of style. Italy operates differently. Instead of the German approach of assembly lines and mass production, the Italians treat industry more like art—one at a time, handmade outputs.*
- On the topic of customers: the largest consumption bases locally are the French and British. Even at the height of the Order, the French liked to keep their market to themselves. The United Kingdom was more open, but in the aftermath of the Brexit drama, count on access to the British market flatout ending. The Dutch will continue to do business with Germany, but the Netherlands is more a flow-through economy than a large final destination. The seventeen million Dutch can absorb only so much.

But looking east, the Germans see opportunity. The former Soviet satellites of Poland, the Czech Republic, Slovakia, and Hungary are not only near, but geographic barriers between them and the Germans are minimal:

- If long-haul energy imports are not an option, Russia is brimming with (relatively) short-haul delivery options, much of which flows via pipes that are already in place.
- Labor costs in the Central European states are one-third (or even less) of those in Germany. Integrating manufacturing systems with them not only makes economic sense, but also provides economies of scale the Germans could never achieve on their own.

---

* When the Italians handcraft a car, it's a Ferrari. When they mass produce, it's a Fiat.

- Poland and its Central European cohorts may be only about one-half as wealthy as Germany, but there are nearly as many people in those countries as there are Germans in Germany. They provide both manufacturing labor *and* a consumption base.
- Just beyond the Central European zone are 44 million Ukrainians and 140ish million Russians who have proven unable to meet their own manufacture needs in *any* era.

What's left that is worth having, fits Germany's needs, and yet remains within reach isn't to the west, but to the east—putting the Germans and the Russians on a collision course. German success puts an integrated German manufacturing system in territory directly adjacent to lands the Russians see as their critical security zone. No more buffer zone. In a post-Order world, everyone starts to get agitated again, with German manufacturing becoming a bit less Volkswageny and a bit more Panzery. And since the Germans will be rearming from a nearly *dis*armed start, the pace of defense improvement will prove terrifying to anyone who has ever crossed swords, bayonets, or artillery with the Germans in the past (which is *everyone* within fifteen hundred miles). On the flip side, Russian success puts Russian occupational forces within a four-hour drive of Berlin on a light traffic day. So where does that take us?

ONE LAST DANCE

Germany's core issues of materials and export access are not new. Neither are Russia's concerns with frontiers and internal rebellions. Nor is it the first time the two countries have found themselves in a mutually antagonistic overlap.

Historically, when such things happen, the leaders of both countries tend to take deep breaths and have a thorough conversation about the lands between them. They then proceed to draw lines around topics and on maps with the intention of avoiding a bloodbath. The problem is always the same. Divvying up influence over the territories between them is the easy part. Once any agreement is executed, however, the buffer zone between them vanishes, swallowed up by the two powers. Direct borders mean direct competitions. Every time the Germans and Russians reach this point, they eventually end up shooting at each other. The only question is how long does the honeymoon last? A generation? A decade? Less?

The last time, the pact held for only twenty-two months.

This time around, four issues argue for, at most, a brief period of coexistence.

First, military technologies are evolving again. The last Russo-German war started with artillery and planes. Now there are cruise missiles and jets. As a rule, increases in speed and reach and lethality compress time frames.

Second, change is not simply coming to Germany and Russia's neighborhoods, but to Germany and Russia themselves. Surviving the coming global breakdown will spawn political, social, and economic revolutions in both countries. The last time these two peoples experienced such rapidly shifting fortunes, the results included Nazism and Leninism. The point is less that we'll see a return to fascism in Germany and fascism with a different name in Russia, but instead that the pressures that generated the last harshly negative government overhaul in both countries are very similar to those of today.

For example, consider what drove the fascist rise during the Weimar period of 1930s Germany: massive dislocation of labor due to new industrial techniques that limited the need for manpower, massive economic dislocations due to disrupted export markets, massive economic inequality, massive population and migration imbalances, and political leaders who weren't so much tone-deaf as trapped by circumstances beyond their control. The Weimar system ultimately found itself overthrown by populists who were far from shy about lurching Germany in a dangerous new direction. Any of that sound familiar? If the Americans are not managing Europe and seeing to the Continent's economic and security needs, then Germany must act like a normal country again. Whatever form the Fifth Reich takes, it cannot possibly be as pacifist and deferential as the Fourth.

For Russia, the possibility of a sharp systemic turnover looms even larger. Fewer than two hundred people remain in Russia's intelligence-officials-dominated political elite. Russia is now firmly in the zone where a single plane crash or assassination campaign or bad flu season could trigger a government breakdown or coup.

The third argument for a quick slide into confrontation has to do with the limited time remaining for the countries themselves. Both countries suffer from demographic weakness so extreme as to be a country-killing issue all by itself. The time horizon is so short that most Germans and Russians alive today will live to see the end.

That end might come sooner than even the countries' abysmal demographic profiles suggest because *not all revolutions succeed*. When a political, economic, or cultural system experiences violent overhaul, there is typically a multiyear period of reset while the old system is being wiped away but the new system hasn't fully replaced it. The last time that happened in Germany, it involved the World War II defeat. The last time for Russia, the post–Cold War breakdown. In each case, simply finding the floor took at least half a decade.

A failed political, social, and economic transformation now would likely gobble up so much of what little time Germany or Russia have remaining that they *never* recover. And that's *before* one considers the bilateral competition. What might a successful Russia do to take advantage of a failed Germany? Or vice versa? A German failure to secure an economic orbit or a Russian failure to secure better borders really does mean the end this time around. The all-or-nothing nature of this competition intensifies pressures, raises stakes, and—again—compresses time frames. Both countries are quite literally "raging against the dying of the light."

Which brings us to the fourth and final reason this is a struggle unlikely to end in a calm co-dominion over Central and Eastern Europe: Germany and Russia are not the only fish in the pond.

At the top of the list of states-of-concern must be the United States. It isn't that the Americans are likely to see anything that overly bothers them in the intensifying German-Russian collaboration-*cum*-confrontation directly, but instead that actions by either Berlin or Moscow may inadvertently rub Washington the wrong way and prompt American action.

A world in which American interests in Europe are *not* grounded in NATO and the Order is one in which American involvement is far more likely to be corporate. Because Germany's primary motivation is to secure an economic sphere of influence, the Americans and Germans could well find themselves tangling over this or that market or resource flow. A failure in Berlin to manage such entanglements effectively could sever what few connections to the wider world Germany would have otherwise managed to maintain. The Americans can (easily) operate beyond Europe while Germany cannot at all.

A falling out between Washington and Moscow would likely have a more visceral outcome. The United States defense and intelligence community has decades of experience in countering all things Russian. That experience—and the information and weapons systems that go with it—can easily be shared with countries or ethnic groups that could do Russia irreparable harm.

British intervention is less of an *if* and more of a *how*. For the better part of the past three centuries, the United Kingdom (and the British Empire before it) was involved in over a dozen wars and near-war competitions with both the Germans and Russians—sometimes at the same time. In part, it was because

London is an inveterate sea power while Germany and Russia are land powers. In part, it was because at the height of empire the British were sort of fighting with *everyone*. But in the years to come, British motivation will follow two themes.

First, it's simply good strategy. British policy since the beginning has been to prevent the formation of a European navy that can threaten Great Britain. The easiest way to achieve that goal is to take preventative measures so no single power can rule all of Continental Europe. A fully united Europe would be a peer not with Britain, but with the United States, and the navy it could float would be able to sweep the British from the seas. Taking preemptive action to prevent the rise of a European hegemon is as central to British identity as a gin&tonic and sturdy rain gear.

The two countries that have come closest to dominating Europe since 1850 are Germany and Russia. For London, the possibility of Germany *and* Russia disappearing from Europe *forever* is too tantalizing a possibility to do anything but encourage.

Second, there will definitely be a revenge play.

When it comes to Russia, the same Russian cybertools that have so inflamed American politics have been used the same way in the United Kingdom on both sides of the Scottish independence issue, the Brexit issue, and the rise and fall of both the Conservatives and Labourites. The Brits would *love* for that sort of shit to stop. For good.

Against Germany, British motivations are more instinctive. There is a powerful strand of thinking among the English that the Germans should have been more understanding of the British divorce from the European Union. While at the time of this writing the Brexit drama remains a work in progress, it is already messy, and the British economy is facing down a multiyear depression. Yet Britain is an experienced sea power that can apply diplomatic, economic, financial, and military pressure nearly anywhere it wants without fear of reprisal—and it has centuries of experience applying that pressure to Europe. Payback's a bitch.

Next comes a family of nations: the Scandinavians. The region's mix of geographies is eclectic in the extreme. Norway is in effect a city-state with a ridiculously long ocean frontage. Finland lost its best territory to the Soviets in World War II and at times has resembled an armed camp. Denmark is an island nation with an almost British worldview. The Balts were Soviet-occupied. Sweden has fought wars with pretty much everyone in the region on multiple occasions. With the Americans gone, no one will stand up for them . . . save one another. Courtesy of the exploits of Danish and Swedish Vikings, all are quite literally family.

None will exactly be enthused about German activism, but ultimately it is Russia that keeps Scandinavian leaders up at night. The only way the Russians can truly be secure is if the Baltic trio once again becomes subordinate to Moscow. That alone will prove sufficient to make the Scandinavians close ranks, and maybe even lead them to cheer the Germans on (through clenched teeth).

Finally, there is one power in particular that will perceive *any* iteration of a German-Russian lovefest as nothing short of horrifying: Poland. Poland has warred with the Germans and Russians more than any other European nation, and on two occasions was divvied up and devoured by its two larger neighbors.

In a straight-up fight, the Poles don't stand a chance. Their population is half that of Germany, and their level of economic development is not all that much better than Russia's. They lack the demographic, economic, or technical prowess to go toe-to-toe with either Germany *or* Russia, much less Germany *and* Russia. But three points come to mind.

First, unlike the World War II carve-up, this time around the Poles are industrialized. They'd still lose, but they'd certainly put up a bigger fight than the three weeks it took the Soviets and Germans to snuff them out in 1939.

Second, they'd get loads of help. Anything that keeps the Germans and Russians focused on another land power is one that takes the pressure off the more maritime British and Scandinavians. In the end, the maritime powers—with their more diversified economies, easier supply lines, better demographics, and superior strategic insulation—really just need to play for time. Poland can buy some for them, but only if it doesn't get steamrolled. Expect to see extensive naval, air force, logistics, matériel, intelligence, and completely disavowable special forces assistance.

Third, nuclear weapons could very well enter the mix.

In the Order, countries have typically sought nuclear weapons to deter countries they feared might prove unbeatable in a conventional war. Israel against the Arab powers. Pakistan against India. North Korea and Iran against the United States. But remove the Americans from the mix, and the list of countries that might find a nuclear deterrent useful changes and expands. Japan and Taiwan against China. South Korea against North Korea and Japan. Saudi Arabia against Iran.

Finland, Sweden, Germany—and especially Poland—against Russia.

There isn't anything secret about *how* to make an atomic weapon; the technologies involved have existed since the 1940s. Back then the Americans did it start-to-finish in about four years at a budget of less than $25 billion in today's dollars, with some 90 percent of that budget going toward the production of fissile material—material available today as a waste product from every existing nuclear power plant. Nukes—especially if the intent is to lob it at your next-door neighbor rather than sling it around the planet—are not hard. And if the Germans and Russians come sniffing around Central Europe after the Americans go home, no one needs nukes more than the Poles.

A nuclear-infused stalemate among Germany, Poland, and Russia would hamstring Germany's economic plans as well as Russia's security expansion, while also ensuring Poland's survival—all without a shot being fired. That might make it the best of all worlds for everyone . . . save Germany and Russia.

Which means that any serious Polish effort to develop a deterrent would likely be met by exactly the sort of assault the Poles would be hoping to forestall.

Can Germany and Russia survive their challenges in this sort of degraded economic and security environment, even if they can somehow avoid coming to blows? The honest answer is probably not. Neither country has enough staying power to make it through this competition, and there are too many countries capable and close that have a vested interest in their mutual failure.

What is certain is, whether it takes the form of collaboration or confrontation on any given day, in their death throes Berlin and Moscow will only have eyes for each other and those unfortunate enough to be caught in the line of their gaze. Such a grand strategic distraction is a wondrous opportunity for many, but for none more than the country that will define Europe for the next century.

GERMANY'S REPORT CARD

BORDERS: There are few significant buffers between Germany and its western, eastern, and northern neighbors.

RESOURCES: The greatest concentration of wealth-generating navigable internal waterways in the world, the most efficient manufacturing and production systems in the world, and the best trained labor force in the world. But jack for actual physical resources.

DEMOGRAPHY: *Old! So very ooooold!* One of the grayest and fastest-graying populations in the world, Germany's population is too old to consume the goods its industrial sector produces, creating a dependency on exports.

MILITARY MIGHT: Germany makes excellent tanks, diesel submarines, and electronic surveillance equipment. Unfortunately for Germany (but not Poland or Belgium or anyone else), decades of reliance on NATO and being hamstrung by the Second World War and the Cold War have left it a paper panzer.

ECONOMY: The entire German economy is predicated on leveraging its manufacturing sector to push high-quality exported goods to a globalized consumer base. In a post-Order world this will not work. At all.

OUTLOOK: Few countries are more dependent on the American-led global Order. Germany's best backup plan—the European Union—is already falling apart. Germany needs a new way of doing things. Or an old one.

IN A WORD: Outdated.

CHAPTER 8

FRANCE

DESPERATELY SEEKING DOMINANCE

France has been the world's second power for a very long time.

After France's emergence from the breakup of Charlemagne's empire, Paris seemingly always found itself on the back foot. Versus the Arabs. The Italians. The Spanish. The English. The Germans. The Americans. With every shift in political and technological trends, France just didn't quite measure up. When the Americans crafted a global system, freezing geopolitical competitions in place, France could only be a cog in the machine.

France never really took advantage of the economic side of the global Order. The French kept most of their production in-house, jealously protecting their statist, domestic market. The French economy barely even Europeanized. The French diplomatic network has been kept jealously separate from EU positions—except in those EU positions that have been led by France. The French military maintains a strong capacity for independent action.

Under the Order, this was all massively inefficient, broadly un-successful, and a huge wasted opportunity.

In contrast, the contemporary German system was custom-ized to operate within the Order's parameters: global economies of scale, bottomless resource and market access, zero security responsibility, a heretofore unheard-of European continuity. Germany has boomed; France has stagnated.

And yet with the Order's end, history thaws. Remove the Order and Germany will find itself either doing without all the economic features that enable it to be a multiparty social-welfare democracy, or it will be forced to fight for everything from market access to resource access, and fight some more to protect its sphere of influence and perhaps even its borders. A competition with Russia over the countries between them beckons. Potentially catastrophic? Certainly. Yet Germany's occupation with economic decay and geostrategic competi-tion positions France at a bay window of opportunity that has *never* occurred before.

For *without* the global Order the *current* French economic model is one of the few systems that will still work. Paris main-tains reasonably good working relationships with the Americans, the British, the Dutch, the Spanish, the Italians, the Germans—even the Russians. French political history oozes with examples of managing and maximizing and manipulating in a shifting geopolitical environment. The mix of positioning and prepara-tion is half of why so many European treaties are signed in Paris in the first place, and *none* of these factors have anything to do with what the Americans are—or are not—up to.

France is emerging as the only significant European power with a sustainable domestic system and no strategic entangle-ments, leaving it free to shape Europe in its own image—and perhaps do the same in lands beyond.

After a history full of coming in second, France's hour has *finally* arrived.

THE POWER OF PARIS

Distilled to its basic elements, France is a power due to a specific region: the Beauce. The Beauce's limestone-rich soils place it among the world's most fertile agricultural regions, and its surpluses have proven sufficient to feed all of northern France for centuries. But that is just the beginning. There is more—far more—to the Beauce than the inputs for Brie, butter, and baguettes.

The river Yonne cuts south to north through the Beauce's eastern half, bringing all the normal benefits of cheap transport to high-quality agricultural lands that are the recipe for success throughout the world. But the Yonne is no standard river. Some sixty-seven miles downstream of the head of navigation at Auxerre, the Yonne empties into the Seine system just above the Paris metropolitan region. The Seine system webworks about 280 miles of river frontage in France's northern lobe, connecting the Beauce's bounty to several population cores, including the capital.

The Seine then drains into the English Channel, making France part of the hustle and bustle of the Northern European Plain. The plain is home to not just the French but also the Belgians, Dutch, Germans, Danes, Poles, Lithuanians, Latvians, Estonians, and Belarusians before it merges seamlessly into the great wide opens of the Eurasian Hordelands, home of the Ukrainians, Russians, Kazakhs, and more. Nowhere in the world are so many different ethnicities with their own countries crammed into a single geographic feature. Paris's location along the Seine—near the far western end of the NEP—guarantees the French a seat at any European table.

Undeniably, it is the Beauce-Seine connection that makes France a quintessential Northern European power, complete with all the wealth and capacity that appertain thereto. But as central to the French identity as Paris and the Seine are, they are not enough. Luckily the French have more. Much more.

Just as the Seine makes France a power of Northern Europe, the Rhône makes France a power of Southern Europe. But unlike the busyness of Northern Europe, there simply isn't very much competition down south. The Mediterranean Basin's extreme aridity limits civilizational options. There are few places where the land rises sharply enough to wring some rain out of the dry air; those locations bloom as a result, but that same sharp rise prevents local rivers from having the calm flatness required for navigability—and from that, national potency. France is the *only* power with Mediterranean frontage that's an exception. Undeniably, it is the Beauce-Rhône connection that makes France a quintessential Mediterranean power, complete with all the wealth and capacity that comes with that stature.

Tickling the Beauce's southern edges flows yet *another* river: the Loire, the river of kings.* While technically still a Northern European river, the Loire actually flows west, linking the cities of the French imagination—Orléans, Tours, Angers, Nantes— before ultimately emptying into the Bay of Biscay.

The flood-prone and rapids-filled Loire may no longer be commercially navigable by the somewhat exacting standards of modern containerized shipping, but it has been an artery for Parisian influence throughout French history—and not just internally. The terminal Loire port of Saint-Nazaire—even today home to France's largest shipyards—is far removed from traditional French competitors in Scandinavia, Great Britain, and the North European Plain. In many ways the Loire Basin serves as France's window out of the Old World, enabling it to be a player on a far larger stage than "merely" Europe.

* And wine!

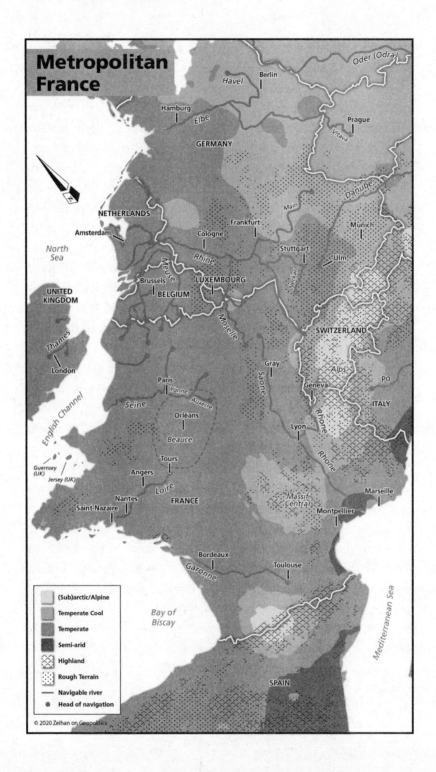

And these aren't even all of France's rivers. The Meuse enables the French to dominate much of Belgium. The Moselle is a tributary of a border river—the Rhine—that enables French influence into both Germany and the Netherlands. Follow the Rhône upriver and one ends up in Switzerland, giving the French a clear path to the heart of Europe's supposedly neutral power. The Garonne is navigable all the way upstream to Toulouse, a mere hundred miles from the Mediterranean coast, increasing French influence over countries south, while also creating a river-and-road shortcut that negates the need for the long sail around Iberia.

Even the arrangement of France's rivers encourages her onward.

The navigable flows of the Seine and Loire are close enough together to cradle all of northwest France. Perhaps a better verb would be *smother*. Somewhat rugged regions, like Normandy and Brittany, would normally be too remote from the river zones to be assimilated easily and thus be likely to turn rebellious. The two rivers' layout enables the direct injection of Parisian influence throughout both. For all intents and purposes, this makes *all* of northwestern France part of an enlarged core territory. The Rhône has a similar effect on the Alps and the Riviera, while the Garonne links in Bordeaux. Even France's most rugged, upland region—the Massif Central—is not immune. The region is so steep, it received its first true multilane roadway only in the 2000s, but because it is bracketed by the Rhône to the east and the Garonne to the west, even here French cultural power eroded meaningful separatist tendencies centuries ago.

Even the shape of France, in essence a knob near the end of the European peninsula, helps. France's frontage on the English Channel, Bay of Biscay, and the Mediterranean gives it a mild, marine climate far more typical of an island. Compared with the bulk of Europe, France is warmer and wetter, while also experiencing less extreme seasonal variations. With the exception of the highlands of the Massif Central, all of France is firmly temperate.

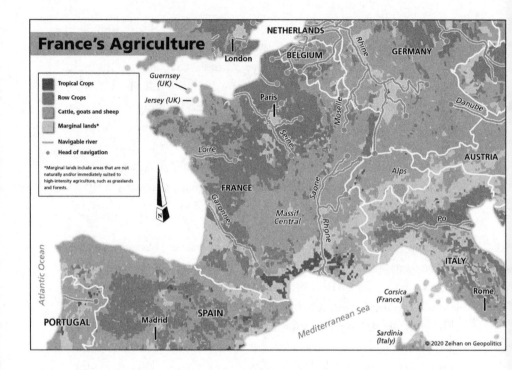

France even has pretty good borders. On France's southern and southeastern borders jut the Pyrenees and Alps, both replete with peaks so sharp, they have permitted little direct land contact with the Iberian and Apennine Peninsulas throughout recorded history.

Even France's borders on the invasion-happy corridor of the Northern European Plain are somewhat circumscribed. To France's northeast, a series of densely forested foothills push from the Alps nearly to the North Sea coast, making this Belgian Gap the Northern European Plain's narrowest point. Unsurprisingly, this chokepoint has been the site of some of the fiercest fighting throughout European history.*

France's many rivers distribute the Beauce's bounty far and

---

*   Most recently the Battle of the Bulge.

wide. They serve as arteries for commercial interests. They entrench the French national identity. They encourage power projection in three wildly different directions into three wildly different regions: the Channel and North Sea, the Mediterranean and the Middle East, and the Bay of Biscay and the New World. They shelter France in the most defensible position possible on the Northern European Plain while still enabling the French to take advantage of all the opportunities the plain can provide.

And as geography doesn't change, those opportunities have existed for a *long* time.

With the Roman Empire's fall, Latin influence merged with—as opposed to overwriting or being expunged from—local norms. The result? A cohesive French identity as early as the fifth century, a full half millennium before the English and twice that before the German. From the French core in the Beauce, this identity seeped into peripheral spheres: remnant Celts in Normandy, Germanic tribes to the east, pockets of Basques in the southwest, even the isolated peoples of the Massif. The French have arguably the longest tradition of operating as a cohesive culture vis-à-vis their location of any people on Earth.

France's unique mix of geographies does more than make it special. The configuration of French lands makes it culturally sophisticated, economically aware, technologically advanced, politically robust, and diplomatically essential. It makes France rich, whether in the sleepy countryside or the City of Light. It makes France involved in places as close as Berne and Brussels and as far away as Montreal and Saigon. It grants the French a deep understanding of how the pieces fit together and what levers need to be pulled to achieve this or that result. It makes France a superpower, and a durable one at that. . . .

. . . which is *not* the same as saying it has always been an easy road.

THE GEOPOLITICS OF LIMITATION, PART I: NATIONALISM

Late-eighteenth-century France was a gooey Petri dish, marinating a riotous mix of destabilizing elements. The early industrial technologies were disrupting traditional craftsmen. The printing press was granting the media huge reach, but it had not yet established standards for accuracy or a commitment to the public good.* The Catholic clergy was experiencing a rise in political influence like that of the Protestants during the Thirty Years' War, and they were wielding it in ways both enthusiastic and inexpert. Halfhearted economic reforms contributed to food shortages. Mounting debt restricted the government's room to maneuver. The reigning monarch—Louis XVI—was kind of a moron.

In short, the place was ready to blow. Think of the virus of American democracy upon this system as less the systemic disruptor and more the proverbial straw.

It got pretty ugly. The French Revolution didn't simply behead the royal family. Its subsequent Terror introduced Western Civilization to populism of the ugliest sort—mobs large and small literally burning down the Ancien Régime, complete with impromptu execution tribunals so prolific that there were occasional *basket shortages* because so many heads needed to be caught. Things like this happen when ossified cultural, military, and economic norms melt away in a few short months.† What emerged from the detritus in the French Terror's aftermath, however, wasn't simply a reborn France, but something fundamentally new.

Nationalist France.

From a geopolitical point of view, nationalism's defining characteristic is its capacity to better harness national power. Medieval

---

* Think social media.
† I'm looking at you, China.

feudalism splits capacity among the royals, the nobles, and the peasantry, with loyalty within and among the three largely determined by a system of bribes and threats. It works, but is wildly inefficient. In contrast, nationalism fuses ethnic identity directly to a centralized governing system, both eliminating the middleman and making loyalty *part* of the governing rationale. Such organizational slimming funnels more resources—financial and labor, in particular—to the one government that rules it all.

Nationalism took root quickly, in no small part because France's rivers enabled the quick transmission of ideas. A single horrific bloodbath to annihilate the economic and political classes of the medieval order and France was off to the races.

The question quickly became what to do with all those newly concentrated resources. Nationalism suggests the answer. Because the nation-state is rooted in ethnicity, not everyone qualifies as "people." Sitting not-so-comfortably on the other side of the twentieth century, you probably see where this is going. Nationalism might make for a much more powerful, capable, inclusive, and accountable state than feudalism, but it also makes it excruciatingly easy to march to war.

Empowered by the social technology of nationalism, France was the most consolidated, stable, mobilized, and potent country of the early 1800s. In contrast, the rest of Europe was politically shattered, emotionally demoralized, and in many cases militarily incompetent. Napoléon Bonaparte wielded the merger of ethnic and government interests against his unsuspecting foes like a flamethrower against soccer hooligans. In a few short years, France's citizen armies had either conquered or forced alliance upon *every* country on the North European Plain, as well as Italy and Iberia, and stood at the gates of Moscow itself.

Yet the French bid didn't simply fail, it failed *catastrophically*. Which brings us to the first lesson of French power: even when France has its ducks rowed up and everyone else has gone fishing, the geography of Europe—the endless Northern European

Plain, the variety of highlands and peninsulas and islands—
means France can *never* win.

THE GEOPOLITICS OF LIMITATION, PART II: INDUSTRIALIZATION

If nationalism gave the French a leg up on the competition, then
industrialization did the same for France's competitors.

Early nineteenth-century German lands were a cartographic
mess made up of dozens of independent and often mutually antagonistic statelets immensely susceptible to the Industrial Revolution's challenges. Chemical fertilizers reduced farm-labor
requirements, impoverishing feudal lords while forcing farmers
off their lands. Imported manufactures blasted away the Germans' sophisticated guild system (i.e., the middle class of the
time). The pressures built to a head with the unrest of 1848, when
many of the statelets flirted with open collapse, civil war, or both.

Yet the Industrial Revolution also provided the tools to solve
the Germans' age-old problem: the disconnected nature of the
German geography. The new road and rail technologies over-
came the physical barriers separating the German principalities.
Simultaneously, the Germans learned about nationalism's invig-
orating capacities from the French.

These twin revolutions—technical and social—had the deepest impact on Prussia, the largest and most powerful of the
German entities. The Prussians extended their writ by rheto-
ric and rail to adjoining statelets, becoming more authoritarian
and aggressive with each expansion. A mere sixteen years after
the revolts of 1848, Prussia absorbed several of its Germanic
neighbors and defeated Denmark in a war. Two years later, it
repeated the process with Austria. A mere four years after that,
the Prussians came for the French.

The French were breathlessly unprepared for industrialized

warfare, not simply because panache and élan are no match for rifles and artillery. Industrialized warfare is about more than simply putting new hardware in soldiers' hands. It's also about everything that backs an army up back home.

Industrialized medicines drastically reduced fatality rates. Industrialized agriculture and fertilizer created food surpluses, eliminating many restrictions on military planning. Famine and disease—factors that in preindustrial warfare were far more likely than bullets to kill soldiers—became more manageable. Railways reduced deployment times from months to days and allowed resupply from hundreds of miles away. The telegraph enabled military planners to adjust attacks, order retreats, reinforce besieged units, or take advantage of gaps in enemy formations within hours.

Early German victories enabled German forces to reach Paris in a mere six *weeks*, where they laid siege and waited for the city to starve. Germany's new long-range cannon underlined with force the thought that any breakout attempt would merely be a very splashy way to commit suicide. When all was said and done, the French suffered five times the troop losses of the Germans.

The French found themselves similarly outclassed in both of the World Wars. Which brings us to the second lesson of French power: the interconnected nature of the European geography means other European countries need *not* be at their peak to defeat France.

The country that matters the most to France's rises and falls will *always* be Germany.

The Germans have the power: as cool as the Beauce is, it is no bigger than the US state of Kentucky. German core lands are *six* times as large, while German rivers have approximately twice the capital-generation capacity of French rivers.

The Germans have the proximity: the distance from Paris to the German border is less than 250 miles, roughly the distance from DC to New York City.

The Germans have the concentration: France may be able to wield power in many places, but Germany is concerned *only* with the Northern European Plain.

If the Germans can unite their disparate regions, there is little the French can do in a straight-up fight but look for white sheets to cut into easy-to-wave rectangles.

Managing the German question *must* be France's top priority.

FRANCE UNDER THE ORDER, PART I: IT WAS THE BEST OF TIMES

The best way to "manage" Germany is to turn the tables and ensure that *Paris* is firmly in charge of the relationship.

Direct occupation sounds sexy but is ultimately counterproductive. If French forces advance through the Belgian Gap, they discover there is *nothing* to hunker behind. The Rhine Valley bleeds into the northern German coast into the Elbe and Oder systems into the Polish Plain into the Hordelands. Napoleon didn't march on Moscow because he liked the food; he did so because moving past Alsace requires breaking *all* powers that live on the plain.

Something savvier and more holistic is required. The trick is to find a way to harness Germany so France can benefit from German economic strength, while also hobbling Germany so Berlin can't use that strength to Paris's detriment. Such puppet-mastering requires the modification of the very environment of Europe.

It is with no small amount of French annoyance that it was the Americans who crafted that perfect alignment. With the imposition of the Bretton Woods Order, the Americans took security planning off the table completely—defanging Germany as a military power. Politically, the Americans and their British (and French!) allies imposed a new constitution upon Germany so the Germans could never threaten the Order.

This neutering was only one piece of Europe's changed circumstances: After World War II, Portugal and Spain remained locked away in their own isolationist dictatorships. Britain was strategically exhausted and could no longer manipulate Continental events. The Iron Curtain cut Europe's eastern third off from all consideration. The war's victors amalgamated Serbia into a unified Yugoslavia that was neither Soviet- nor Western-aligned. The still-independent Turks were on the far side of the Soviet-occupied Balkan Peninsula and so could no longer participate in European affairs. The Finns and Swedes retreated into armed neutrality to limit the rationale for a Soviet attack.

The Netherlands, Belgium, and Luxembourg—three countries that had been ground under the Nazi boot for a half-decade—needed whatever association was on offer to facilitate their economic and political rehabilitation. As a defeated Axis power, Italy had foreign affairs fully removed from its remit, leaving it to someone else to . . . tell it what to do. Germany wasn't simply occupied; it was partitioned. The original German heartland between the Elbe and the Oder was carved off into the Soviet sphere of influence. The most significant bit that was still part of the west? The Rhineland, hard on France's border.

France could not win in a European battle royale, even when the deck was stacked for it, but if what was left of Europe was only one-third the size, if the Northern European Plain was no longer an endless road, if the Germans were economically, strategically, and politically neutered . . . well, that's a completely different sort of competition.

Facing a very different European geography, Paris got down to the nitty-gritty of crafting a French order for Europe. The result was a series of institutions to lash German power tightly to French goals: the European Coal and Steel Community, the European Atomic Energy Community, the European Economic Community—all precursors to what would in time evolve into the European Union. Technically, all were joint Franco-German efforts, but with the Americans conveniently smothering German decision making, the French didn't feel obligated to issue Berlin an opinion.

The money was good too. The new European institutions regularly pulled funds out of the German budget, transferring them to French needs. Call it exploitation. Call it war reparations. Call it agricultural subsidies or development funds. The French called it *fantastique*! France's centuries-long goal of being Europe's first power had finally been achieved.

But it wasn't all champagne and cheese.

FRANCE UNDER THE ORDER, PART II: IT WAS THE WORST OF TIMES

Because the new French-led European system was possible only because the Americans ran the table in World War II, the French were forced to follow the Americans' new global lead. Yes, when it came to European political and economic affairs, the French were large and in charge, but in the military sphere as well as every facet of the global environment, the Americans dominated. France may have been on the rise in Europe, but—like all the other imperial systems—it was in screaming retreat everywhere else:

- In Vietnam local insurgents managed to gut French forces at Dien Bien Phu in 1954, collapsing the government back in Paris and forcing a French withdrawal from Indochina.
- When the Algerian Revolution began later that same year, the Americans largely sat on their hands, leaving the French to fight an increasingly bitter war that, in the end, went so badly that it contributed to Charles de Gaulle leading a military rebellion that successfully overthrew France's constitutional order.

- Attempts to collude with the British and Israelis in 1956 to seize control of the Suez Canal were met with direct, public, angry American opposition. Not only did the effort fail; the global humiliation shattered the image of France as a major power, spawning a raft of ultimately successful independence movements throughout France's remaining overseas colonies.

For the country that had been at or near the top of the human experience for the previous millennium, to have the Americans simply blanket the world in their power—to know that France's recent successes in Europe were possible only because of the American-led Order—on occasion made Parisians break out in hives.

The core issue was, while the Order altered European geopolitics in France's favor, it also negated many of France's geographic advantages. Being protected by the Belgian Gap didn't mean much if Germany was contained within NATO. Having a politically unified system didn't mean much if the Americans' barring of imperial predation the world over enabled even shattered systems like China to unify. What *had* made France special was a geography that made it unified and wealthy and stable and secure. What the Americans offered the world was a system that made *everyone* unified and wealthy and stable and secure.

In no circumstance was the Order's elimination of geographic weakness more worrying to the French than in the case of Germany.

The most positive feature of Germany's geography is its dense webwork of navigable rivers, granting it the best topography on the planet for economic development. The most negative feature is that those same rivers are on Germany's outer edges rather than in its middle, making it easy for Germany's neighbors to hamstring German unity. The Order magnified the benefits of the former while eliminating the weaknesses of the latter. Now expressly forbidden from having a military policy, the Germans poured all their efforts into economic and financial growth. It would have been impossible for France to keep up under any circumstances. With France attempting to maintain a military capability independent of NATO and the United States—complete with a fully functional nuclear deterrent—it became laughably so. Had Germany not been lashed to the French will, the disconnect would have been terrifying.

Soon it became so. In 1989 the Soviet Empire cracked apart, and on November 9, throngs of Germans—Eastern and Western—surged onto and over the Berlin Wall. The Americans, sensing a once-in-a-generation opportunity to *end* the Soviet Union, directly and publicly threw their support behind the budding German reunification discussions. It didn't take much encouragement for the Germans to, well, plan. Within a year the Germans had legally knitted their country back together. With the Soviet Union's entire western periphery of satellite states now unrelentingly hostile to Moscow, the Soviet Union itself shattered only two years later.

On the day reunification occurred, Germany immediately became France's demographic and economic superior. Even worse, German economic linkages quickly started spreading not simply from West Germany into East Germany, but into *all* the former Soviet satellites. In parallel, the German government started to experiment with—*gasp!*—*independent* stances in foreign affairs, flexing its political muscles in the former Yugoslavia. France was, in a word, horrified.

The French strategy for dealing with this unfortunate reality had three pieces.

First, Paris sought to build an anti-German alliance with the intent of surrounding and submerging Germany in the new European geopolitical context. Countries that had sat out the Cold War. Countries that had been on the other side. Anyone and everyone who might prove to be a counterweight to German political and economic power was slavishly wooed. In each and every case, the French government pushed for rapid accession to the European Union, the institution the French still controlled. Between 1995 and 2008, the European Union admitted fifteen new members, including all from the former Warsaw Pact.

Second, the French turned European diplomacy into a co-dominion effort. Rather than asserting that Paris spoke for the entire EU, the French leadership would first court the German leadership and then present joint statements to the world on the EU's behalf. Such coordination ranged from the purely cere-monial (annual joint meetings of the two countries' cabinets for some well-choreographed ego stroking) to the deeply substantive (collaborating with the Russian leadership to oppose American efforts in Iraq).

Third, the French sought to constrain the Germans within an economic web to complement the political cage. Paris became an unabashed supporter of the common currency. The French became founding members of the Economic and Monetary Union with the Maastricht Treaty in 1992, and the euro became a reality at the turn of the millennium.

Unfortunately for the French, instead of these three "achieve-ments" ensuring France's grip on Germany, the strategy instead gave Germany a grip on *all* of Europe.

First, France's envisioned alliance against Germany proved to be nothing of the sort.

Throughout France's long history, it had at one time or another found reason to abandon and/or betray *all* its would-be allies. Such actions are a derivative of French strength. Unlike the rest of Europe, France could typically feed itself, trade with itself, and unify with itself. The French—and especially the Parisians—see themselves as eternal. As better. In many ways, the French are Europe's answer to China's "Middle Kingdom" mind-set. And just like the Chinese, the French tend to see their neighbors as inferior. Disposable. The moment a country is no longer useful as an ally, it becomes useful as an item to be traded away.

People remember stuff like that. Most of the new EU members at least in part blame the French for their fall to the Nazis in World War II, and by extension their domination by the Soviets for the half century that followed. Instead of bowing to French political needs, they proved infuriatingly uppity. Now that all were in the European Union, they had the means, motive, and opportunity to snarl French plans.

Second, while Paris went to great lengths to cultivate German involvement in great-power diplomacy, the Germans had lots of topics they found need to issue statements on that had next to nothing to do with French interests. France couldn't care less about clearing bridge debris from the Danube in the aftermath of the American bombing of Serbia during the Kosovo War, but as a littoral state, the Germans were keenly interested. France had left the military side of NATO in a fit of pique after the Suez crisis, and so wasn't a party to military–military talks with the Americans and British as the Germans were—and since NATO expanded right along with the EU, united Germany had a voice in the regional security grouping while France did not. France certainly didn't want to become guilty by association when this or that Central European or Balkan leader brought up German actions in World War II. In the end, all the French effort did was get the Germans used to being in the spotlight again.

That proved particularly problematic when the third pillar of France's Germany-containment effort snapped. The euro didn't generate the straitjacket the French thought it would. One early outcome of the common currency was that all European countries could borrow at interest rates appropriate for the *German* economic system—a system predicated upon careful planning, measured growth, and fiscal responsibility. Those are not terms normally associated with places like Italy. As soon as Europe's poorer members gained access to the new currency, they engaged in massive spending programs, funded with cheap borrowing. Social spending boomed, retirement ages dropped, and pension rolls grew as governments tossed cash at citizens and public works projects. For the first seven years, the eurozone economies had growth rates that felt a bit Chinese . . . until it came time to pay for things. For countries like Greece, who were burning cash like a Saudi prince on Instagram, reality hit hard—and it brought receipts.

At first the French didn't see the euro crisis as a big deal. They had forced the Germans to pay for European integration before: first in the post–World War II settlements, second in transfer payments to bring the weaker parts of Europe up to snuff in the 1970s and 1980s, and third in the expansion of European Union structures to all those new members in the 1990s and 2000s. In inflation-indexed terms, it amounted to some $3 trillion (with France happily carving out the meatiest share for itself). What's wrong—so the thinking in Paris went—with a fourth tranche? Instead, the unthinkable happened.

Germany said no.

The German government made clear there would be no bailouts . . . unless the countries seeking funds agreed to a series of strict budgetary controls to remake their government spending systems along the lines of the German model. Germany led the negotiations. Germany set the terms. Germany wrote the checks. Germany enforced the deals.

Courtesy of France seeking new allies to constrain Germany, the Germans now economically dominated all of Central Europe. Courtesy of France's efforts to make the European system a Franco-German duet, the Germans lost their hesitance to stake out public stances. Courtesy of France's effort to constrain Germany within a new European institution, the Germans ended up dominating the entire eurozone. Paris had not just lost control over Germany. Paris had not just lost control over Europe. The country that France had built the EU to contain now ran the show. By 2016 Germany was more powerful in Europe through political and economic control than it had been at the Nazi zenith, all without firing a shot. All because of French miscalculations.

For the French who can swallow their pride and admit they have lost control of their continent-size creation to the Germans, this living nightmare of the Order cannot end soon enough.

Luckily (if that is the right word), it *is* ending. The European Union faces any number of internal challenges: demography, exports, currency, migration, banking, financial, consumption— any one of which is sufficient to end it. Hit with all more or less at once in a time of international turbulence, the EU lacks the capacity to cope. About the only way the EU *might* linger on is if the Germans relent and agree to provide a massive financial backstop of nearly *all* EU members. Even *that* is possible only until the Germans hit mass retirement around 2026.

In a world beyond the EU, however, France's capacity to adapt is robust. The French's statist preferences discourage the sort of multinational supply chains so prevalent in Germany or China, which in turn slims France's international exposures compared with its European peers. Of all the Euro-area states, only Portugal and Luxembourg are less dependent upon trade with the outside world. Only Greece trades less intra-EU. Germany, in contrast, is near the list's top in relative terms, and it is absolutely in the top spot in absolute terms. Post-Order France is the European country that will suffer the least and rebound most strongly.

Even better, France's strategic freedom and capacity make the French integral to German survival.

France's expeditionary-themed military is fully capable of reaching places like Libya, Algeria, Nigeria, Chad, Equatorial Guinea, Gabon, Congo (Brazzaville), and Angola. Most are former French colonies. All are oil exporters, and post-Order none have external security providers. France is one of very few coun-

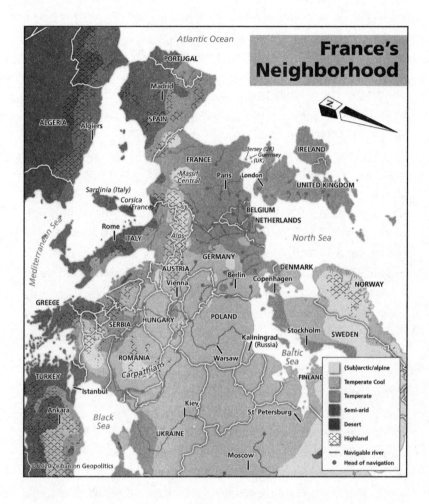

France's Neighborhood

tries that can provide technical assistance for their energy sectors as well as security for their commerce. For the energy-poor Germans, such French middlemanning could well prove the factor separating grim success from catastrophic failure.

And considering Germany's demographic plunge into oblivion, the Disorder's "solving" of the German question might actually be . . . permanent.

And *that* frees up French resources to focus on issues beyond Continental Europe.

ISSUE ONE: FENDING OFF BRITAIN

No matter the era, no matter what French politics look like, no matter who has been in charge in London, the French have always considered the English to be a pain in the ass. The French geography is obviously superior in its ability to generate political unity and economic strength. Great Britain is but one-third the land area of France, roughly half the island is broadly useless, and the highlands of Scotland are easily triple the political challenge of the Massif.* While Great Britain is littered with rivers, only one of them—the Thames—is navigable for a long enough stretch to generate a meaningful economic core on its own. Positioning hurts too: the Thames enters the North Sea facing Northern Europe, so should any mainland European power prove capable of floating a meaningful navy, the existence of England itself falls into doubt.

But all that is of secondary relevance to the simple fact that Great Britain is an island. Britain *must* have a navy, and any navy that France can float will *never* be able to compete.

It's an issue of the regional northwestern European geography. Great Britain isn't simply an island; there are no strategic competitors whatsoever in a roughly 300-degree arc from the island's south-southwest rolling clockwise to its east-southeast. Denmark and Norway's interests broadly mirror the United Kingdom's as regards the European mainland, and Ireland oscillates among satellite, protectorate, and cowering quietly. Any British naval vessel faces minimal challenges in returning home in a crisis and forming up with its sister ships into a single, massive, mailed fist.

The French don't have that option.

First, the French navy will typically be smaller because, as insulated as France is from its mainland neighbors, the country is

* *Freedom!*

still a mainland power with land borders. Once the English conquered those damn Scots, London never required ground forces
for defense ever again.

Second, France doesn't have one coast, but two. It has an Atlantic coast on the Bay of Biscay and the English Channel, and
a second one on the Mediterranean. While France's border with
Spain is only 375 miles, the *sailing* distance between the Atlantic
and Mediterranean points on the Franco-Spanish border is over
six times that. In a strategic crisis when France needs its navy all
in one place for maximum hitting power, the British have traditionally smashed the French fleets in detail with almost contemp-
tuous ease. Even worse, the British have gone out of their way
to make Portugal—an otherwise unremarkable Atlantic-facing
country on Iberia's southwest end—its oldest ally *expressly* to constrain the French from even trying.

Third, the Brits *must* have a powerful regional navy both for
defense and trade, which in the deepwater age puts the Brits a
single imperial baby step away from having a globe-spanning
navy. Not only does Britain's naval strength tend to enable it to
outmaneuver the French in European waters *and* well beyond
Europe, but it also grants the Brits power and wealth on a global
scale—while constraining everyone else. At the British Empire's
height, its rulership of the waves enabled London to dictate terms
to any other power that sought to use the ocean for trade—often
even in that country's own coastal waters. That proved triply true
for European waters, including *French* waters. It was *galling*.

France may boast the best continuity of mainland Europe, but
even it suffers a revolution, loses a war, or throws itself a coup
every generation or so. France's governing system at the time of
this writing is the fifth constitutional iteration since the Revolution in 1789, ergo the country's formal name: the Fifth Republic
of France. In contrast, Britain has suffered only two continuity
breaks since the penning of the Magna Carta in 1215: the War
of the Roses and the Protectorate under Oliver Cromwell. It isn't

so much that the Brits have less drama, but that no matter how good a day, year, or century the French have, the Brits pretty much always have it together and they are *always . . . right . . . there.*

In the coming era, the British will *not* be in a good mood. The French have gone out of their way to make the Brexit process as painful and punitive as possible, and whoever lives at 10 Downing Street is going to be itching for payback. Under the Order, the Americans would instinctively tamp down such activity to maintain alliance coherence, but those days are gone. In the Disorder, the Brits can draw upon a long imperial experience, rich with examples of how to bring economic, political, and strategic pressure to bear. Trade deals with the Netherlands to stymie French economic primacy. Military support to Greece to hamstring French ambitions in the Balkans. A partnership with Turkey or Denmark to complicate French actions in the Mediterranean and North Seas. Kingmaking in Italy or Spain to limit French penetration into the Apennine or Iberian Peninsulas. Maybe even toying with Germany a little to give Parisians some late-night panic attacks.

London's strategic options will always be more flexible and have greater reach, and even in those cases where the British suffer massive geopolitical defeats—such as the loss of the American colonies—such losses affect neither the British mainland nor the core of British power: its navy. France, as a continental economy whose power is measured by its ranking and penetration in the European hierarchy, has no such buffer.

But now, as the Order slides into Disorder, for the first time in a *millennium*, that might not be true. The British navy has been significantly downsized since the Cold War's end, and at the same time the Brits are attempting to float their first pair of supercarriers. Add recent financial crises and the rising costs of the Brexit drama, and the British navy is so thin that it lacks the ships required to provide an escort ring for those new carriers. The only way the British can field them safely is to do so hand-in-glove with the Americans, and the Americans will undoubtedly think nursing Brexit-related angst requires a tool somewhat less gargantuan than an American-augmented aircraft carrier battle-group.

In the two decades it will take the Brits to build up and shake down their own rings of escort ships, the French have a window. "All" the French need to do is ensure that any ill will between the Americans and themselves doesn't rise to the level of ammunition exchanges. Considering that the Americans and French haven't been on opposite sides of a war for the entire history of the American system, that is *probably* as safe a bet as can be made as the world ends.

ISSUE TWO: MATTERS OF IDENTITY

Both Germany's convulsions and the United Kingdom's temporary impotence present Paris with historically unprecedented opportunities. France's next issue has no such silver lining. Like nearly everyone else on the planet, France faces a demographic imbroglio.

First, the good. Yes, France's birthrate *has* decreased, and yes, the French demography on the whole *is* aging, but overall the French demography is very similar to the American: relatively young, relatively stable, aging relatively slowly, and in far, *far* better shape than the rest of Europe. Post-Order France is the only significant consumption base in Continental Europe, which means post-EU France is far more than highly likely to keep its consumer market to itself. That's wonderful for French businesses, disastrous for the neighbors.

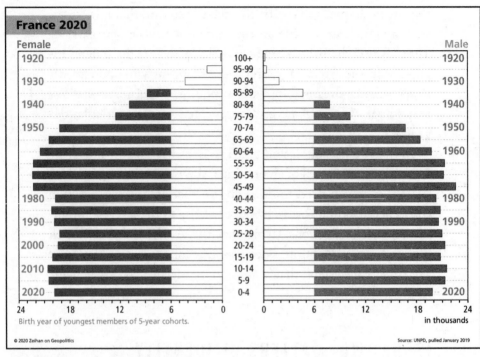

France 2020

Female / Male

Birth year of youngest members of 5-year cohorts.

in thousands

© 2020 Zeihan on Geopolitics

Source: UNPD, pulled January 2019

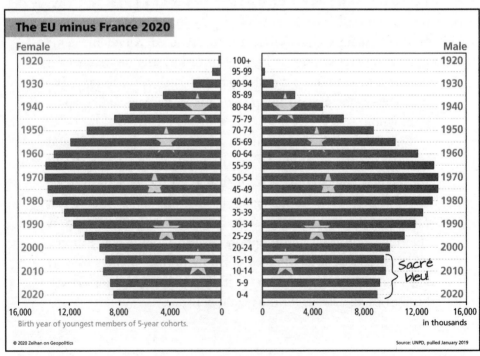

The EU minus France 2020

Female / Male

Sacré bleu!

Birth year of youngest members of 5-year cohorts.

in thousands

© 2020 Zeihan on Geopolitics

Source: UNPD, pulled January 2019

Now, the bad. France's demographic problem isn't with the headline data, but instead a level down, buried in the French national consciousness. Bretons, Basques, Alsatians, Catalans, and more didn't simply dissolve into the French milieu in the late 1700s. To encourage merger, the First French Republic adopted the motto "*liberté, égalité, fraternité,*" a declaration similar to the defining American phrase "all men are created equal."

The differences between the two phrases are telling. The American version proclaims that all are equal *before the law*, while the French version is a declaration of ethnicity—asserting that all *peoples* within the borders of France are now part of the French family, and family looks out for one another. With the horrors of the Terror still large in their rearview mirror, the founders of the First Republic took this commitment to fraternal values seriously and strove to make it permanent.

For contemporary France, this has become a problem. There are more people in France than just the French, and you cannot just invite yourself to join a family. You must be invited. Outsiders are rarely welcomed. There's some truly ugly stuff under this overturned rock.

France was one of the old empires, and in some places the French had ruled not for decades but for centuries. The merger of imperial thinking with *liberté, égalité, fraternité* meant that the French saw their overseas territories less as colonies and more as extensions of the homeland. That meant including the extended "family" in France's economic and political integration regardless of where the flag flew: Guyana, New Caledonia, Tahiti, Vietnam, Senegal, Algeria. When the French Empire collapsed, many of those people—many of these *citizens*—relocated to the imperial center. As years turned into decades, none of the former colonies did particularly well. Trickles of migrants making hay of past imperial commitments followed that first batch to France.

Many French (grudgingly) acknowledge that these imperial imports and their descendants are legal citizens, but they steadfastly, allergically reject the idea they could ever be *French*. Even if they were born in France. Even if their *grandparents* were born in France. Resentment runs hot on both sides of the racial divide.

In a world of Disorder, this wedge of disunity in the French system is France's biggest weakness. It stratifies French society between Frenchmen of Caucasian background and those of the colonial imports. It weakens France's appeal to African nations. It provides outside powers with a fifth column to exploit. It diverts scarce resources to maintaining civil stability. It creates alienated communities in every major French city. And in the country that *invented* nationalism—the concept that a country and its ethnicity are inexorably linked—a lack of civic unity on the issue of identity is potentially catastrophic.

Perhaps the most critical aspect of this disunity is that the French have at best only a vague idea as to the scope of the problem. *Liberté, égalité, fraternité* is so entrenched into French political culture that it is unconstitutional even to collect data on people's ethnic background. This was done to help merge all the various late-1700s ethnicities into a homogenized system. But now it means France doesn't know how many of its people are second-class citizens. The figure is *probably* about six million people, or 10 percent of the population, and it is *probably* skewed toward the younger end of the population scale. But the nitty-gritty details of the truth are anyone's guess.

How the French deal with this issue cuts to the core of the very definition of the French nation in the near future. Option one would be for the French to renegotiate their core cultural definitions, to repeat the social experiment of the late 1700s and once again broaden the definition of "French" by adopting the liberal dream: full multiculturalism.

For a country that has bent over backward to make itself as unwelcoming as possible to Polish workers (fellow Caucasians from a fellow Northern European country), as well as to block the immigration of refugees from Syria (*a former French colony*), it is difficult to envision the multicultural approach happening. What progress the French have experienced in opening their hearts is largely limited to former imperial subjects who are Christian and who speak (perfect) French. Southeast Asians and especially Algerians most definitely are not feeling the love.

Option two is the populist approach. Instead of widening the definition of "French," narrow it to only include the ethnically French, and maybe a scant handful of other Caucasians who can convince the preexisting French of their enthusiasm for all things France. This requires a logistical pretzel: how does the state act as the social arbiter and strip citizenship from the millions of people who don't pass the visual or auditory tests for being suitably French? It is with a dark irony that it is France's late-1700s experience with just this sort of horrific purge that frightened the French into embracing the dream of *liberté, égalité, fraternité* in the first place.

Between these two extremes, contemporary France pinballs its way forward. Non-French ethnics can have and keep citizenship, but they are de facto forced by the French government and French culture alike to live in the country's infamous suburb-slums, the *banlieues.*

In many ways, the French system takes the two types of racism most prevalent in the United States and applies the worst of both. In the American South, racism takes the form of, "We will mingle, but we are not equal." In the American North, it is in the vein of, "We are equal, but we will not mingle." In France, the targets of racism are out of sight *and* out of mind, consigned to ghettos and at the back of the line as regards government services.

This is a combustible and unsustainable mix, tantamount to occupying a sizeable chunk of one's own population. In a democracy,

---

* The French urban model is different from that of the United States. In France the affluent and middle class live throughout the city cores, while the suburbs are the low-cost, high-crime housing.

even in the best-case scenario, it is a great way to get a lot of people killed.

Never forget that one outcome of *liberté, égalité, fraternité* is that France is the European country with the highest percentage of ethnic Arabs who are not merely isolated residents, but permanently disadvantaged citizens. Such ghettoization provides terror groups with communities throughout France that have already proved useful for recuperation, recruitment, planning, and refuge from the authorities. Add a disintegrating Middle East, heavier migrant flows, and general European dislocation, and France—already having suffered from more deaths from terror attacks than any other European state—can now fold terrorism into the heart of its identity crisis.

ISSUE THREE: A SOUTHERLY APPROACH

In the Disorder, Northern Europe is certain to be a busy place. Germany and Russia will be locked into struggle until (mutual?) oblivion, while the British will once again be manipulating events and trends European whenever and however they can.

In contrast, Southern Europe is downright sleepy. It is largely an issue of capacity. While Spain and Italy's extensive ocean frontage necessitates naval forces, both have largely deferred to the United States on security issues. Their current naval orders of battle have a next-to-bottom-shelf feel. On the actual bottom shelf are the naval pigmy states of North Africa; all combined couldn't fight Spain on the waves. For France, at the Disorder's beginning the Western Mediterranean is already a French lake.

That hardly means there is no work to do or competitions to wage. France's Mediterranean challenges will come in three waves, each building upon the one before.

First, there will be issues with the neighbors. Like France, both Spain and Italy import roughly 90 percent of their oil and natural gas from beyond the European continent. Conflict in the former Soviet space will remove what used to be one of the Continent's three sources. An embittered Britain is unlikely to want to share much that comes from the North Sea and is not above twisting some arms in Norway to keep the sea's supplies local.

That leaves just North Africa. After years of civil war in Libya and physical destruction and industry mismanagement in Egypt, there isn't enough to go around. Algeria is the only remaining large-scale producer, but even it cannot quite supply *one* of the three Southern European powers, never mind all of them. France certainly could defeat both its European neighbors in any contest—France has more money and a bigger market and a better manufacturing base and more guns and a better navy— but what would be the point? Wrecking the Spanish and Italian systems would generate scads of refugees while also ruining some low-hanging strategic fruit.

Better to supply them with energy rather than deny it to them.

To that end, the French must look to West Africa. The headline figures look good: the slew of states from Libya to Angola could supply Western Europe with about 6.7 million barrels of crude and product per day, almost precisely the combined consumption of Italy, France, Spain, and Germany.

If France can keep those flows operational and focused on Europe, France will far and away be Europe's first power. Spain and Italy will be satellites, and Germany can keep focused on its eastern frontier. It should work. There should be enough to go around . . . but only if nothing—absolutely *nothing*—goes wrong.

That would be a bad bet. Just as there's just barely enough oil in North and West Africa for France's would-be sphere of influence, there's just barely enough in the North Sea for the United Kingdom's would-be family in Scandinavia. The United Kingdom's backup plan will also be West Africa, so right from the start France's perennial German and British complications will tangle themselves up in whatever else Paris tries to do.

With the margin of error so thin, France has no choice but to expand a second plank into its southern approach. France must get involved up to its elbows in the political guts of North and West Africa simply to maintain stability.

Nigeria's 8 major and 250 minor ethnicities make it a seething stew of racial violence that killed at least 4,000 people a year since 2013 with a peak of nearly 16,000 people in 2014. Angola is run by a quite literally genocidal family and cannot operate its offshore energy sector without extensive outside help. It doesn't take an overactive imagination to conceive a Libya-style civil war occurring in Gabon or Equatorial Guinea or Chad that has a similar impact upon oil exports.

For France, that means political and technical support. That means humanitarian assistance and trade. That means bribes. That means maritime convoy systems and troop deployments to ensure smooth operations. It means dealing with or disposing of inconvenient personalities. It means direct military occupations when other means fail. And even that neo-imperial approach assumes that any competition with the British remains more of the good-natured type over tone and approach and accolades rather than a Thunderdome-style melee over who gets to have gasoline next Thursday.

The only way any of this can work is to ensure that additional supplies of crude oil can reach far Western Europe. And that means shifting attention east to a third front.

In any world where the Americans are not providing global maritime security, the Suez connection is the best way to move goods and people between Europe and Asia. Sending fat, slow boats around the African continent is simply asking for them to be stolen by pirates, privateers, or the actual navies of African states. Whoever controls Suez will be able to both make exorbitant transit income and access directly the Red Sea oil exports of Saudi Arabia—and in so doing shape countries on both sides of the link. Success would make France the broker of Europe. And the Mediterranean. And the Middle East. And give France a seat at tables in South and East Asia. What France wants—what France has always wanted—is total control of Suez itself.

But unlike the Western Mediterranean, which is France's to lose, the Eastern Mediterranean has a lot going on. Lebanon and Syria oscillate between dictatorship and anarchy. Israel always has a firm opinion on security matters. Cyprus hosts a pair of large British naval bases. Turkey is firmly, unapologetically on the rise. Perennially unstable Egypt is . . . perennially unstable. And of course the nexus of Saudi and Iranian interests has been enthusiastically messy since the eighth century.

France, put simply, needs to seize the Suez Canal and a reasonable buffer of territory on both sides so French forces can ensure smooth operation. Luckily, no plan to invade the Middle East has ever gone awry. . . .

FRANCE'S REPORT CARD

BORDERS: France is the sole country on the Northern European Plain to have meaningful geographic boundaries: the Pyrenees, the Alps, the North and Mediterranean Seas, and the Belgian Gap.

RESOURCES: The French have almost as good a riverine transport network as Germany, but it is in agriculture where France truly shines. The variety of microclimates makes France a world-class agricultural producer and exporter.

DEMOGRAPHY: The French boast the healthiest demography of industrialized Europe, but unassimilated immigration from sub-Saharan Africa and the Arab world is a growing issue that will soon reach critical mass.

MILITARY MIGHT: Paris learned its lessons during the World Wars, primarily to trust no one. The French maintain a military independent of NATO, an economy separate from the EU, and their very own nuclear deterrent.

ECONOMY: France is a significant producer both agricultural and industrial, making it perhaps the most difficult EU member to negotiate trade deals with. While not an industrial producer on par with Germany (who is?), the French can make or grow most anything they need at home.

OUTLOOK: France is rarely number one, but it is almost always in the top five. When France's neighbors struggle—as they are now—French power naturally rises.

IN A WORD: Finally!

CHAPTER 9

IRAN

THE ANCIENT SUPERPOWER

Few regions have confounded, confused, or captivated the world as much or as long as the Middle East. Sitting at the crux of Eurasian trade for thousands of years, Persia was ancient Greece's existential threat. Carthage and Phoenicia traded, raided, and warred with Rome. The Moorish conquest of Spain. The Crusades. The Silk Road. The Spice Trade. The Ottoman Empire. The Age of Empire. The Scramble for Energy. The Global War on Terror.

Part of the reason the Middle East features so prominently and so often in Western lore is that it is right *there*. Athens and Rome had more contact and arguably more in common with the world of the Eastern Mediterranean—aka the Levant—than they ever did or would with Berlin, Paris, or London. Like it or not, this region is a central piece of the West's collective cultural history.

Another reason is money. Before deepwater navigation, much of the trade between Europe and Asia relied on routes through the Middle East. For the relatively short historical period of steam travel and coal, the region languished—until the discovery of oil. The dissolution of empires and the massive injection of wealth into countries built literally on sand made everyone who relied on fossil fuels a party to the sometimes esoteric and often violent disputes that still define the region today.

As hard as it is to believe, the American-led Order has kept most of these disputes limited to a slow burn. As the United States started to step back from the world stage, the Middle East was the first to descend into Disorder, and the region has precious little in the way of stability to look forward to.

Part of why the Middle East always seems so chaotic is that it's next to impossible to establish continuity. It comes down to water—or, more accurately, the lack thereof.

What rain the region receives occurs seasonally in a scant handful of highlands, most of which are too steep to be particularly useful as living space. The resulting streams and rivers follow narrow courses with minimal downstream inflows; miss those, and oases are the only other water source. Nearly all the region's agriculture must be irrigated, and most of its population clusters tightly around whatever water is available: on the Levantine coast, in Mesopotamia, along the Nile, at Mecca, at the Damascus oasis, and so on.

Dictated by that most essential of resources, this settlement pattern forces a base, visceral competition upon the region. One seasonal hiccup in the delicate water supply, and local civilizations are forced into a blizzard of raiding and counter-raiding. Civilization falls apart.

The nature of the region's terrain doesn't help. Nearly all of the Middle East is flat, open arid lands or hard desert. Throughout, there are painfully few geographic barriers a culture can hide behind. Desert communities can fall with a single raid. Coastal communities can be subjugated—or wiped out—by a single foreign vessel. Oasis cities can be besieged and starved. It doesn't take a horde to sweep across the flats of Mesopotamia and force a flag-change. Chokepoints, like Aleppo or Suez, sit at maritime and/or land-based crossroads, attract a lot of attention, and tend to change hands from time to time.

THE POWER OF WATER

Iran is different in every conceivable way, but it all boils down to one factor: populated Iran is a fused, sprawling *mountain* system. Iran's Zagros mountain chain fills the country's entire southwestern third, while the Elburz dominates the northern third. The contemporary capital of Tehran sits on a plateau where the two chains meet. With an average peak elevation of ten thousand feet, the two chains not only force out fairly reliable rainfall, but their valley floors tend to be above three thousand feet. Unlike nearly everywhere else in the region, it actually *rains* in Iran where people *live*. And that changes everything.

Direct rainfall enables agriculture without necessarily requiring irrigation. Lower labor requirements free workers to do other things, like going to school or practicing a trade or composing a poem or waging war. Culture here has roots stretching back five thousand years.

Mountain living has other advantages. Anyone wanting to invade Iran must fight their way uphill into the Persian core and batter through every. Single. Mountain. Line. This defensibility shapes Persia's participation in international affairs. Iran isn't a destination, but instead a knot of difficult territory that must be bypassed by those in Asia or Europe who want to trade with each other. Since passing through Persia is so difficult, trade in all areas tends to flow *around* Persia whenever possible, preferring either the dangerous, often pirate-ridden seas of the Persian Gulf and the Red Sea, or the flatland routes through brigand-rich Central Asia.

The only exception is the Silk Road. Persia did play host to one major route: from the Central Asian steppes and deserts, traders would enter Persia from the northeast and drop down to Persia's coastal strip on the southern shore of the Caspian Sea.*

---

* Take a look at Google Earth. It's the only flat, green place in the country.

Such trade came into Persia's sphere of influence—enabling the Persians to take a big fat bite out of it—but did so without unduly undermining local culture or establishing economic dependency or strategic vulnerability.

Unlike the dozens of city-states and empires that have risen and fallen throughout the Middle East, the Persians have art and history and culture that isn't short-lived or incidental or fused with foreign practice but instead anchored in millennia of continuity. The Persian language and Persian customs—conservatively— date back *hundreds* of generations. Persian continuity enabled a local economy far more sophisticated than the pearl harvesters of the Persian Gulf's southwestern shore or the caravan towns of the Syrian interior or the grunt farmers of Mesopotamia or the Nile. No one-horse town, the Persian economic system is thick with variety, literacy, and technical skill.

The Persians don't simply look down on their neighbors—both literally and figuratively—from their mountain fastness; from time to time the Persians boil out and conquer them all. In past incarnations, the Persian empires have expanded as far as Thrace, the Northern Caucasus, the Aral Sea, the Hindu Kush, the Indus, the Nile, and Cyrenaica.* The Persians have had their highs and lows just like everyone else, but most notable is the fact that the lack of local competition meant that the Persian empires of the past didn't so much rise and fall as expand and contract. Only on rare occasions did the Persian core itself fall. Today the Persians are under only their seventh governing system since the rise of the Achaemenid Empire. Americans—who are still on their *first* governing system—might feel a bit smug at that, but that Achaemenid Empire first rose twenty-six *centuries* ago. These folks have some serious staying power.

---

* To put that into contemporary names, at Persia's maximum historical extent it absorbed most or all of contemporary Azerbaijan, Armenia, Georgia, Turkmenistan, Tajikistan, Afghanistan, Pakistan, Iraq, Kuwait, Jordan, Syria, Israel, Lebanon, Egypt, Turkey, and Bulgaria, along with big bites of Uzbekistan, Libya, and Greece.

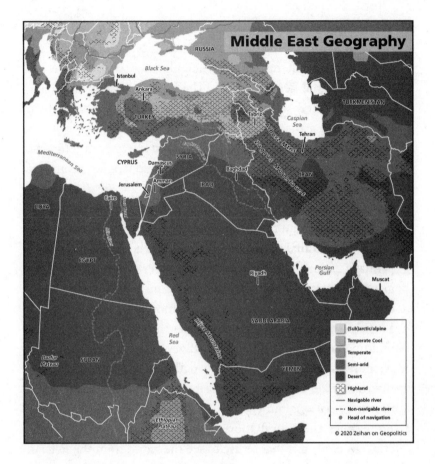

Backstopping Persian sophistication was the need to deal with the heterogeneous nature of the Persian population. Everywhere else in the world, successful, cosmopolitan, thriving cultures *never* emerge in mountainous terrain. Persia is an oddity largely because the lands around it are even worse. Tiny patches of useful land in the inter-Caucasus zone to the north, hard desert to the northeast and east, the invasion corridors of Mesopotamia to the west, the desiccated canyon lands of inner Anatolia to the northwest. There isn't much competition. Being an exception forced the Persians to develop tools to mitigate the fact that each and every one of their thousand mountain valleys had its own cultural identity:

- The early Persians were the original multiculturalists in that they maintained a relatively loose definition of what "Persian" meant, making it possible for neighboring groups to be admitted into the Persian whole.* The relative success of Persian lands compared with the desert communities around them added to the attraction.
- The Persians aggressively promote interaction among the mountain valleys they control. Internal trade, economic dependence relationships, and intermarriage all deliberately and successfully blur the differences among the various valleys into a fluid Persian norm.
- Persia always maintains a huge intelligence apparatus whose primary rationale is to keep tabs on all its myriad sectarian groups: Shia Muslims, Sunni Muslims, Balochis, Azeris, Arabs, Jews, Turkomans. All and more live somewhere in Persia—with even more groups under the Persian roof in times of imperial expansion. The intelligence authorities' primary purpose is to provide the state with insight as to where an internal trade route or bribe or arranged marriage might be useful for national unity.
- Sometimes the information gleaned necessitates beating the crap out of groups that refuse to see the world from the Persian point of view. Since antiquity, the Persians have maintained a massive army relative to their population. In essence, the Persian army occupies Persia itself in order to preserve the ideal of national unity. That such an army can be used for other things has enabled Persia to be one of the world's largest empires at multiple times throughout recorded history.

---

* The most recent evolution of this concept is today's term "Iranian." Persian is an ethnicity, while Iranian is the nationality. Arabs, Kurds, Turkomans, and more consider themselves Iranian despite being of different ethnic background.

It has proven a wildly successful mix. Persia is the *only* component of the ancient world to survive through the classical era and on to the postclassical era with any real continuity.

In modern times, it all went to pot so painfully quickly.

GOOD, BUT NOT GOOD ENOUGH

Persia's geography may be top-notch for the Middle Eastern region, but by global standards, it's just a bunch of mountains. They may provide life-giving water and significant protection, but from an economic point of view, the biggest drawback is the old bugaboo of internal transport. The country has no navigable rivers and precious few flat areas—none of which are connected. Building and maintaining infrastructure in such a topography is especially difficult. In a world in which all interaction is regional, Persia can manage such shortcomings fairly easily.

Times change.

The twin technological revolutions of deepwater navigation and industrialization transformed how humans interact with one another—and how the Persians interact with the outside world. Part of this was military: industrialized Western armies and navies could reach out and touch anyone—Persia included—in ways heretofore unimaginable. But the deeper transformation was economic.

Successful industrialization requires a lot of capital—typically generated by navigable waterways—to fund industrial development, and flatlands to keep the costs down. Persia has neither. What Persia had was a loose, fractured, valley-by-valley economic system based heavily upon local craftspeople and on-again, off-again internal trade. It was spectacularly advanced compared with everyone else in the preindustrial Middle East, but it couldn't hope to keep up with the endless flood of high-quality, cheap goods out of capital- and river-rich industrialized Europe.

For the Japanese, the Industrial Age meant a crushing end to the Shogunate-era system and mass dislocation *followed by* the full leveraging of the industrial technologies to become Asia's first power. Iran got all of the first part but lacked the concentrated population centers and naval capacity for the rebound. The country quickly became dependent upon external supplies of everything, including the capital borrowed to purchase European goods. In a few short decades, the Persians went from being the richest, most sophisticated economy in their region to having the Western nations influence every aspect of their lives—just like everyone else. The Industrial Revolution crushed Persian grandeur and spirits in a way no war could.

It's worse than it sounds. The industrial era robbed Persia of much of what made it successful, wealthy, and unique. Yet the state still had to wrestle with the consequences of the country's messy, mountainous, fractured political system. That massive army couldn't be downsized, but now it could no longer be afforded. Persia slid from a land of cosmopolitan wealth and security into an inward-looking, poverty-stricken police state. In 1600 Persia ranked among the world's richest cultures, but as the rest of the world broke loose in the Age of Discovery and in time the Industrial Revolution, Persia stagnated.

Then there's oil.

Yes, oil guaranteed Persia income like nothing it had had before, but the Persia of the oil era was not the Persia of old. Oil-era Persia had *already* been broken. It was no longer a sprawling empire. It didn't control its own economy. Its educational system was stunted. It was surrounded by European colonies. Worst of all, Iran had the disadvantage of being one of the first places oil could be commercially produced. Even before World War I, oil may have made Iran globally relevant, but it did so in the worst possible way: by making it a target.

The British and Americans were the biggest players in Persian oil, and when the Iranian people tried to exercise more authority over their natural resources and experiment with democracy early in the twentieth century, the British and Americans chucked out the prime minister and reinstated an absolutist monarchy headed by the last shah. The fact that the Iranians got caught flirting with the Nazis to balance against the British, Americans, and Soviets didn't help. The Tehran Conference, famous for its photo of Roosevelt, Stalin, and Churchill together, is infamous in some Iranian circles as being the end of any authority the Iranians would have over their place in the world or over their resources. Irony of ironies, gaining access to the world's most valuable resource *destroyed* Iran's continuity.

World War II and the subsequent global Order gave the Iranians a fresh start. The new Order forcibly ejected the colonial powers from the region, while the Americans included oil-rich Iran in their circle of allies. The Americans deemed Iran's ongoing existence as essential to the functioning of the global economy, their global alliance, and, at the end of the day, American security.

Iran boomed, and it was about (a lot) more than simply safe oil exports. The economic plank of the Americans' Order was about enabling countries to develop, many for the first time. That meant industrialization, and industrialization meant energy access. *Global* industrialization demanded easy, safe, global access to Middle Eastern oil. Oil demand didn't simply rise; it gushed— benefiting every oil producer everywhere. But it also meant the Americans had no choice but to wade hip-deep into the ebb and flow of Middle Eastern politics.

Right from the start, the American-Iranian relationship under the Order was a frustrating one. Iran's political, business, and religious elite were not fused into a single family or city or region. Iran was a complex, cosmopolitan tableau of competing power centers and ideologies. Assuming the Americans got the right person on the phone, that was only the beginning of the process. The Americans experienced quite a few teething pains in their early efforts to manage their Iranian ally. In 1953 this manifested as an American-sponsored coup. In 1979 it took the form of a popular revolution against the American-sponsored Iranian leadership. The shah was out, the ayatollahs were in, and the American embassy staff was held hostage for 444 days.

The bad blood set in all at once.

For the Iranians' revolutionary clerical regime, it was simple: The shah was a creation of the Americans; the shah was fabulously corrupt and gloriously violent; therefore, the American-backed Order was to blame for Iran's troubles. Anything that would harm American-brokered, American-supported, or American-imposed stability would by extension weaken the Order that had kept Iran in check.

For the Americans it was simple: the Order necessitated global access to Persian Gulf oil. Anything that threatened that access was therefore a threat to the global Order and American security. Therefore, American power must be used to contain and ultimately eliminate threats to either oil production or regional stability.

So long as the Americans remained committed to the Order—no matter the reasons or thought processes—Iran was a foe. So long as the Iranians remained committed to breaking out of their strategic straitjacket—no matter the reasons or thought processes—the United States was a foe.

Since 1989 the road has only gotten rockier, largely due to American actions in Iraq—the only country in the region that was a military peer to Iran. In 1991 American military forces ejected Iraqi forces from Kuwait, clipping Iraq's wings. In 2003 the Americans returned and liquidated Iraq's ruling Baath system. The Americans proceeded to directly occupy Iraq for a decade, but when it became apparent to Washington that it could not force Iraq to act like Wisconsin, the Americans decamped.

With Iran's only regional check eliminated, the Iranians proceeded to stretch forth to create a contemporary incarnation of their empires of old.

For Iran, things may have gone a little *too* well.

OVERTURNING THE ORDER

If the goal was to disrupt the Order, the Iranians were well prepared. Modern-day Iran, as heir of ancient Persia, commands an intelligence system expressly designed to manage its polyglot, multiethnic hodgepodge. Just as Iran's oversize military can serve a foreign as well as domestic need, so too is its intelligence expertise hideously applicable in inflaming and exploiting cleavages across the broader region (while also protecting the homeland from any blowback).

Within months of the shah's fall, Iran began deploying intelligence assets throughout the Middle East to play the divide-and-conquer game—no wheel invention required. The game has worked well nearly everywhere the Iranians have played it.

Lebanon is where it all started. The small Levantine country had fallen—again—into civil war in the 1970s with Christians and Sunni Arabs at each other's throats. But there was a third group in play: Iran's Shia co-religionists dot much of the Middle East, and one of their densest concentrations is in southern Lebanon. The local Shia not only share the same religious beliefs as Iran and make up approximately 30 percent of Lebanon's population, but they were *already* organizing a militia in 1982 to resist Israel's ongoing occupation of southern Lebanon. A culture as ancient as Persia recognizes serendipity when it sees it, and so Tehran poured in weapons, money, technical support, and leadership resources. Voilà! The Middle East's best-equipped, best-trained, most-capable, most-(in)famous militant group: Hezbollah.

As the years rolled on, Hezbollah played a central role in antagonizing Israeli forces, became a political power broker within Lebanon itself, and even became the most powerful foreign force in the subsequent Syrian Civil War. Hezbollah's Shiite affiliation meant their success was Iran's success, but the greatest gift of all came from the militia's geographic proximity to Israel. Shiites in Lebanon had the positioning and reach to directly threaten the Israelis on both sides of the border. Iran's issue with Israel was not so much that it's a Jewish state (although that helps), but rather Israel's status as the most obvious local ally of the global hegemon, which gave Iran more leverage over the United States than it could have ever hoped to achieve otherwise.

The **Syrian Arab Republic** was one of the ayatollahs' earliest allies. During the French colonial period, Paris found the Sunni Arab supermajority to be a bit too spunky. The French solution was to elevate the Alawite minority to leadership. As the Alawites had often been the target of Sunni Arab crackdowns, the Alawites took to the role with violent enthusiasm. When the French left, the Alawites remained in charge, turning Syria into the most repressive country in a repressive region. More than a little bit afraid of the uppity Sunni masses, the ruling Assad family seized upon the rise of the Iranian ayatollahs to secure an ally. As the Iranians love a good sectarian play, it was an easy alliance.

Iranian support for the Assads has been a central pillar of Syrian security ever since, but it comes at a cost. Syria has to not only support but often directly operationalize Iranian policies on Lebanon and Israel—policies that often require no small commitment of troops and money. All in all, it has paid off for Syria's Alawites. Without direct Iranian military support in the form of Iranian troops and Hezbollah fighters, the Alawites would have long ago lost the Syrian Civil War.

The Iranians have had the best and worst luck as regards **Iraq**: worst in that in 1980 Iraq invaded Iran, starting an eight-year-long war that inflicted a million casualties on the Iranians; best in that in 2003 the Americans did the heavy lifting and destroyed the regime that had launched the 1980 war. With Iraq's government shattered and the American occupation government inexplicitly obsessed with reshaping the country in its own image, Iran wasted no time and spared no effort in extending its influence deep into Iraq's Shia community. Within two years of the American invasion, Iraq's Shia were treating Iran as the conquering hero and actively cooperating with Iran in deepening its connections throughout the Iraqi system. Iran was so successful that when Iraq's Sunni community began to ally with an al Qaeda faction that would eventually morph into the Islamic State (ISIS), the Americans had little choice but to informally ally with pro-Iranian groups. By 2012 even the Americans had to admit that they couldn't find an Iraqi Shia to serve as prime minister who was not at least in part beholden to Tehran.

The internal political fractures in **Yemen** make Lebanon look well thought out. The rugged, mountainous topography of Yemen has at times housed competing kingdoms in mountain valleys, enclaves of religious minorities, Marxist rebels, a socialist government, foreign occupiers—and often a variety of the above simultaneously. Also, unlike Lebanon, in Yemen, the Iranians lack a true sectarian ally. The closest group they've found are Yemen's Houthis, a Shiite-ish group that many Iranians see as borderline heretical.* Between the distance and this ideological disconnect, the Iranians know control of Yemen isn't an option, but that's OK. Anything that wrecks Yemen's balance of power is bad for Saudi Arabia, Iran's regional rival. Encouraging a Yemeni brush fire in Saudi Arabia's backyard is reward enough.

Another mountainous hotbed of internal competition, **Afghanistan** is a less advanced, landlocked, poorer, formerly

---

* Sunni Islam is a bit like Catholicism, in that it is theologically united, while Shia Islam is more like Protestantism, in that every sect has its own traditions and hierarchies. Iranians are of the Twelver denomination, the largest Shia sect. Yemen's Houthis are predominantly Zaidi Muslims.

Soviet-occupied version of Pakistan. Roughly half the country is Persian-speaking (Dari is a close cousin, perhaps half brother, of Farsi), while 15 to 20 percent of the population is Shia, giving Iran a stable foothold in the country. The sad reality is that, beyond providing recruits to throw into Iran's other projects and heroin to smuggle to Europe, Afghanistan doesn't have much value for Tehran. It was a lovely launching pad for attacks against US troops during the decade Washington spent patrolling the entire country, but since the large-scale American withdrawal in 2014 Afghanistan's trademark rough, dry terrain has prevented Iran from implementing any kind of direct control as it has enjoyed in Lebanon or Iraq.

The broader regional geopolitics set the Iranians up for success:

- America's late–Cold War crusade against the Soviet Union shattered Russian power to Iran's north while removing Soviet forces from Afghanistan.
- America's subsequent wars in the region smashed Iraq, historically the biggest local check on Iranian power.
- America's strategic sponsorship of the Arab states of the Persian Gulf enabled the Gulfies to avoid investing in their defense or acquiring any meaningful military expertise; America's regional retreat leaves them all vulnerable.

It's a well-set table, and the Iranians' sectarian plays have paid off handsomely:

- Iraq is one step from being a client state, even smuggling a trickle of Iranian oil past the Western sanctions on Iran.
- Syria is too weak and dependent to resist orders from Tehran.
- Hezbollah is far and away the most powerful military and political faction within Lebanon and has more than proven itself as a useful expeditionary force in both Syria and Israel.
- Afghanistan provides a steady stream of bodies to die in Syria.

- Iranian influence in Yemen has embroiled the Saudis in a lovely strategic distraction.

The Iranians have won the Middle East in large part *because* of American security commitments there. Now with the Americans leaving, it would seem the Iranians just have some mopping up to do before consolidating themselves as the regional superpower.

Not so fast. This is the Middle East.

It is *never* that easy.

IRAN'S REPORT CARD

BORDERS: Iran is an arid, mountainous country surrounded by deserts and even more mountains. Invading Iran is therefore damnably difficult, but the same things that protect Iran also tend to stymie expansion.

RESOURCES: Iran's mountains hold mining potential, but for over a century it's been all about an overdependence upon oil, oil, oil.

DEMOGRAPHY: Iran's population is the largest in the Persian Gulf region, but a crash in birthrates after the 1979 revolution makes Iranians much older than their peer Gulf Arab populations.

MILITARY MIGHT: Iran has not one, but two armies. Both are fiercely loyal to the country's ruling elite, outfitted with primarily re-worked Soviet matériel, and are as likely to fire on domestic agitators as external armies.

ECONOMY: Oil exports form the country's export basis, distantly followed by low-grade industrial and agricultural products such as cheap steel, dates, pistachios, and carpets. All are subject to one of the strongest sanctions regimes in history. The economy is . . . struggling.

OUTLOOK: Cast in the role of troublemaker for four decades, Iran has recently experienced mammoth success in disrupting its foes. Now that Iran has more or less won regional leadership, it is woefully ill prepared to protect its gains.

IN A WORD: Winner?

CHAPTER 10

SAUDI ARABIA

THE ANTI-POWER AND
THE DESTRUCTION OF
THE MIDDLE EAST

Right up until the early decades of the twentieth century, the territory of today's Saudi Arabia was crap for development. The Arabian Peninsula is among the world's least habitable areas. Temperatures regularly breach 110 degrees Fahrenheit, even in what passes as winter. The peninsula's interior is almost entirely rock and sand. The Saudi desert doesn't benefit from a subterranean aquifer like the Sahara, nor is it speckled with oases like the Karakum in Central Asia.

Historically, the vast bulk of the region's population has clustered near the mountains along the peninsula's southeastern and western flanks, where elevation wrings a few drops out of the sky.

Even these somewhat less arid subregions are bereft of trees, generating a weird dissonance. The nearly contained nature of the Persian Gulf makes it a sailor's delight, but the lack of trees anywhere on the peninsula grants the locals zero capacity for developing naval cultures. Any outsiders who can reach the Gulf can dictate most terms to most of the locals.

Modern Saudi Arabia is even worse, because it didn't even get the Arabian Peninsula's best land. The greenest lands in the region, on the peninsula's southern flanks, were chunked together into Oman (an entity with centuries of expertise as a trade way station) and Yemen (an entity with centuries of expertise as a failed state). Most of the Persian Gulf ports were carved out first for the British Empire and then granted to local loyalist tribes in the 1970s, becoming the United Arab Emirates, Qatar, and Kuwait. The only non-useless bit of land that eventually made it into the modern Saudi kingdom is the thin western mountain fringe of the Hejaz, where a handful of oases serve as the region's historical hub and home to the Holy Cities of Islam, Mecca and Medina. The world knows well the state that commands those cities: the wildly wealthy, wildly violent, oil-drenched Kingdom of Saudi Arabia.

But before there was oil, before there was Saudi Arabia, there was only desert. Before we can tease out Saudi Arabia's future (and from that the rest of the Middle East), we need to understand the transformations that brought us the House of Saud. That requires a bit of history to set the stage.

UNLIKELY ORIGINS

Until a little over a century ago, the royal family in Arabia wasn't the House of Saud, but instead the Hashemites, an Arab clan that could trace its lineage back to the family of the prophet Mohammad himself. The Hashemites controlled the most culturally and economically significant territory the Arabian Peninsula had to offer—the Hejaz on the peninsula's Red Sea coast, home to most of Arabia's cities, infrastructure, agricultural capacity, and trading opportunities.

The Hashemites of the time were a colorful, cosmopolitan people. The Hejaz wasn't simply a trading zone; it was also a tourism center. Yes, the Muslim Holy Cities of Mecca and Medina were the end-destination for pilgrimages, but with all the toing-and-froing, the cities were also well known as good locations to swap goods, ideas, and body fluids. It was as if Jerusalem somehow merged with Las Vegas.

The Hashemites' allies-*cum*-sponsors-*cum*-rulers were the Ottoman Turks. From the mid-eighteenth century, the north-eastern and western coasts of the Arabian Peninsula—which is to say, most of the useful bits—were the southernmost extremities of an ailing Ottoman Empire. In exchange for a cut of the Hejaz profits and pledges of Hashemite loyalty, the Ottomans tended to let the locals run things, providing only enough of an imperial presence to maintain security.

None of this sat well with a local pair of tribes. The first, the Saudis, were dirt-, er, sand-poor, consigned to the Nejd region in the depths of the barren Arabian interior, where they subsisted largely by raiding wealthier groups, such as those on the edge of the Hejaz. In essence, the Saudis cherished a stalkery relationship with the Hashemites, jealous of their wealth and power. The second tribe was that of one Muhammad ibn Abd al-Wahhab, a throwback theologian who advocated a strict interpretation of Islam as it existed in the seventh century—social and technological progress in the intervening millennium be damned.

It was a perfect match. The Saudis wanted the power of this world. Al-Wahhab's tribe sought a religious reset to access the next. The jealous Saudis wanted to wallow in all that sinful wealth themselves and resented the Ottomans' protecting the Hashemites. The hyperviolent, raid-happy, camel-riding, fanatic Wahhabis saw the Hashemites as sinners and the Ottomans as usurpers of the Holy Cities.

To seal their alliance, the two families agreed to a truly epic series of intermarriages, which would undoubtedly make for a wild Netflix series. All they needed was someone to give them some modern weapons, tactical intelligence, and a bit of logistical support to move against their common foes. As luck would have it, someone dripping with such weapons, intelligence, and support was sniffing around for just such a goon squad.

By the early 1900s, the entire Middle East was primed for conflict. Ottoman power had already failed in North Africa and the Balkans, and the restiveness of Arabia was matched by Russian agitation throughout the perennially agitated Caucasus. While the Ottomans were the power to please in the Red Sea and the Persian Gulf and so controlled regional trade, the British held sway in the broad open of the Arabian Sea and controlled transcontinental trade. When the two great empires clashed in the First World War, life in Arabia got decidedly lively decidedly fast.

The Brits dispatched one Lawrence of Arabia to the peninsula to recruit forces and coordinate opposition to the Ottomans. Larry provided the weapons, intel, and the occasional car pool required to fuse the Saudi and Wahhabi tribes into a semi-competent fighting force.

By global standards—hell, by *regional* standards—this tripartite marriage of convenience wasn't militarily impressive. It didn't have to be, because it wasn't unique. The Brits' slapdash Arabia strategy wasn't a one-off, but merely a single example of the sort of thing the manpower-light Brits did *whenever* they found themselves in a fight with a land power. London stirred up similar hodgepodges of disposable troublemakers in Mesopotamia, the Persian Gulf coast, Kurdistan, the Jewish territories, Transjordan, and the northern Levant. The Turks found themselves suppressing Brit-stoked rebellions throughout their southern territories even as they came under more-conventional military pressure from Serbia, Romania, and Greece on their Balkan front and from Russia in the Caucasus.

One big war later, the Ottoman Empire was on history's ash heap. The British left Arabia as quickly as they had arrived, but not before dismissively anointing their local allies the commanders of some of—even by Arabian standards—the world's most useless territories. The tribal Saudis immediately rebranded themselves the ruling House of Saud, while their in-laws—the Wahhabis—gained control of the Muslim world's Vatican City.

Modern Saudi Arabia emerged from the Ottoman Empire's wreckage in 1932 . . . and six years later some Americans poking around in the new Kingdom of Saudi Arabia discovered oil.*

The result was a country that hasn't exactly restored anyone's faith in humanity.

THE POLITICS OF BRUTALITY

The state of Saudi Arabia is first and foremost a medieval-style monarchy—a tyrant-king, multi-wife, family-murdering, crush-the-peasants, rich-get-richer, poor-get-poorer, off-with-her-head *monarchy*. Power is concentrated wholly within the ruling family. Political dissent is routinely punished by torture and execution.

Modern Saudi Arabia is quite literally the marriage of the Saudis' Comanche-style raiding history to the Wahhabis' Seventh-Day Adventist–*cum*-Amish worldview on issues of technology, topped off with both tribes' mafia-esque opinions about earthly punishment. Such political inflexibility and deliberately maintained economic inequality are unremittingly fused with one of the most repressive, sexist, and sadistic interpretations of any religion, ever, with the repression of individual rights being a

---

* Today's Saudi Arabia isn't built only on oil and religious fanaticism; it's built on a far from insignificant amount of luck. Had oil been discovered thirty years earlier (as it was in Iran), Saudi Arabia would likely *still* be a British colony.

cornerstone of national identity.* Such brutality is also the first pillar of the Saudi state.

The result has been the establishment of some truly hateful norms.

At home, the population lives under lockdown. Despite some recent, glitzy, and completely insincere promises of economic and political reform, any dissent is met with clubs and bullets. Dissenting members of the royal family itself are treated a bit better: they're held under house arrest at the Riyadh Four Seasons until they recant their views, re-pledge their loyalty to the crown, and sign over the bulk of their assets to the state. That is, of course, assuming they're not taken out for a one-way August picnic in the desert.

Saudi cruelty isn't casual or incidental, but instead starkly deliberate.

Saudi Arabia's wretched geography drains its people of any hope for economic or political advancement. With social stability a fragile thing, the last thing the ruling family wants is a bunch of college graduates. Anyone with a technical degree would have no reason to remain in the country.

Rather than go through the effort, expense, and risk of educating the Saudi population, Riyadh imports millions of guest workers as needed—Americans and Europeans for highly skilled work, and South Asians, Filipinos, and Indonesians for mid-skilled and unskilled labor. The country's manufacturing base is

---

* I cannot emphasize enough how much Saudi brutality and general lack of civility is a product of its geography and history, and *not* a result of Islam. Civility plays a weird role among the deeply religious Bedouin. Granting shelter to strangers is the ultimate act of propriety, so long as those strangers are not competing for resources. Should *that* happen, the result is a combination of righteous indignation and a visceral need to ensure the tribe's survival. Reactions tend toward extermination. Any people in such a geography would likely twist any religion into such a perverted interpretation. That was certainly the case with the various animist religions popular in the region before the advent of Islam in the seventh century. Since the Saudis took control of the Holy Cities, most associate their version of Islam with the religion writ large. In reality, the Saudis represent but a tiny splinter faction representing only a few percent of all Muslims.

miniscule outside of oil-related stuff such as refining and petro-chemicals.

For an idea of just how distorted this makes the Saudi work-force, consider the following: the hajj is a major boon for the Saudi economy. Religious tourism in general employs nearly one million people of the population and attracts 17 million visitors annually. In the United States, Disney World alone draws 55 million people per year yet employs fewer than 70,000 people.

Similarly, the Saudis are hesitant to build an actual, functional army. Giving a large, organized group of men competence with firearms in a state ruled by a family that's a cross between *Game of Thrones* and *The Beverly Hillbillies* would guarantee coups.

Consequently, Saudi Arabia's "military" is quite a bit different from the norm. For actual military purposes, the Saudis maintain a foreign mercenary force that draws heavily from Pakistan for its air force, Egypt for its army, the United Kingdom for trainers, and the United States for equipment. This model has two prob-lems. First, the foreigners are as hypercompetent as the Saudi rank-and-file below them are incompetent. The military has no skill depth. Second, it is unclear that the foreigners would fight for Saudi Arabia when the rubber hits the road. Pakistan, Egypt, and the United Kingdom have all taken sharp issue with recent Saudi policy toward Yemen, Qatar, Iraq, Syria, Egypt, and Iran, while the Americans are looking for excuses to burn their maps of the entire region.

Parallel to—or, depending on the family politics of the day, in competition with—the military is the Saudi Arabian National Guard. Unlike the Guard system in the United States, the Saudi version isn't charged with guarding the nation, but rather the House of Saud. That requires a skill set that doesn't use artillery and jets so much as truncheons and waterboarding equipment.

The Guard also serves a political purpose. Most Guard mem-bers are from rival tribes the Saudis have elevated in an attempt to secure their buy-in to the Saudi state. It's a combination of keep-your-enemies-closer while also diverting the violent tendencies of preindustrial tribal social management systems into channels the Saudis find of use, most notably the state-sponsored lynching of anyone who doesn't do what the House of Saud says.

Social management follows a simple process: use free food and housing to keep the population quiet, and beat the crap out of anyone who steps out of line. None of the population hold actual economic value to the ruling family, and so all the population is utterly expendable.

Such attitudes of disposability don't drive only the country's domestic management; they form half of its external strategy as well.

Not all Saudis are sheep. As befits a cultural elite that evolved from camelback desert raiders, a far from insignificant slice of the male Saudi population is aggressive and gets a bit restless sitting in air conditioning all day. Those with a track record limited to domestic violence are ignored. Those whose violence leaks out of the home a bit are typically imprisoned and beaten into sheep. Those who are a bit more vehement are brought into the security services and become responsible for beating their countrymen. And for those special men who demonstrate a penchant for more intense and sustained violence, a special future awaits.

The government gives its sociopaths some weapons training, packs their wallets with oil cash, puts Korans in their pockets, straps bombs on their backs, and ships them abroad to fight for greater glory of the House of Sau-, er Saudi Ara-, um, Islam. Exporting such malcontents acts as a pressure valve to ease social management at home while also generating endless headaches for Saudi Arabia's foe-of-the-day. Iran, Iraq, Syria, Libya, Lebanon, Afghanistan, Pakistan, and Russia have all found themselves inadvertently importing Saudi militants at one time or another.

The second pillar of the Saudi state is, unsurprisingly, oil. Saudi Arabia is home to several of the world's "biggests": the biggest conventional oil reserves (298 billion barrels); the biggest producing oil field (Ghawar); the biggest oil-processing facility (Abqaiq); the biggest oil pipeline (the Petroline); the biggest oil-loading platform (Ras Tanura); and for nearly all of the post–Cold War era, the world's largest oil production and exports.*

Such vast assets and the income that comes from them would enable *any* state to buy pretty much anything it feels it needs: a national airline, modern cities, influence, leverage the world over, and a population so fat it becomes uninterested in rebellion.

But there is more at play here than "merely" a metric butt ton of money. Saudi income comes from *oil*—the material at the heart of the modern economy. Oil is the source of some 95 percent of the world's transport fuels and 85 percent of the world's various petrochemical products—a list that includes everything from wallpaper to tires to asphalt to safety glass to nylons to insulation to herbicide. Getting their wealth through this vector grants the Saudis far more influence than if they had a printing press that could churn out a bottomless supply of American twenties. Oil doesn't simply make them rich; it makes them *essential*. To *everyone*.

Which brings us to the final pillar of Saudi success: outside sponsorship.

Among the world's countries past and present, Saudi Arabia is a very odd cat. It cannot risk having a normal military or a professional class for fear of overthrow. It must import vast amounts of labor on temporary contracts to run its economy, applying liberal doses of violence to the unskilled and a steady stream of bribes to the skilled to keep its foreign workers in line. The one economic asset the country has that is worth anything is something that most countries would do *anything* to obtain. And Saudi Arabia wholly lacks the sort of geography that would provide them with any natural shielding from out-of-region powers. A high-value,

---

* In recent years, the Saudis lost out on both measures to the Russians (whose production has finally recovered from the post-Soviet collapse) and the Americans (whose shale revolution has overturned what passes for "normal" in the world of oil).

easy-access target with zero ability to operate its own economy, much less defend itself? It seems that more capable countries would be lining up to relieve Saudi Arabia of its independence.

BUT . . . THERE'S THE ORDER

Since the end of World War II, far from being a nowhere, Saudi Arabia became the everywhere—especially for a certain superpower whose global strategic alliance was predicated upon global economic growth. Regardless of what the Americans thought of Saudi greed or Wahhabi extremism or the country's preference of dismemberment as a media-management strategy, the Americans had no choice but to ally with one of the world's least functional and most repressive regimes. Saudi oil powered global growth, enabled global trade, solidified the American alliance, granted American security.

No matter how economically distorted or militarily inept or politically hateful Saudi Arabia might be, the Americans *had to* have the Saudis in their camp. The Saudis lucked into a world in which their absolute security was a prerequisite for European security and East Asian security and American security. Completely as a side effect of American efforts, Saudi Arabia—a country whose internal discordance means that by all rights it shouldn't even exist—was elevated to the status of regional political and economic power.

With the Order providing the Saudis with near-blanket immunity to any and all geopolitical trends, the Saudis gave pretty much everyone—ally or not—solid, lasting reasons to hate them:

- Their formation of OPEC in 1960 and the subsequent oil embargos in the aftermath of the 1973 Arab-Israeli War quadrupled the price of crude oil and triggered a global recession.

- Their public (and private) beheadings, casual enslavement of women, and enthusiastic participation as an "end-user" in the global sex-slave trade keep them on the radar of anyone with a conscience.
- The Saudis' militant-export strategy proved instrumental in inflaming conflicts in places like Afghanistan, Chechnya, Iraq, Libya, and Syria—meaning the Saudis have deliberately deployed assets to fight *every* Middle Eastern power, as well as the Russians, as well as the principals of the NATO alliance. That includes their de facto security guarantor, the United States.
- Militants whom the Saudis have "misplaced" have gone on to form groups such as the Taliban, al Qaeda, ISIS, and other groups that have carried out attacks targeting the interests of America, Russia, Iran, Pakistan, India, Nigeria, Sudan, Israel, France, the United Kingdom, Germany, Morocco, Mauritania, Libya, Malaysia, Thailand, Australia, the Philippines. . . . You get the idea.

The Americans were so tangled up in the need to guarantee oil flows that even in the post–Cold War era, even when the Saudis "lost control" of a group of terrorists who flew passenger jets into iconic American buildings, killing some 3,000 American citizens—even then the Americans dared not walk away from their alliance.

Saudi Arabia is a brutal, dictatorial monarchy ruled by a small clique of violent men who brook opposition about as well as the medieval tyrants they explicitly model their government on, and who view the murder of people a hemisphere away as simply the cost of doing business.

There are only two reasons why anyone in the contemporary age deals with them at all: oil and a façade of civility. Between the shale revolution and their retreat from global management, the Americans no longer have an interest in the former. And once it sinks into the heads of the militarily weak, economically malformed, and strategically vulnerable Saudis that their security guarantor has abandoned them, the latter loses all usefulness.

The mask has come off quickly, and what lies beneath is truly hideous. In October 2018 the Saudi government disposed of a dissident journalist—one Jamal Khashoggi—in Istanbul by baiting him to visit the Saudi consulate, torturing and strangling him in a back room, hacking his corpse to pieces, and smuggling the remains to the consul general's house for cremation in a custom-built barbeque pit. The consul general then proceeded to host a big party that involved the roasting of a few hundred pounds of meat to muddle any forensic evidence. It isn't that this sort of thing is new in Saudi Arabia. What's new is that the hit was carried out abroad, under cover of diplomatic immunity. The Saudis are not so much becoming emboldened as they are adapting to a world in which the Americans neither defend them directly at home nor protect Saudi commercial interests abroad. They're taking matters into their own hands.

Broader Saudi strategy is evolving as well. After all, if the House of Saud and their Wahhabi allies are to survive, they *must* find a new way to operate.

OPTION 1: FIND A NEW SECURITY GUARANTOR

This is harder than it sounds.

Security guarantors are tricky things. They need to be powerful enough to reliably project power into the Saudis' neighborhood, possess a burning strategic need to keep Saudi Arabia independent, yet be far enough away to not care what the Saudis do in their backyard. In this, the Americans were perfect. Those super-handy supercarriers enabled the Americans to be *the* determinant in Persian Gulf military affairs whenever they wanted to, but because the Americans live in a different hemisphere, they didn't care what the Saudis did at home so long as the oil kept flowing. It's a magic mix that is impossible to replicate.

Of the European states, only the British and French boast even the theoretical capacity to project power as far as the Red Sea and the Persian Gulf. The British get their energy from locations far closer to home, so they're likely out of the picture. The French might see a partnership as being in their interests, but it is far from clear that France could be the sufficiently reliable or powerful partner the Saudis need to hold off a country as large and determined as Iran. We are, after all, talking about ground troops. Russia—on the wild chance the Saudis would seek Russian protection—is on the wrong side of Iran and Turkey.

China would love to be the Saudi guarantor, and it certainly has an intense need for Saudi oil and a complete lack of concern for how the Saudis treat their people, but Beijing lacks the naval power projection to break out of East Asia, much less get past Singapore, Indonesia, and India. Japan's projection capabilities are more impressive, but it has bigger issues to stress about closer to home, while also lacking the army manpower required to defend Saudi frontiers.

That leaves only two potential players who *might* be able to help.

The first is Turkey. The Turks command the largest ground forces in the NATO theater and come in third in terms of regional naval forces after the Brits and French.* Yet the Saudis would never seek Turkish assistance except perhaps to establish a regional balance of power. For reasons we'll dive into more deeply in the next chapter, the Saudis view the Turks as less a savior and more a problem on the horizon.

---

* The Americans obviously would rank first overall, but the vast bulk of their forces are nowhere near Europe. Haven't been since before the early 1990s.

That leaves only . . . Israel. At first glance, this seems a few steps past ludicrous: the keepers of the Holy Cities of Islam under the protection of the imperialist Zionists? Please!

Take another look.

Israel has a long history of working to turn the various Muslim powers against one another. It has de facto allied with Egypt, once the "leader" of the Arab world. It has turned Jordan into a de facto satellite. It's stuck its chocolate into Syria's peanut butter on multiple occasions to help keep the civil war there burning bright. Once upon a time, there was even a fairly firm and active alliance between Israel and *Iran*. For the Israelis, co-opting a pint-size military like Saudi Arabia would be a continuation of a long-running theme.

For the Saudis, the strategic math is much simpler. The House of Saud's commitment to the precepts of Islam is, shall we say, somewhat fluid. Besides, the Wahhabi interpretation of Islamic law enables you to waive your ethical and moral obligations if it helps you defeat an enemy of Islam. Saudi Arabia's premier cultural, economic, political, and strategic foe is Iran, and because Iran subscribes to *Shia* Islam—a sect the Wahhabis view as a heretical abomination—doing anything up to and including allying with heathens is fair game.* Israel—thinking the Egyptians, Jordanians, Iraqis, Libyans, and Saudis neutered—has also turned its attention to Tehran.

So Israel provides the Saudis with training on American weapons systems and a steady flow of useful intelligence and munitions. The Saudis surreptitiously "misplace" the odd cargo of crude oil or refined product in the general vicinity of an Israeli port. Toss in some mutual political and military intelligence sharing, and

---

* I'm really not trying to judge the Saudis too harshly on this specific point. The Americans cozied up with Stalin to fight Hitler, and shortly thereafter got into bed with Mao to box in Stalin. Geopolitics makes for consistently strange bedfellows. The only angle unique to the Wahhabis is their assertion such flip-flopping and backstabbing is expressly Allah-approved.

you have yourself a bona fide friends-with-benefits relationship. About the only sticking point is that Israel would certainly have some opinions as to which specific countries should *not* get Saudi oil. Considering the House of Saud's willingness to use its energy sector for political purposes, it is hard to see the crown getting worked up about needing to sell its in-high-demand exports to a slightly shorter list of clients.

Yet as tidy as an alliance with Israel is, it simply cannot replace one with the Americans. Israel lacks the naval reach to purge Iranian forces from the Persian Gulf, nor can it deploy sufficient troops to secure Saudi Arabia's northern border. Riyadh will have to do something more.

OPTION 2: BUILD UP A (MORE) INDIGENOUS DEFENSE CAPABILITY

The problems the Saudis have with building a potentially coup-prone army haven't gone away. The current cult of leadership understands this instinctively. From 1932 onward, the country's ultimate leader has been the king—an office that passes from father to son, and then through said sons from oldest to youngest. That is, unless someone breaks the system.

The current king, Salman bin Abdulaziz Al Saud, controlled the country's security services before he became crown prince. That enabled him to wrestle his way into becoming crown prince out of order, skipping over two of his older brothers. Upon becoming king in 2015, Salman then went on to push his son, Mohammad bin Salman Al Saud, ahead of some seventy other princes who held age-based seniority. MBS became crown prince shortly after his father's coronation, and as his wily father was already advanced in age and facing a set of health issues,* MBS

---

* Including rumors of dementia.

became the de facto monarch at the tender age of thirty-three. Barring illness, an unplanned meeting with an assassin's blade, or slipping in the shower and falling on some bullets, MBS will be the chief decision maker for all things Saudi for a half century to come.

That takes us a few gritty places.

MBS is his father's son, so he fully understands what happens when someone who isn't king controls the men with guns, and he has maintained direct control of both the defense and interior ministries rather than share them with potential successors (as had been the Saudi norm until his father's tenure).

MBS has already forced the royal family's ranks closed. A not insignificant portion of the royal family—some supported by previous monarchs, some who skirted the edge of the Saudi system—was subject to a mass arrest in mid-2018. None were released until they (a) transferred huge portions of their personal wealth back to the state, and (b) publicly pledged loyalty to the person of MBS. Shortly thereafter it was undoubtedly on MBS's orders that Khashoggi was tortured, suffocated, dismembered, and cremated. Many abroad criticized both actions as being bad for the country's image.

That's missing the point. Love him or loathe him, Mohammad bin "Bonesaw" Salman is the country's absolute ruler, and *he is acting like one*. No one within the country—royal family or otherwise—dare act against his wishes, and the Khashoggi hit vividly demonstrates that he casts a shadow well beyond Saudi Arabia's borders. That shadow is getting long. In 2017 MBS de facto kidnapped Lebanese prime minister Saad Hariri for twelve days for his unwillingness to toe the Saudi line, going so far as to force Hariri to publicly resign while under house arrest. Throughout the Sunni Arab world, nary a peep of protest was uttered. It ultimately took the French president personally flying to Riyadh to strong-arm the Saudis into releasing the Lebanese leader (who promptly rescinded his resignation).

The young regent fully understands just how weak Saudi Arabia is as a country, so he's trying to buff up bits of the country's security apparatus for an actual war. In 2015 on MBS's order, the Saudi military launched an invasion of Yemen. Officially, the war is about preventing the overthrow of President Abdrabbuh Mansur Hadi in the face of a Houthi insurgency . . . or some such nonsense.

MBS has two goals. First, to get Saudi commanders and diplomats some (limited) experience in coordinating multinational Arab efforts against Iranian proxies like the Houthis. Second, to get the Saudi air force as much target practice as possible. The two most likely war scenarios the country faces are either a slow Iranian infantry thrust across Saudi Arabia's northeastern desert or a naval invasion via the Persian Gulf ports. In either case, the oil fields of the Ghawar region would be the target. In either case, a strong air force would be the most useful tool to repel the assault. And unlike the army, the air force doesn't have the sort of ground forces that would be useful in storming the royal residence, granting MBS a bit of coup insulation.

But it isn't enough.

Riyadh has lost its security guarantor, and any possible replacements are, at best, imperfect. MBS must get the country ready not just for war, but for a broad-scale competition with a much more stable and powerful foe. Success requires him to be bold. Brash. Dangerous. The communities and countries he chooses to target to build his Iranian firebreak will determine the shape of his own country, the broader Middle East, and the global energy market for decades to come.

And that is because the Saudis need a *third* leg if they are to stick around.

OPTION 3: BURN IT DOWN

As good as the Iranians have proven themselves at exploiting sectarian divisions, the Saudis are better at making sectarian divisions *burn*. Six reasons:

First, in no place where the Iranians have worked to sow discord and overthrow the local orders are Iran's allies the "natural" rulers:

- The Shia of Lebanon come closest at perhaps 30 percent of the population, but Lebanon is (by far) the least consequential country in the region.*
- The Alawites of Syria, used by the French as a collaboration group during colonial rule, comprise (at best) 15 percent of the population. The Syrian Civil War sorely tested their capabilities, and Alawite military performance would have been insufficient to the task without extensive Iranian and Russian support. Even with their Druze and Christian allies, the Alawites fall short of 30 percent of the population.
- Yemen is not only 60 percent Sunni, but the non-Sunni population is split among a variety of groups whose only connection to Iran is *not* particularly liking the Saudis. It is an alliance Iran can wield only awkwardly, and it takes specific orders from Tehran only if the pay is good.
- Iran has linguistic, cultural, and a small but vibrant sectarian linkage to Afghanistan. What it lacks is a track record. In the 1990s, the Taliban loved to display Sunni dominance by publicly executing Afghan Shia. Sometimes dozens at a time. Sometimes in *stadiums*. Iran utterly lacked the military capacity to stop such atrocities, yet didn't want to accept Afghan Shia

---

* A quick FYI: demographic statistics in the Middle East—especially where minority groups are concerned—are, at best, guesses. None of the governments want anyone to know exactly how big the governance problems they face are.

as refugees because they weren't Persian.* So Tehran had to sit and quietly seethe until the Americans pushed the Taliban out in 2001–2. The Saudis, for their part, helped *form* the Taliban and retain strong religious, intelligence, and military links throughout Afghanistan's underbelly despite their nominal alliance with the Americans.

- From a sectarian perspective, Iraqi Shiites are the closest to Iran of all these groups, but by virtue of being the most established and most likely to be able to fend for themselves in a post-America Middle East, they are the most resentful at being subjugated to Tehran's ambitions. That, and their history of fighting to the death of the last Shiite to determine which Shiite is in charge is so legendary that the Iraqi Shia have *never ruled Iraq except under direct American sponsorship.* As soon as the American withdrawal was completed in 2015, the Shia fell back into their habitual infighting, providing openings for the Sunni and Kurds to expand their hold on the Iraqi state.

In contrast, the Saudis, for the most part, are not tapping into restive minorities. Instead, in places where the Iranians have experienced success live bitter, radicalized *majorities.* With the unnotable exception of Lebanon, the dominant force in all these places has been Sunni Arabs—a group that has similar ethnic and religious proclivities as the Saudis. Mobs with pitchforks are always more effective when it is the majority lynching the minority.

Second, exports of terrorists are part of the Saudis' domestic management system. So long as the Saudis maintain their system with a preference toward a mono-religious, culturally stunted, economically backward tyranny, they'll have no end of aggroed young men to throw into this or that meat grinder.

---

* Iran still hosts nearly a million poorly integrated refugees from the Afghan wars of the past four decades.

Third, the Saudis have a great deal more money to burn than Iran ever will. On any average day, Saudi Arabia produces roughly two and a half times more oil, which is also higher quality and so fetches a higher price per barrel then Iran's. Saudi Arabia has humongous volumes of excess refining capacity, enabling it to export high-value fuels rather than low-value crude. Add in that the Saudi population is less than half that of Iran, and Riyadh's per-capita export income is septuple that of its rival in Tehran. Bribing others to do things your way is ultimately about how much scratch you can throw. Iran will never win that fight.

Fourth, the Saudis are not micromanagers. In contrast, when the Iranians work to shift a regional strategic or political system, they have a clear idea of how they want things to end up: with Iran calling the shots. One of the region's more famous half jokes is to call Lebanon's Hezbollah "Syrian managed, Iranian-owned." That's absolutely how things have played out in the Syrian Civil War, where Iranian Revolutionary Guard Corps generals have directly *led* Hezbollah assets into battle within Syria. In part that direct puppet-mastering generates some ill will. It certainly shortens the list of folks willing to put themselves under Iran's aegis, as Iran has discovered in every country in which it dabbles.

While the Iranians believe that they should—and eventually will once again—control the entire Middle East, the Saudis operate under no such illusions. They are perfectly happy to provide cash, recruits, weapons, explosives, and intelligence to various anti-Iranian groups, stir the pot a bit, and then step back and see what happens. In some ways, this gives the Iranians the advantage. Iran's allies and proxies operate far more cohesively, but the Saudis' grand strategy makes the most of their more disassociated approach, which brings us to the fifth factor:

The Saudis *don't care if they win the Middle East so long as Iran loses it.*

The Saudis are not trying to twist the region's populations into being on their side. They know they lack the institutional bandwidth and expertise to rule the broader region even if somehow they could work up a little cultural magnetism. They can't even ride herd on the many extremist groups they've spawned and hurled willy-nilly throughout the region.

The Saudis have something else in mind. Iran's successes throughout the region mean the Iranians are now supporting proxies and allies far beyond their mountain fortress. They are overextended. Vulnerable. The Saudis aren't trying to seize control, or even win a war. The Saudis' goal is to burn it down. Burn it *all* down.

There are a few hundred miles of empty desert between the populated Levant, the Syrian interior, and Mesopotamia on one side and populated Saudi Arabia on the other. If Iranian forces are desperately engaged battling or containing the fallout from groups like ISIS, they aren't marching south across that buffer. The Saudis even see complete civilizational collapse of the sort that parts of Syria have experienced as an unmitigated *positive*. A wrecked Iranian client state is one that cannot be a springboard for future Iranian action, and any refugee flows are going to head north into Turkey and on to Europe rather than south to a certain horrible death in the Arabian Desert. The Saudis do not need to control the aftermath so long as the aftermath is ruins.

This will likely be more horrific than it sounds. The oil revolution injected the Middle East with vast wealth, and a big slice of that wealth is used to import food of quantities and qualities heretofore unimaginable. Since World War II, the Middle East's population has *sextupled*, putting it far beyond the region's meager carrying capacity. Anything that disrupts oil flows disrupts income disrupts food supply. Some countries now—literally—face the apocalyptic combination of war, famine, pestilence, and death. The House of Saud is hardly oblivious to the threat. Resource competition gets ugly fast when you *know* you will die of starvation if you lose. What truly sets the Saudis apart from their neighbors is that they see mass starvation throughout the region as less a threat to avoid and more an event to *encourage*. After all, *they* will still be able to afford food imports.

It gets worse. The sixth and final factor relates to Iran itself. As useful as the sectarian tool kit has been to Iran's recent geopolitical successes, it isn't as if Iran is immune. The Saudis can play the sectarian card *within* Iran. For a country with Iran's multicultural underpinnings, the potential list of fifth columns is lengthy. Azeris of Turkic descent in the northwest, Kurds in the northwest, Baluchi in the southeast, Ahwazi Arabs in the southwest, Mazandarani in the north, and a freckle pattern of tiny groups throughout.

The point isn't that this or that group will succeed in breaking away (although I'd watch the Ahwazi Arabs very carefully, as they live atop Iran's oil-production capacity), but instead that for an anti-Iranian power like Saudi Arabia, it is a bit of a target-rich environment. Every solider the Iranian army has to assign to deal with local unrest is a soldier who's not available to march on the House of Saud.

Iran is now struggling with the same question the Americans faced after their initial lightning wins in Vietnam, Afghanistan, and Iraq. Tearing down the previous governments is the fun and easy part. Ruling territories in the aftermath is not nearly as rewarding. Occupying territory absorbs a lot of troops, and propping up local authorities that are manned by sectarian minorities absorbs a lot of money.

Normally, one would deal with strategic overextension by consolidating gains. Establishing governments. Involving civil society. Building infrastructure. Increasing educational levels. Building a health-care network. Winning hearts and minds. Nation-building.

Nation-building sucks. It's hard. *Really* hard. As in it is difficult to name *any* country that has *ever* been good at it. Nation-building is even harder if your country is not only the cause of the previous government's collapse, but if you also inflamed the country's religious and ethnic cleavages to do so and explicitly sided with a specific group. The Americans have experience in nation-building dating back to the Spanish-American War, most of which hasn't worked out very well. The Iranians—stretching back to the Persian empires of old—have never really tried it out. Nation-building in Iran/Persia itself is several thousand years in the making, and yet only half of Iran's people consider themselves ethnically Persian. The old Persian empires fully realized they'd never be able to win over the Arabs they conquered, so they treated their conquered territories as just that—conquered territories administered from Persia.

Now that the Iranians have caught the car, they're finding they don't know what to do with it, and worse, that they've set themselves up for a fall far more impressive and violent than the collapses they triggered. This time it won't be a minority rebelling against the American-maintained Order, but instead a *majority* rebelling against the Iranian-imposed order, a majority that shares an ethnic and religious affiliation with the well-funded Saudis. And *that* the Iranians have no chance of resisting.

Iran faces one other monumental problem.

Tehran has been so obsessed with opposing American interests in the Middle East for so long that it views the American withdrawal from the region as an unmitigated win. In reality, it is an unmitigated disaster. Every drop of crude oil the Iranians export travels by tanker ship through the Strait of Hormuz. Historically, the same has been true of crude originating in Iraq, Kuwait, Saudi Arabia, Qatar, Bahrain, and the United Arab Emirates.

Things have changed. The Iraqis now have a pipe that runs north into Turkey and out to the Mediterranean, while the Saudis and Emiratis have built a Hormuz bypass line. The Arab states of the Persian Gulf still use Hormuz because—in a time of Order—it is cheaper, but they are no longer wholly dependent upon it. Of the 12.3 million bpd of crude the Arab states of the Gulf exported in 2018, fully 6.6 million *could* be transferred via pipe bypass options. The Iranian volumes that could similarly avoid Hormuz? Zero.

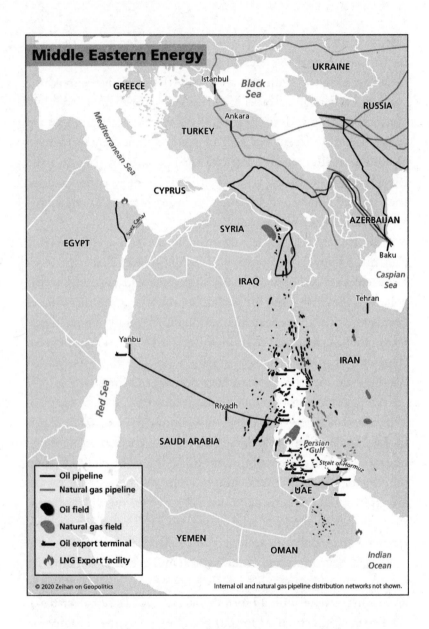

Even beyond Hormuz, the Iranians are in a tight spot. Saudi Arabia can buy friends. It has already done so to a large degree in Israel, Egypt, Pakistan—even India! That ensures Saudi shipments easy passage in the Red Sea, Suez Canal, Eastern Mediterranean, and Indian Ocean . . . while perhaps doing the opposite for oil from Iran. In this aspect of the game, the Iranians don't even have pawns.

In a bizarre turn of events, easily defensible, militarily potent, politically stable Iran now needs the global Order *more* than indefensible, militarily weak, politically unstable Saudi Arabia—just as that Order bleeds away.

THE GEOPOLITICS OF ARSON

In this fight, the initiative is certainly with the Saudis. They have more financial resources and more modern military equipment. They can be more aggressive, and they have institutionalized instability as a governing strategy. In a world in which global transport is curtailed, the value of their primary export will soar—encouraging foreign powers to partner with them just as the Americans did during the Cold War.

That hardly means the Saudi position is without risk. In a straight-up land war, a coalition of the kids from *Stranger Things* and *It* would rip them apart. The bulk of the Saudi national defense strategy is to *avoid* having to fight a real war but instead to burn down civilization anywhere Iranian power touches, working from the reasonable understanding that an Iran under siege from all points is one that cannot march across the desert to confront the House of Saud.

Faced with an opponent who views civilizational arson of its neighbors as a *positive*, Iran is ill suited to a protracted fight. But losing isn't the same thing as being lost. The inconvenient truth most foes of Iran ignore is that Iran's mountainous nature ensures that it cannot be destroyed with anything short of a World War III–style massed nuclear assault. Persia has existed since very nearly the dawn of human civilization, and while it will be the worse—far worse—for wear, it will survive the Order's end as well.

The most likely outcome is a sort of horrific stalemate, as the foundations of civilization throughout the Middle East burn away, leading to outbreaks of famine, civil collapse, vast refugee flows, and unprecedented depopulation—most notably in Afghanistan, Iraq, Jordan, Lebanon, Syria, and Yemen. The only reason the Palestinians might be spared is because a functional state largely divorced from the region's politics and traumas—Israel—is responsible for feeding them. On average, this entire region imports over half the grains and soy it consumes, and most local production is possible only because of imported agricultural inputs and industrialized irrigation systems. It won't take much more than a strong breeze to knock these places over, and the brewing Saudi storm will be terrible indeed.

The Saudis will achieve their goal: a mass uprooting of Iranian influence across large swathes of the Middle East. The cost—the civilizational collapse of the entire region as the relative stability of the Order burns away—is one they are willing to force others to pay.

It will be very loud and very messy, yet with the whole region either on fire or smoldering, the region as a whole becomes surprisingly *un*dynamic. Saudi Arabia and Iran will only have eyes for each other for some time to come, which means neither is the *real* power broker.

It is someone else's turn to take up the baton.

SAUDI ARABIA'S REPORT CARD

BORDERS: Vast expanses of sand, some scraggly mountains, and shallow local seas . . . but mostly just the sand. Moving tanks or soldiers or supplies in and out of most of the Arabian Peninsula is simply a bitch of a task.

RESOURCES: Oil. Jihad. Sand. And oh so very much of all three.

DEMOGRAPHY: Like much of the Persian Gulf, the explosion of oil income in the latter half of the twentieth century enabled mass subsidies and a flood of foreign workers to do all the work, freeing the Saudis to reproduce in air conditioning. During the past two decades, however, the economy has stalled somewhat, taking birthrates with it.

MILITARY MIGHT: Saudi Arabia is one of the largest buyers of military equipment in the world, but much of it is operated by veterans of foreign militaries such as those of Pakistan and Egypt.

ECONOMY: Heavily oil-dependent. Oil sales make up 70 percent of state revenues.

OUTLOOK: Saudi Arabia is in the rare position of having the money, military equipment, and the will to position itself as a legitimate counterweight to Iran in a region long defined by American (mis)management.

IN A WORD: Arsonist.

CHAPTER 11

TURKEY

THE AWAKENING SUPERPOWER

Turkey's position where West meets East made it the world's connective tissue, not so much of the world before the American Order, but instead of the world before *that*; in the pre-deepwater-transport era, when the Silk Road acted as interstate, railroad, container ship, and passenger jet all in one. The Turks' dominance of that era made them so wealthy and expanded their territory so much that they were able to linger as a major power for a full four centuries after the era of the Silk Road ended. Turkey is a living relic of an age long forgotten.

But the Turks are about to come roaring back.

The Turks are ancient, dating back to at least the sixth century, making them one of the oldest consolidated identities of the Middle East and Europe. But that doesn't mean they're locals. The original Turks were nomadic horsemen, wandering far and wide across northern China and the Eurasian Hordelands, on occasion raiding, wrecking, ruling, or trading with communities they happened upon, based on personality, season, opportunity, and whim.

Around 1000 AD, one tribe of Turks—the Seljuks—ventured

very far from home, pushing through the Caucasus Mountains into the Anatolian Peninsula. What followed was the weakening of the Byzantines, a series of crusades, and the continent-spanning Mongolian tide. It wasn't until three and a half centuries later that the Seljuks' successors, the Ottomans, dismounted for good at the gates of Constantinople. The Ottomans put an end to the vestiges of the Roman Empire and Christian Europe's sense of security all in one go. The Turks renamed the city Istanbul. Within a century, the Turks had become the most powerful nation on Earth.

MARMARA ON LAND, MARMARA AT SEA

Pre-Columbian sailing was a dangerous and limited affair. Sails weren't yet mastered, so it was slow and required lots of big guys with oars. The ships couldn't carry much and couldn't go far at a stretch. In fact, they had to stay close to the shore, stop at night, and hope. Pirates, storms, enterprising locals, and sudden changes in both actual and geopolitical winds could all kill you. Hell, loading the wine wrong could sink a ship. Traders and sailors took such risks regularly, but . . . well, there weren't many retirement parties for old sailors, and there was a reason a silk outfit was the privilege of emperors.

Well-positioned locations that could also offer some semblance of security and shelter became crossroads. And Istanbul was the ultimate example of a secure crossroads.

Istanbul sits at the eastern end of the Sea of Marmara. While Marmara is technically not a river, its expanse is so enclosed and calm, it might as well as be one. That's less the end of the story than the beginning. Sail east from Marmara through the Bosporus and on through the Black Sea and one reaches the Eastern Balkans, the Northern European Plain, and the endless flats of Slavic lands. Sail west through the Dardanelles and one reaches

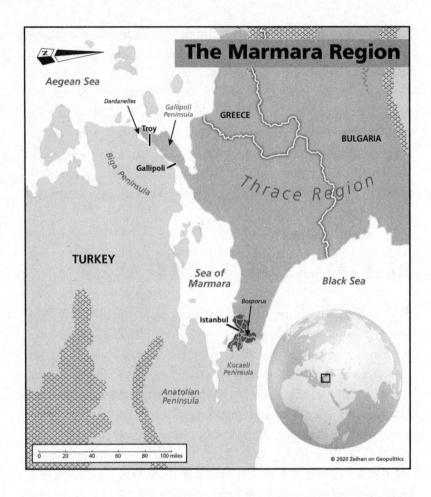

The Marmara Region

Aegean Sea

Dardanelles

Gallipoli Peninsula

GREECE

BULGARIA

Troy

Biga Peninsula

Gallipoli

Thrace Region

TURKEY

Sea of Marmara

Black Sea

Bosporus

Istanbul

Kocaeli Peninsula

Anatolian Peninsula

0   20   40   60   80   100 miles

© 2020 Zeihan on Geopolitics

the Aegean, the Mediterranean, the Nile, the Adriatic, and ultimately the Atlantic Ocean.

And it wasn't just sea routes. During periods of continuity, Istanbul was a safe, central location between Africa, Europe, and Asia. Most Silk Road caravan traffic found its way to and through both the Anatolian Peninsula and Istanbul itself.

Nowhere else on the planet has there ever been as dense a focal point for regional and intercontinental trade. And nowhere else has geography seen fit to enable such a focal point to be *defended* so easily. To Marmara's east, the lands of eastern Anatolia form a barrier so severe that they have only ever been crossed by those with Turkish or Persian determination. To the northwest, the rich lands of Thrace give way to the Balkan Mountains. The city has fallen to hostile forces only twice in the past thousand years—once when the Crusaders sacked it in 1204, practically burning it to the ground, and again when the Turks conquered it somewhat more gently in 1453.

In the Marmara region, the Turks hold the most useful and powerful geography in their interconnected region. Marmara's mild climate, good lands, and peninsular and mountain enclosures make it the perfect site for a wildly successful city-state. But it is more than that. Add omnipresent trade connections to the entire Old World, and that city-state is almost doomed to rule an empire.

The Turks leveraged that geography to dominate everything within a thousand miles. It was a natural progression: secure a physical barrier, metabolize the land it protects, break past the barrier to the next piece of good land, repeat. This took the Turks from Marmara to Thrace to the Lower Danube to the Pannonian Plain and the Crimean Peninsula. From Hatay to Beirut to Jerusalem to the Nile, the Shirak Province, the Red Sea, and the Hejaz. From the Pontic Coast to Zangezur to Mesopotamia, the Persian Gulf, and the shores of the Caspian Sea.

The sprawling empire became the largest on Earth of its time, and if a European coalition had not stopped the Turks at the gates of Vienna during the Ottoman surge of the sixteenth and seventeenth centuries, one power would have dominated *all* of Europe and *all* of the Middle East. History would have looked very different.

But the Turks failed. What went wrong?

In a word, technology.

About the same time the Turks were laying the groundwork for their first attack on Vienna, the technical limitations that blocked everyone from sailing the ocean blue evaporated. Sailors and scholars and tinkerers in far Western Europe learned to construct bigger, stronger ships that could survive on the high seas. They developed the means to sail not just with the winds, but also against them. They figured out how to discern their location, day or night—*enabling them to navigate and sail at night*. The deepwater era began.

For the Turks, this was an unmitigated disaster.

All the transport limitations that made the Ottoman territories the most powerful in the world—that granted the Ottomans a de facto monopoly on global trade—bled away in a matter of two centuries. Instead of being empowered by their geography, the Turks found themselves imprisoned by it. For centuries the Turks had pushed their borders back until they rubbed up against a barrier—geographic or political—that gave them pause. They would then use their superior economic position to break through that barrier, expanding bit by bit. Unfortunately for the Turks, this also worked in reverse. Maintaining their gains depended upon the Turks' superior trade-income-fueled economic position.

Since the Ottomans were a river-and-land empire, none of the new maritime technologies helped them one iota. The rise of the deepwater powers gutted the Turks' trade-based income while also presenting the Turks with new threats. Habsburg Spain's fleet of the newfangled ships challenged the Turks not just in the wider Mediterranean, but even in the Adriatic and Aegean. Tiny Portugal provided a naval means of transporting Asian goods to Europe that not only bypassed the Ottomans' land routes completely but enabled the Portuguese to take over production of the fabled Asian goods *in Asia*, establishing full imperial control over the entire supply chain. And all the while the Turks' land-based rivals in Europe kept the pressure on.

Stripped of what made it special, facing more foes while garnering less income and suffering greater exposure, the end was inevitable. The Ottoman Empire, the greatest empire of its era, slowly collapsed over three painful centuries. The Northern European

powers gutted the Turks in the Pannonian Plain, while the Iberi-
ans and English gutted the Turks' trade income, and the Russians
harassed the Turks on the Black Sea and in the Caucasus. By the
early twentieth century, the Turks had lost all their Danubian, African, and Caucasian territories. Their World War I defeat ripped
away everything but Anatolia and Marmara.

But history wasn't done. There were more humiliations to
come.

First, a condition of the post–World War I settlement forced
Marmara and the Turkish Straits open to international traffic.
Not only had land-based global trade between Asia and Europe
evaporated because of the better logistics of deepwater transport, but now the Turks couldn't even charge duties on the regional trade passing through the Bosporus and Dardanelles, even
though all that trade sailed through downtown Istanbul.

Second, the Soviet rise in the 1920s and domination of Central
Europe at the end of the Second World War not only completely
locked the Turks out of what once had been some of their richest
territories, but also out of what once had been lucrative markets
of the Black Sea littoral. The Danube, Dniester, Dnieper, Don,
and Volga systems were now all internal waterways of the newly
risen Soviet Empire—and Soviet ideology frowned on trading
with outsiders.

Third, the creation of Israel and the subsequent Arab-Israeli
standoff ended nearly all trade within—much less through—the
final economically interesting bit of the former Ottoman Empire:
the Levant.

Fourth, and most damning, was the Americans' new Order.
In dismantling the empires and making the oceans safe for
everyone, the Americans extinguished any hope of the Turks'
geography mattering at all to global trade. The Americans
forced *all* the world's waterways to be part of the global commons. If goods can go from any port to any other port without
needing to be concerned about either safety on the high seas

or persnickety local naval powers, then the specific location of this or that port or coast isn't all that important. Traders certainly didn't need a central land-based clearinghouse. What had made the Turks special—*critical*—evaporated.

Marmara was still nice, and it ensured the Turks remained the most powerful people in their (suddenly much smaller) neighborhood, but no longer was it a kernel of empire. With the world's long-range commerce now moving on the seas and oceans instead of rivers and roads, Marmara shifted from being the center of the universe to a forgotten backwater. It went from being a bridge to everywhere to a bridge from nowhere to nowhere.

Yet with the Americans' departure and the Order's end, Turkey cannot help but thrive.

RESETTING HISTORY

Many—particularly Europeans—like to compare Turkey to the advanced economies to illustrate why the Turks are animals and therefore ignorable. The idea is that the Turkish economy is substandard compared with Northern European norms, with a far heavier emphasis on lower-skilled industries like textiles and basic manufacturing.

That's true, but it's also idiotic. Turkey may have (strong) European tendencies and influences, but it isn't part of the Northern European Plain. Any comparisons must evaluate Turkey within its neighborhood: the Black Sea Basin, the Eastern Balkans, the Western Mediterranean, and the Middle East. In *that* neighborhood Turkey— isolated, insular—is *already* the regional superpower.

Consider the major consequences of the Order's collapse. *All* bode well for Turkey.

THE END OF SAFE OCEANS AND THUS OF GLOBAL COMMERCE. Remove the Order, and there is no longer an integrated system of global trade. Instead, the world devolves into a series of national and, in some lucky areas, regional systems. Turkey is among the few countries that have *already* adapted to the new reality. The Turkish economy is heavily regionalized by global standards, with most trade exposure limited to Europe and its immediate neighbors. Better yet, in a world where the Turks once again assert direct military control over the straits, they can charge transit fees for any through trade. As Turkey is the linchpin in any connections among its neighboring regions, the possibility of making some mad bank looms large.

THE RETURN OF LAND TRADE. It is undeniable that the Order provided the wealth, investment capital, and stability for many countries to build out their physical infrastructure to facilitate local economic development. But when it comes to trade, the vast majority of countries focused instead upon port expansions. Why wouldn't they? Under the Order, everyone has global reach.

The Order's end will do more than gut shipping potential and volumes; it will make regional land-based geographies matter a great deal more. Trade that must travel only a few dozen miles is less likely to be disrupted than trade that must cross an ocean. In any system where global maritime transport becomes constrained, *land*-based transport becomes more viable, particularly within the borders of any particular political authority. Turkey's physical location makes it the obvious connection point for such land-based transport between Europe and the Middle East, and with roads and rail lines snaking in all directions, Turkey is *already* prepared for the switchover.

A GLOBAL ENERGY CRISIS. Turkish energy demand weighs in at about 1 million barrels of oil and 4.6 billion cubic feet of natural gas daily, about average for a country of its size and economic structure. Also like its peers, the Turks import almost all of it. But *unlike* its peers, the Turks don't have to bend over backward to secure energy. Two of the world's most important oil transshipment pipes connect Turkey's Ceyhan port to oil fields in Azerbaijan and Iraqi Kurdistan. Either of them has enough pipeline throughput capacity to fuel *all* Turkish needs, and with a bit of . . . encouragement from Ankara, neither would face too heavy of a lift to provide the appropriate volumes. Should the Turks build one more midsize refinery anywhere along their lengthy coastline, their oil-products problem vanishes without taking a drop of Russian or Saudi Arabian crude (which could well remain available to the Turks). Natural gas is a bit more complicated, Turkey having an unhealthy dependence upon Russian imports. Yet new lines to Azerbaijan, the possibility of a short-haul line from Iraq, and a burning Russian desire to not piss off the Turks while Moscow is jousting with Stockholm and London and Warsaw and Berlin takes the sting out of the question.

As a fallback, the Turks have just enough crappy coal to bridge themselves to more geopolitically reliable (i.e., non-Russian) energy supplies.*

THE FALL OF THE RUSSIANS. The Soviets didn't want to trade with the Order, and the Americans didn't want the Order trading with the Soviets. The ideological and strategic conflict generated extreme economic spillover, producing consequences that sucked for the Turks (much of their former empire fell on the wrong side of the Iron Curtain). But with the Cold War's end and the reopening of the former Soviet space, trade from Europe, the Balkans, the Caucasus, and the Hordelands is once again flowing to Turkish ports, into Marmara and beyond to the Mediterranean. Even in the unfortunate circumstance that Turkey ends up crossing swords with the Russians, there will be far more trade on the Black and through Marmara than there ever was under the Order. Without American-enforced freedom of the seas, the Turks can return the Bosporus and Dardanelles to their normal status of being internal waterways. Everyone will have to pay the

---

*   As a rule Turkey has limited green energy potential. While there are some moderately-sloped zones for wind in extreme western Anatolia, the bulk of the country is simply too undulating for either large-scale wind or solar.

Turks to sail through. At least in part, the income stream that made the Ottomans so wealthy is coming back.

A CONFLICT IN THE PERSIAN GULF. The Middle East is replete with examples of two local powers rising simultaneously, clashing, and containing each other's ambitions and opportunities to the point that they massively weaken each other. Then a third power comes in and sweeps the board. Saudi Arabia and Iran both fancy themselves regional powers. As the Turks have far more military and economic capacity than the pair *combined*, the Turks find the Saudi-Iranian battle for influence rather adorable. What Turkey does *not* find cute is the fact that the Saudi-Iranian battles in Syria and Iraq are starting to harm Turkish interests: militants in Syria, threats to energy supplies from Iraq, swarms of refugees from both. The Turks reserve the option of interfering in the Saudi-Iranian fight and settling it however they choose. The real fireworks of the Iranian-Saudi confrontation will happen in densely populated Mesopotamia—an area not of core concern to Ankara. The Turks could push into Iraq as far south as Baghdad within a couple of months. Similarly, with a week's worth of effort, the Turks could shatter the entire Iranian and Saudi position in Syria, withdraw the next day, and evaluate the smoking wreckage from a distance.*

END TO THE BAR ON IMPERIAL PREDATION. Deepwater navigation turned Turkey into a backwater, and the Order turned Turkey into a nowhere, but an era of Disorder means Turkey matters again for all the same reasons that it mattered at the Ottoman Empire's height. Turkey is transformed from a nothing power into, if not the center of the world, at least the center of its region. And while Turkey was never one of the grand, globe-spanning, naval-based empires the Americans were thinking about when

---

* Such a smash-and-leave strategy is broadly how the Ottomans "maintained" what is today's western Iraq and eastern Syria whenever a local group of crazies (think ISIS) would rise up. There isn't anything of value to Turkey there now (or then), so the strategy has no strategic downside.

they crafted the Order, an empire it certainly once was. Once again, Turkey will be able to expand outward from Marmara, seeking territories and opportunities.

BACK TO THE FUTURE

The Turks' post–World War I shattering was just as traumatic for the Turks as any empire's ruin is to its core citizens. The military crumbled, the economy collapsed, the political system splintered, and the Turks' sophisticated culture—rich with hospitality and tolerance—splintered inward. Reformers attempted to overhaul a sclerotic bureaucracy that had little to do but absorb resources. Minorities facing starkly reduced economic security rebelled, sparking violent responses from the Turks, culminating in, among other things, the Armenian Genocide. Foreign agents were able to tap into the resentments to achieve military goals, the most romanticized one being Lawrence of Arabia. When it was all over, the empire had done more than die; it had dissolved into chaos. At one point, it was so bad that pieces of Anatolia weren't controlled just by the French and British, but by the *Greeks*.

Normally when an empire collapses, it is either occupied and restructured by outsiders (like Germany and Japan after World War II) or so wholly immersed in a new reality that the transformation occurs very quickly, often with extraordinarily painful impacts upon the citizenry of the home state (such as the British Empire's collapse in World War II, or Russia after the Soviet Union's dissolution). A truly destroyed empire can rarely dwell on its defeats in privacy, shielded from the world, to map its way forward.

However, that's precisely what happened to the Turks after World War I. The Soviet rise barred commerce and contact with former imperial possessions to the northwest, north, and northeast. The hostile, arid geography of the Middle East combined with the rise of mutually hostile and/or totalitarian governments that cared little about economic development or trade walled off the south. The only "open" border the Turks had was with Iran, the adjoining territories being physically rugged and far removed from the two countries' capital regions, both in terms of distance and culture. Turkey was sequestered, and in its isolation, it obsessed over what it meant to be a Turk.

Two competing visions emerged.

On one side were the secularists, who sought a modern Turkey that would be a completely separate, independent pole in international politics. They rejected the cosmopolitanism for which the Ottomans were known in favor of a new ethnic Turkish ideal. Islam played no role in the secularists' identity, at home or abroad. The wealth of Istanbul was embraced, and economic and cultural interaction with Europe was encouraged, but no one confused such acceptance with multiculturalism or a celebration of "lesser" peoples, like Greeks or Armenians. It went so far as to assert that Kurds, who today comprise one-fifth of the Turkish population, never even existed in the first place.

Opposing the secularists were the Anatolians, who sought to retain many of the Ottoman system's cultural norms— particularly its religious identifiers—and use those characteristics to bind the new Turkey tightly to their co-religionists throughout the Arab world. The Anatolians saw themselves as the logical heirs to Ottoman grandeur and viewed the shift of the capital to the interior city of Ankara as less a (necessary) strategic move to insulate the government from outside powers than a means of transforming Turkey's political life into something more akin to the countryside. Economic links were a means to a cultural end, rather than ties that bind Turkey to Europe. Deep, unrelenting efforts were made to transform liberal Istanbul into an extension of the culturally conservative interior.

It was the greatest cultural fight the Turks had wrestled with since swapping their saddles for houses after the conquest of Constantinople a half millennium before. From the close of the First World War until century's end, the argument over "who are we?" raged. But history has a way of overtaking events. Between the Cold War's end and the Order's impending demise, the Turks' context evolved, and their existential identity crisis evolved with it:

- You didn't have to agree with Osama bid Laden's politics or policies to understand that the 9/11 attacks vividly demonstrated that Islam was relevant in global affairs. For the Turks who used to *lead* global Islam, but who had been divorced from the world for three generations, the attacks pressed home the idea there was a mantle out there waiting to be picked up.
- The Soviet Union's collapse opened up Turkey's entire northern horizon. Trade began flowing through Marmara again. The Turks realized the world was re-interfacing with them once again whether they were ready or not.
- The Soviet collapse was a partial motivation for Turkey's effort to join the European Union, as several of the former Soviet satellites and republics in the membership queue were also former Ottoman provinces. The secularists saw membership as a way to entrench Turkish modernity. The Anatolians saw it as a means of guaranteeing democracy. Once it became apparent that the EU would never admit Turkey, both Turkish factions reconsidered what it was they were after.
- For the Turks, the United States' 2003 war in Iraq was notable not for its outcome or its nearness to Turkey's borders, but instead for the fact that the Turkish parliament voted to refuse the use of Turkish territory by American forces. Turkish nationalism surged, now flavored by a weird cocktail of local militarism, pro-Islamism, pro-Arabism, and anti-American feeling. The secularists and Anatolians discovered themselves . . . agreeing on things.

- With the Turks sitting out the Iraq War, the Americans found a new ally in northern Iraq's Kurdish groups. This was a problem for Ankara. Not only were there more Kurds in Turkey than Iraq, but the Turks had just recently finished fighting a thirty-year civil war over Kurdish separatism. Even worse, the Kurds in time rose to be the Americans' ally of choice in Syria as well. Such betrayal (in the Turks' eyes) functionally ended the American-Turkish alliance, while provoking both secularists and Anatolians into seeing Turkey as beset by an ever-more-distant superpower. America's abandonment of the Kurds in late-2019 hardly restored the US-Turkish alliance, although it did result in a bit of an upgrade from seething hostility to cold distrust.

After seventy years of culture war, the two factions found themselves intermingling and eventually merging in the personality of Turkish president Recep Tayyip Erdogan. Many of Erdogan's foes see him as the quintessential Anatolian—wanting to use Islam as a pillar of Turkish identity and policy, and disdainful of connections to the West. That's true. But what is also true is Erdogan's embracing of many of the secularists' traits: a rejection of multiculturalism, a willingness to bash heads when elections don't go the way they are "supposed to," the idea that Turkey is a strong and independent power.

The new Turkey combines the cultural grandeur, muscle tone, and arrogance of the Ottoman Empire with the religious leanings, disdain for secularism, and distrust of the Western world of the Islamists, and with the authoritarian, chauvinistic, and ethnic parameters of the secularists. If the goal is a peaceful, multicultural, globalist society, it's not quite the worst of all worlds, but it isn't far off.

Most significant of all, the Turks are late to the game of becoming an ethnically defined nation-state. That's extraordinarily dangerous. When a group starts defining itself in terms of ethnic purity, sketchy things tend to happen. The French, who started us all down the road toward fusing our ethnic identities with state power, also started us down the road of the various consequences: the Terror and the Napoleonic Wars. When the Germans and Japanese followed, we got the World Wars, the Nazis, the Holocaust, the Rape of Nanking, and the Bataan Death March. Mentally and culturally, Turkey today is in a condition very similar to that of late-1700s and early-1800s France, and early-1900s Germany and Japan—flaunting a brash, bold, unapologetic nationalist identity like a teenager on a meth high. Time will grind such thinking down into a more pragmatic form, but time takes, well, time.

Erdogan's party—the AKP—rose to power in 2002, granting Turkey a government willing to look beyond the parameters of the American relationship just as the Americans were starting to view their strategic picture from new angles. Fast-forward to 2020, and the Americans are unwinding their global strategic commitments at the same time the Turks are getting over their shyness in international affairs.

The Turks are excited about exploring what it means to be a nation-state just as the world is shifting back into empire mode, and the Turks have the perfect topography to be one of those empires. That all but hardwires in Turkish conflict—strategically, economically, politically, and *racially*. What with a fracturing Europe, a collapsed Middle East order, and Russians on a warpath paved with desperation, Turkey's entire periphery is in wild motion, and the Turks are about to leap into the storm.

Which begs the question: which way will the Turks jump? There are opportunities at every point of the compass.

THE FRONT YARD: BULGARIA AND ROMANIA

For the Turks, the brightest spot in this emerging world is the Eastern Balkans. The reasons are legion.

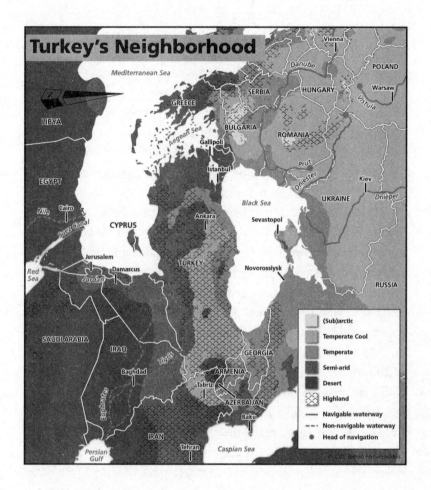

Without the Americans anchoring NATO and the European Union, most of the Continent becomes a free-for-all as once-imperial powers compete for regional influence over matters political, economic, cultural, and military. But the portions of Europe closest to Turkey—the Eastern Balkans—are a world away from such competitions.

The European Union did not admit Bulgaria and Romania to their membership on the same timetable as their wealthier former Soviet satellites for good reasons. Their infrastructure is clearly inferior. Organized crime penetrates throughout their societies. Corruption bleeds through into much of public life. Their cultures are not as entrepreneurial.

But mostly it was about physical access.

The Lower Danube's littoral zone is largely isolated from everyone else in Europe. To the west of the Lower Danube, a series of knotty peaks where the Balkan and Carpathian ranges meet blocks nearly all transport options aside from the Danube itself. The point where the Danube forces its way through those knots is a particularly rugged series of cliffs. These Iron Gates inhibited *all* transport between the Eastern and Western Balkans since antiquity, and it was only with deeply expensive Soviet reengineering of the river, completed in the 1970s, that the Danube became safely navigable year-round. Even that advance was set back two decades when the Americans destroyed most of the Danubian bridges in Serbia during the Kosovo War. The Eastern Balkan pair is not as integrated into Europe *now*—a decade *after* EU and NATO membership—as closer countries, like Poland, Latvia, and Hungary, were *before* EU and NATO membership.

The relative isolation has its perks. Bulgaria and Romania's relative poverty compared with the rest of Europe enables the Bulgarians and Romanians to switch partners with relative ease.* Because the Turkish Straits are the best maritime connection between them and the wider world, there's not much wiggle room in choosing a post-Order partner.

The pair bring a lot to the table. Both Bulgaria and Romania are significant agricultural exporters, and between the two and Turkey, nearly every climate zone that generates foodstuffs is represented. The result will be a regional supermarket for nearly every type of food humans can grow: wheat out of Turkey's Mediterranean coastal climate, corn and soy from continental Bulgaria and Romania, oilseeds from Romania's drier east, fruits

---

* *Relative* being the operative world. The two Eastern Balkan countries may be minor European players, but they are heavily dependent upon Europe.

of various types from the three countries' uplands, citrus from the Black Sea coasts, grapes from the three's mountains, and so on. In a world where global trade breaks down, and once everyday luxuries become exotics, the extended Turkish family will have all it needs close at hand—and complete food security to boot.

Above all else, the Bulgarians and Romanians will be *willing*. Both Sofia and Bucharest realize that their chances for charting their own destiny—even if the two allied—are zero. The Balkan and Carpathian Mountains box them in, the Russians dominate their northeast, the Turks their southeast, the Germans their northwest, and any access to the wider world requires the purposeful and ongoing permission of multiple other powers. They are quintessential examples of the sort of countries that have no long-term hope of survival outside of the global management structures of the Order.

That is, they have no chance without a sponsor. On their own, they are prey, but partner them with the Turks, and their position shifts from pathetic to enviable. They gain entrée to a country with energy security, a market twice the size of their own combined, and access to the Mediterranean Basin—something otherwise impossible. Defense guarantees from Turkey may not have the gold-star value of those from the United States, but the Turkish military is both competent and close by. Best be on its good side. *Especially* if there's a risk in the Disorder that the Russians start acting like Russians again.

GO FOR THE THROAT: UKRAINE AND BEYOND

The former Soviet Union abounds with threats and opportunities for Turkey, but none glow so bright as the Russian death throes. They raise the distinct possibility that the Turks will seek to nudge their old rival toward oblivion. Such a strategy is *far* from risk-free—the Russians have one of the world's most powerful air forces and a hefty nuclear arsenal—but the benefits would be legion.

Let's start with the benefits:

- During the Cold War, the Russians controlled all the territories to Turkey's north, contributing to Turkey's slide into strategic oblivion and economic mediocrity. Even now the Russians' presence dampens the region's economic horizons. Their penchant for hijacking trains in and shooting down passenger jets over Ukraine severely limits the potential for trade. Removing the Russians would drastically change the fate of those living along the Dnieper and Dniester rivers. Unlike Russian rivers, the Dnieper/Dniester pair flows south instead of north and lies in more temperate zones. There is no ice-dam danger, and both remain navigable nearly year-round.
- In Ottoman times the Ukrainian rivers were robust commercial arteries that pushed Ottoman interests deep into the western Hordelands, but under Russian domination both have become anemic, corrupt, circumscribed, and dreadfully poor. About the biggest complication is that Ukraine and Turkey have similar steel industries and are traditional competitors. (In a world of shortfalls, however, today's competitor could be tomorrow's oligopoly.) Bulgaria and Romania might be solid choices for a Turkey seeking economic opportunity, but Ukraine—with more people, a huge metals industry, and a larger agricultural production than Bulgaria and Romania combined—would be a close third.
- Azerbaijan would be fourth. While there isn't a lot that happens of economic interest anywhere in the Caucasus, there is one large exception: the Caspian oil fields of Azerbaijan. By themselves they export over five hundred thousand barrels per day, and there is sufficient installed infrastructure for double that amount to transit through Georgia to Turkey. Russia is

most certainly the dominant military, economic, and cultural power in the Caucasus. Armenia is in essence a satellite state while Russian forces maintain secessionist enclaves on the Georgian side of the Greater Caucasus Mountains. Yet Turkic peoples throughout the region still look to the Turks for succor. Removing Russia from the board raises the tantalizing probability that Turkey would be the first power of the Caucasus. Again.

- Russia isn't all that . . . nice. The same intelligence services that are so good at pacifying Russia itself and Russian-occupied territories are just as skilled at stirring up trouble in other places. In ages past the Russians encouraged Iranian belligerence against not just the Americans, but also the Turks. They've put flies in the Azerbaijani ointment. They've formally backed Armenia with weapons and intelligence. They've worked to hinder—even sabotage—pipeline projects that would bring Azerbaijani energy to Turkey. They've deliberately spawned humanitarian crises in Syria, deliberately flooding the Turkish border with refugees. And they've stoked Kurdish rebellions in Turkey proper in order to limit Turkish options and keep Turkish power pinned down. Admitting all that might be ideologically inconvenient to this or that faction of the current Turkish government, but for those aware of imperial, Cold War, and recent history, calling Russia a friend to Turkey requires substantial mental creativity.

The kicker is, Turkey doesn't have to win for Russia to lose. The only outcome of the brewing contest between Russia and Europe that extends the life of the Russian state is one in which Russian forces consolidate control of the Baltic Coast and the Polish Gap as well as the Bessarabian Gap, all of Ukraine, Moldova, and the Caucasus republics. Failure to achieve all these goals leaves the Russians unanchored and engaged in a war of numbers and/or movement that they cannot possibly win.

Anything shy of total success would also gut the Russian state's income. Russia's competition with Europe will end meaningful oil and natural gas exports via the Baltic Sea and North European Plain. The only other large-scale route is southwest via the Black Sea and Turkish Straits. Even a minor military conflict with Turkey would utterly end Russia's ability to export oil and natural gas to the west, removing it from the list of significant energy exporters (it currently ranks number one for combined oil, natural gas, and petroleum products, with oil and natural gas sales being the government's top two sources of income).

While predicting tactical moves in a strategic conflict that has yet to begin is a bit like playing darts blindfolded while doing tequila shots, the optimal time for Turkish action would be once the Russians become fully committed against the Northern Europeans. At that point the Russians would have fewer forces to spare to a front in the south, vastly improving the success rate of what would have to be a sizable amphibious assault.

The optimal place would be the Crimean Peninsula, Turkish control of which would eliminate the only meaningful Russian naval presence on the Black Sea, turning it into a Turkish lake. Because the Crimea's link to mainland Ukraine is only three miles wide, defending it from a mainland assault would be easy. While Turkey's air force couldn't hold its own against Russia in a one-on-one fight, Russian forces would already be engaged against every country that borders the Baltic Sea. It wouldn't take much Turkish strike capacity to sever most of the Russian army's supply lines into western Ukraine and even Belarus. And the Turks would have local help: the Crimea's Ukrainian and Tatar minorities would likely view the Turks as liberators from Russian occupation.

The problem is, all this is very all-or-nothing. Russia faces nothing less than an existential crisis; it is fighting for its very survival. The parts of the region the Turks see as the most lucrative are those that the Russians view as the most strategically essential. Even distracted and off-balance, the Russians have the capacity to strike back. Russia already has thousands of airmen and soldiers in not just the Crimea, but also Armenia and in secessionist regions in Georgia. Turkish intrusion into the Crimea or Azerbaijan would so change the facts on the ground that the Russians would feel they have no choice but to use every tool at their disposal, from prompting a Kurdish insurrection in eastern Turkey to a sponsoring an Armenian assault into Georgia to bombing Istanbul itself.

High reward brings high risk.

THE BACKYARD: IRAQ AND SYRIA

The Turks' first foray into the Middle East *predates* their calling themselves Turks. The Seljuks toured real estate in Mesopotamia and the Levant before they ultimately settled on the better neighborhood of Marmara. Later, after the failed assaults on Vienna, the Ottomans found it easier to expand into areas their forebears had explored in centuries previous. The empire shifted from its Occidental orientation to an Oriental one during the fifteenth and sixteenth centuries, earning it territories that comprise contemporary Egypt, Israel, Lebanon, Jordan, and Iraq.

The problem with contemporary Turkey returning to the region is that it isn't clear where it can stop. There is no equivalent of the Crimean isthmus or the Iron Gates where Turkey could stop to digest a small bite of territory. The fact is that much of the imperial debris in the region *is from the Turks' own empire*, and any sane planner in Ankara must look south with at least a concerned wince. This isn't a region you expand into because you want to; you go in because you fear what will happen if you *don't*. Which is precisely why the lands to Turkey's south are on Ankara's radar.

Ultimately, it's an issue of security. The vast majority of Turkish territories today are not the rich lands of the Lower Danube, but instead the arid, rocky uplands of Anatolia. The farther east one travels on the peninsula, the higher, drier, sharper, and less productive the terrain becomes. But while this argues against the Turks settling in it, it both provides an excellent block against invasion *and* provides within Turkey precisely the sort of territories that house rebellious, persnickety ethnic groups. Russia has its Chechens, Lebanon the Druze, Spain the Basques, America has West Virginia, and Turkey has the Kurds. The entire southeastern quadrant of Anatolia is a Kurd-majority region, home to half that ethnic group's total population, with the rest distributed in the border regions of Syria, Iraq, and Iran.

The Kurds of Turkey have been fighting an on-again, off-again insurrection against Turkish power since the end of World War I. Its hottest flare—claiming over thirty thousand lives—was in the 1990s. One of the most effective means the Kurds have of fighting is to seek succor and basing with their co-ethnics in Syria, Iraq, and Iran. And because the Turkish, Syrian, Iraqi, and Iranian governments are all run by different ethnoreligious-linguistic groups, they don't care much for one another and have often been willing to encourage Kurds in the others' lands to rebel—so long as they do so on the other side of the border.

Turkey often finds itself flat-out invading Syria and especially Iraq to pursue fleeing Kurdish militants and rip up Kurdish bases. A more permanent assault that left Turkey in control of Syria and Iraq's northern territories would bring over three-quarters of the Kurdish population within the Turkish system. There would still be unrest and violence, but direct, permanent occupation would enable the Turkish state to bring all its many tools to bear.

Conquering Iraqi Kurdistan would come with some notable fringe benefits. Permanently stationing Turkish troops within a short commute of Baghdad would sharply refocus minds in both Tehran and Riyadh. Neither Middle Eastern power can hold a candle to Turkish military might, and seeing Turkish troops much closer than the horizon would encourage both to take Turkish preferences into account throughout the region.

There's also the energy question. Iraqi's Kirkuk oil region is less than two hundred miles by road from the Turkish border. An improved security environment complemented by some engineers who know what they're doing could easily coax enough oil out of the ground to meet Turkish demand in full. Even better, the Kirkuk region *already* sports preexisting pipeline capacity to Turkey capable of carrying nearly twice that much. Technically and logistically, a Turkish southern expansion could easily bring the entire producing infrastructure under Turkish control. The "only" complication is what the Kurds might do.

Syria is messier, in large part because it encapsulates everything that makes the Middle East dysfunctional: A coastful of collaborating minorities. A vast interior of impoverished Sunni Arabs. Sharp, forested mountains full of hidey-holes. It's the perfect recipe for political instability and economic dislocation. And it's on fire.

The territory that is today's Syria has suffered hugely under the global Order. Pre-Order interior "Syria" was a caravan route throughout lightly populated terrain, dotted with a quartet of ancient oasis cities—Aleppo, Homs, Hama, and Damascus. But the Order's freeing of the global ocean combined with its shattering of the empires ended the caravan trade while erecting hard political borders. The people of Syria could no longer trade for food; they had to find something else.

On the one hand, the Syrians exported crude oil and used the subsequent currency earnings to import food. On the other, they transformed the desert to a degree possible only in the Industrial Age, drawing water from the Euphrates to grow wheat. It worked for a while. But the Syrian population expanded to the point that the oil exports stopped, and a drought (among other factors) in 2011 wrecked nationwide irrigation. The Syrian Civil

War, first and foremost, is a civilizational collapse with its roots in national starvation.

Which is damnably inconvenient for a newly emergent Turkey. Even if the Syrian Civil War ended today, there isn't enough water and oil to feed the population. Syria will not—will never—recover. Rivers of Syrian refugees into Turkey are the new normal, for they've nowhere else to flow. Lebanon already has more refugees per capita than any country in human history. The Israeli border is a wall of mines and barbed wire. Jordan is even dryer than Syria (and already has over seven hundred thousand refugees, about four hundred thousand more than the poor country could be expected to support without its fragile political system cracking). The Saudi and Iraqi borders are hard desert. That leaves Turkey. And the only way the Turks can truly manage the situation and keep the migration flows manageable is to go in and take over the whole damn place.

THE LONG PLAY: THE EASTERN MEDITERRANEAN

A successful Turkey needs a navy nearly as much as it needs an army, and on no side is a navy more relevant than to Turkey's west. In part, this is because the magic of sea travel made it easier for the Ottoman Empire to integrate with Europe and North Africa than the Caucasus, but mostly it is because of the countries most pathologically paranoid about a Turkish return.

Consider the Aegean Sea and northeast Mediterranean to be Turkey's Gulf of Mexico. The islands of Greece—islands the Europeans transferred forcibly from the Ottomans to the Greeks as part of the post–World War I settlements—are like the various islands of the Caribbean. Cyprus a sort of Cuba. Greece and its myriad islands have to be brought under Turkish control, or they will always threaten Istanbul and the trade that flows in and out of Marmara. Similarly, Cyprus is too big and too close to Turkey for Ankara to tolerate the possibility of a foreign power on the island. So the Turks invaded in 1974, conquered the island's northern third, and have been there ever since.

Conquest of Greece and securing the Aegean Sea would insulate mainland Turkey from regional naval powers, even those as powerful as the United Kingdom or France. And since those two European nations are likely to have bigger fish to fry closer to home, odds are the Turks wouldn't face many extra-regional challenges if they try.

Turkey's problem is not military. On paper, Greece maintains a huge air force expressly to keep the Turks at bay, but with Athens' financial implosion, the Greeks haven't maintained their jets for a decade, much less managed to get their pilots much flight time.

Indeed, both Greece and Cyprus are likely to fall without any help from the Turks. Both import all their energy and nearly all their food. Both are economic basket cases whose existence continues only because of an ongoing—and resentful—financial drip-feed from the European Union.

The problem is the last thing the Turks would want responsibility for is another Syria. With the fall of the European Union and the general breakdown of maritime economies the world over, Cyprus and Greece are about to decivilize, albeit likely (hopefully) not so violently as Syria has managed to.

If the Turks move west, they're likely to attempt to split the baby: seize Cyprus outright along with the bulk of the Greek islands but leave mainland Greece to wither on the vine. That makes Turkey responsible for a manageable 1.5 million occupied Greeks and Greek Cypriots rather than 12 million.

From the Turkish point of view, such territorial gains would be more than enough motivation to move west. Both the Aegean Islands and Cyprus were some of the Ottomans' oldest territories. The Turks lost control of them only because the Europeans—

primarily the British, French, and Greeks—were so enthusiastic about carving up the Ottoman corpse at the end of the First World War.

But there is more at stake here than simple historical grievance.

In any post-Order world, maritime shipping will be more difficult and less safe, while ships will have no choice but to travel faster and do so with smaller cargos. Add instability and conflict in the world's two largest oil-exporting regions—the former Soviet Union and the Persian Gulf—and the result is lower oil supplies combined with sharply higher and far more erratic oil prices.

Turkey is on a short list of countries that enjoy some insulation from that environment. It is hardwired via pipelines to *both* former Soviet *and* Persian Gulf supplies, providing the Turks with more than enough for their needs. Pretty much everyone else needs to keep importing crude oil by tanker. Traditionally, one of the world's three biggest oil shipping routes is from the Persian Gulf states to the Red Sea, through the Suez Canal and Suez bypass pipelines, into the Eastern Mediterranean and on to Europe. In a post-Order world, the Mediterranean transforms from being a safe and unified European shipping channel to a fragmented and highly contested maritime environment—just as it was from the dawn of recorded history to 1945.

The Turks are not blind. If they succeed in resecuring control of the Aegean Sea and Cyprus, they will de facto control the entire Eastern Mediterranean Sea.* Expect to see Turkish military vessels escorting oil tankers that didn't request escorts. Think of it as an echo of the old Silk Roads. The Turks command the route's middle sections and so can take a bite out of the commerce as they see fit.

And *that* will trigger a response. With increased Turkish naval activity in the Eastern Med, the likelihood of Egypt jacking up transit fees for Suez, and the increased difficulty of sourcing crude

---

* The aforenoted pipeline supplies from the former Soviet world and Persian Gulf to Turkey terminate at the Turkish port of Ceyhan, also on the Eastern Mediterranean.

oil in general, many European nations are likely to view Suez and its oil flows with covetous eyes. France is by far the country with the greatest propensity and capacity to move in force.

That puts Egypt firmly and permanently in the Turkish camp, while also putting Israel on notice. A marriage of convenience with Turkey gives the Israelis local security under a regional hegemon and all the crude they could ever need. The irony of the Jewish state being under Muslim protection and using Muslim oil is not likely to be lost. The other option would be for Israel to side with France. That introduces a great deal of risk but also gives the Israelis a hedge against the local Middle Eastern super-power. Luckily for the Israelis, they are used to difficult choices.

SWING FOR THE FENCES: IRANIAN AZERBAIJAN

Russia and Turkey are not the only major powers with their fingers in the Caucasus. Any increasing Turkish presence there would be met with hostility in Iran. Such hostility would be far more visceral than even Russian opposition, for while the Russians see the Caucasus Mountains as an ideal border and buffer, the Iranians see the region as a twofold threat.

First, the Lesser (Southern) Caucasus are not nearly as geographically impressive as the Greater (Northern) Caucasus. The Greaters are, as the name suggests, pretty great. Tall, steep, imposing, and pushing east and west to the very shores of the Caspian and Black seas.* The Lessers are somewhat less great. More of a knotty plateau than true mountains, the Lessers sport several fairly easily negotiated access points that link the intra-Caucasus region of Georgia and Azerbaijan to the highlands of Anatolia and the Zagros Mountains.

---

* If it weren't for the locals' tendency to flay outsiders, I'd go backpacking there.

Several valleys connect throughout the region, but the most important by far is the Zangezur Corridor, which accesses Turkey *and* Armenia *and* Azerbaijan *and* Iran. Should a single power prove able to control the Zangezur in its entirety, it could march on the other three with relative ease. Therefore none other than Joseph Stalin himself redrew the region's borders in the 1920s, splitting the Zangezur roughly equally among Turkey *and* Armenia *and* Azerbaijan *and* Iran, with the express intent of maximizing strife among the four. Stalin was pretty good at this sort

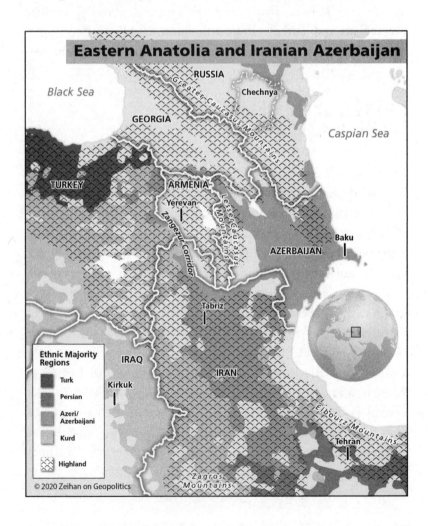

Eastern Anatolia and Iranian Azerbaijan

© 2020 Zeihan on Geopolitics

of thing. Any serious push by Turkey into the Caucasus means putting Turkish troops in the Zangezur—near the Armenian capital of Yerevan as well as within striking distance of the Iranian heartland.

As if that weren't bad enough for the Iranians, the second factor is even worse for them.

The second most numerous people in Iran—next to the Persians themselves—are the Turkic-speaking Iranian Azeris, ethnically identical to the Azerbaijanis of the namesake former Soviet republic, who live almost exclusively near Iran's borders with Azerbaijan and Turkey. There are *more* Iranian Azeris living in Iran than there are Azerbaijanis in Azerbaijan itself. The geographic center of their population is the city of Tabriz, less than a hundred miles from the Zangezur access point.

A Turkish move against Iran would be far from simple. It would require a flat-out occupation of northwestern Iran despite the direct, unrelenting military conflagration that would come from it. While Iran's jetless, tankless, outdated, sanction-starved military is laughable compared with Turkey's modernized, NATO-armed, and increasingly self-supported complex, this is about more than order-of-battle comparisons.

First, the terrain is mountainous, not only making northwestern Iran easier to defend but also providing precisely the sort of advantages that infantry would require in a battle against mainline tanks enjoying air support. The Turks could still carry the day, but not quickly or without considerable bloodshed.

Second, there's the "home-field advantage" defenders have when their homeland is under attack. Moreover, Iran's Persianification efforts are literally millennia old. Because the Azeris are Iran's largest minority group, extra effort has gone into grinding as much of the Turk out of the Azerbaijanis as possible. While most Iranian Azeris certainly still consider themselves of Turk*ic* descent, that is not the same as saying they consider themselves anti-Persian or fully Turk*ish*. For example, the dominant religion among Iranian Azeris is Shia Islam, the same as the Persians, and not the Sunni Islam practiced in Turkey. The Turks will certainly enjoy the support of a large fifth column in any invasion, but they will not be categorically welcomed. There will be plenty of resistance, and not only from the Persians.

Third, Iran has tools beyond its military. Iranian intelligence assets at any given time are active in Azerbaijan, Afghanistan, Iraq, Saudi Arabia, Jordan, Syria, Lebanon, the Palestinian Territories, . . . and Turkey. In times of Turkish-Iranian hostility, Iran has worked overtime to stoke tensions and militancy among Turkey's many mountain minorities and political factions, most notably the Kurdish population. And because any Turkish assault on Iran would need to pass through Turkey's own Kurdish region, Iran would undoubtedly work to set that entire region on fire. Even if Turkey could seize northwestern Iran, holding on to it would be a whole other problem.

This whole scenario seems like more trouble than it's worth, but it is worth considering, for two reasons.

First, most of Turkey's border regions have interlocking issues. Going into Romania can lead to confrontation with Russia, which can lead to intervention in Ukraine, which naturally leads to competition in Azerbaijan, which means Turkish forces in Zangezur, which means Tabriz is in play. Whoever has controlled Marmara, their top problem has *always* been the spiraling combination of seemingly unrelated topics and theaters into a royal, unentangleable mess.

Second, Turkey is still figuring out not only what it wants, but who it is. In 1995 Turkey was a hard-core authoritarian military regime with a healthy civil society but accountants who might as well have been Italian. In 2005 Turkey was a mildly Islamic democracy run by economic technocrats who might as well have been German. In 2020 Turkey's democracy is dead, and the country is on a rapid slide into authoritarian, ethnic-based populism that looks practically Russian. Considering that trajectory, the emergence of an ethnically driven strategic policy that takes Turkey directly into Azerbaijan—both independent and Iranian—isn't so far-fetched.

THE VOTES OF OTHERS

None of these is an obvious choice. Economically, the Balkan advance makes the most sense. Ethnically, it is difficult to argue against grabbing the Azerbaijans. Going southeast solves both an internal security issue as well as an energy dependency. Taking Cyprus would be a strategic coup and give the Turks leverage against all of Europe. Grabbing Crimea would condemn the Turks' most powerful historical foe to dissolution. What is clear is that Turkey has options—and that includes options for the fights it will pick.

Options make the foreign policy of contemporary Turkey seem erratic. Within the past decade, the Turks have offered Armenia peace and threatened it with invasion, funded some infrastructure in (independent) Azerbaijan while also haranguing Baku on other projects, cozied up to Russia economically but shot down a Russian jet, alternately let Syrian migrants flow through its territories to Europe and stopped them, encouraged Islamic militants to flow into Syria and then invaded Syria to kill them, encouraged Iraqi Kurds to ship their oil through Turkish territory while invading Iraqi Kurdistan, and competed with the Iranians for influence in Azerbaijan and Iraq while also helping the Iranians bust US sanctions.

If this seems a bit schizophrenic, that's because it *is*. Turkey's geography is complex. What works for one border doesn't work for others. What might have worked during the Cold War is different from what works during the Order's dying days is different from what will work in the Disorder. And once Turkey chooses a path, the calculus shifts again.

It will get messier.

Unfortunately for the Turks, they are not the only people with a say in the matter. While Turkey easily has the power to go *any* direction it wishes, it has nowhere near the power to go *every* direction it wishes. It will have to choose, and others have the opportunity to shape the decision.

The Europeans and Russians wouldn't like the Turks moving into the Eastern Balkans. The Russians and Iranians wouldn't like the Turks moving into the Caucasus. The Iranians obviously wouldn't be keen on the Turks moving into Iranian Azerbaijan, while neither the Iranians nor the Saudis would like to see Turkish troops moving into Iraq *at all*. Turkish forces there would be able to cut off any Iranian assault on Saudi Arabia in a day, while the last thing the Saudis want to see on their northern border is a functional state and military.

What's a bit weird is, *everyone* would like to see the Turks in Syria.

If Turkey is forced to deploy a hundred thousand troops or so to stabilize Syria, it will largely occupy Turkish strategic attention for years, absorbing any military bandwidth that might have been used to venture into Greece or Romania or Crimea or Azerbaijan or Iran. It would also enable Russia, Iran, *and* Saudi Arabia to pin Turkey down more firmly by spawning violence in the occupied areas. (The Saudis, in particular, have a vested interest in nudging Syria's decivilization process along.) Even the Europeans would see this as a good thing: if Turkey is forced to occupy Syria, the Syrians are more likely to *stay in Syria*.

The Turks know this, but that's not the same thing as saying they won't fall for it. The Turkish institutions that managed foreign and strategic affairs died with the Ottoman Empire in 1922. That's a long time to be out of the game. Since starting to wake up from their century-long slumber a few years ago, the Turks have made some horrid mistakes:

- A botched attempt to patch things up with Armenia handed the Caucasus lock, stock, and barrel to the Russians.
- Thinking they could crack open the Palestinian issue, the Turks instead managed to rupture relations not only with the Israelis, but with nearly all the Arab states.
- Efforts to manipulate the Syrian Civil War instead slimmed relations with the Americans and led to international embarrassment by the Russians, Iranians, and Saudis.
- Turkish use of the refugee crisis as a cudgel to force concessions out of the European Union soured relations with all Europeans at all levels.

In all cases, the Turks made a typical freshman mistake. They assumed everyone would do everything the Turks said, simply because the Turks are so awesome. Such narcissistic self-aggrandizement is a natural outcome of only recently settling upon an ethnic identity after not having deep conversation with anyone on the outside for a century.

Emerging into the heart of the region is the Turkish Question: No one alive has *any* experience in a world in which the Turks are outward looking. No one knows how Turkey defines its interests, and so no one has a clue as to how Turkey will prioritize.

The Turks included.

TURKEY'S REPORT CARD

BORDERS: Turkey itself is a mountainous peninsula that typifies an ugly house in a gorgeous neighborhood: Turkey bridges Europe and Asia, the Mediterranean and the Middle East, Russia and the West.

RESOURCES: Bad coal, the best agricultural lands in the Middle East, but above all else the ability to control maritime trade between the Mediterranean and the Black Sea, and land trade between Europe and southwestern Asia.

DEMOGRAPHY: Turkey has a stable, relatively young population. Great for consumption-led growth and a future capital-and-trade-driven move up the value chain. Also good for political protests.

MILITARY MIGHT: Turkey has *the* most capable army in Europe and the Middle East. Despite the military's recent political weakening under the imposition of civilian authority, it retains plenty of pep to deal with any immediate neighbors.

ECONOMY: The Turkish economy has languished in recent years under extremely inexpert political intrusion into the country's economic systems, but the country's position and population structure make it a growing manufacturing hub as well as *the* bright spot for the region's future.

OUTLOOK: Turkey will always be smack dab in the middle of everything. It's relationships with outside powers may wax and wane, but it will *always* be the economic and military heavyweight of its region.

IN A WORD: Returned.

CHAPTER 12

BRAZIL

SUNSET APPROACHES

With an unofficial motto like "country of the future," it's no surprise Brazilians like to make some callouts:

- Brazil's land area is the largest on the South American continent and the fifth-largest in the world.
- Brazil's economy is the largest of all the New World economies except the United States and ranks solidly in the top ten globally.
- Economically, Brazil has taken the world by storm. It ranks in the top three in multiple production categories: soy, corn, beef, iron ore, coffee, orange juice, and sugar.
- Brazil's natural bounty is unassailable: the world's largest river by volume, the largest tropical rainforest, the largest area of undeveloped agricultural lands.
- Brazilian manufacturing is recognized globally as excellent. The national oil company, Petrobras, regularly demonstrates a command of offshore operations that compares favorably to industry leader ExxonMobil. In aerospace, the dominant

Brazilian firm Embraer sells jets pretty much everywhere and is easily among the top four globally.*

Such snappy factoids are great—and they certainly sizzle when shared over copious *caipirinhas*—but they all ignore Brazil's extreme geopolitical weaknesses. It all comes down to difficult transport combined with Brazil's omnipresent tropics.

BRAZIL'S GEOGRAPHIC STRAITJACKET

Brazil has no navigable rivers. Well, that's not *quite* true. Technically the Amazon is navigable a thousand miles inland. However, the Amazon Basin's extreme tropical nature and muddy banks make it broadly uninhabitable and impervious to development. Except for a couple (extremely) subsidized outposts, such as the interior city of Manaus, civilization barely touches the entire region in general or the northeastern coast in particular, even with modern technological advancements.

Everything needs to be shipped via artificial infrastructure, especially roads. That would be expensive enough, but Brazil isn't flat. Think of populated Brazil as a table that's lost two legs, with the missing legs in the *interior* of the country. The bulk of populated Brazil's coastal zones is a series of cliffs known as the Grand Escarpment. Ever tried to run a train up a cliff?

In most cases the Escarpment descends directly to the ocean, blocking Brazil's urban footprints from one another. Its cities cannot integrate. Even today Brazil lacks a four-lane coastal highway. Brazilian economies of scale are few and far between.

Brazil's roads are poor. In part, it is a materials issue: concrete doesn't set properly under high humidity, introducing warps into the

---

* Embraer's technical prowess proved so attractive that America's Boeing acquired an 80 percent stake in the operation in 2019.

finished road, while asphalt often re-liquifies under high heat and normal traffic tends to squeeze the road material to the shoulders.

Brazil's only truly viable territories are in the cooler southeast, but even there sharp limitations abound. The country's rugged coastal terrain ensures that Brazil's southeastern cities are atrociously crowded—far more densely populated than even Japan's trademark lack-of-elbow-room urban zones. But Brazil lacks Japan's capital richness and its First World health-care system, *and* it's in the tropics. The results? Some of the world's most expansive slums—the infamous favelas—and the communicable-disease rates one would expect in a densely urbanized, tropical environment.

The opposite side of Brazil, on the eastern slopes of the Andes, has a different problem. Mostly tropical forest, the entire region is a no-man's-land for roads *and* the perfect sort of climate for growing cocaine. The result? A swathe of territory larger than Texas and California combined in Brazil's northwest is beyond the reach of Brazilian law enforcement and thus a playpen and breeding ground for many of South America's drug groups.

About the only reliable exception to Brazil's problem with economies of scale is its export-driven agricultural sector. Even that has a litany of problems, beginning with the sort of land Brazil has to work with. Vast swathes of interior Brazil are a tropical savanna known as the *cerrado*, which is wildly *un*like the rest of the world's farmland. The *cerrado* must be first stripped of its native vegetation, then repeatedly treated with lime for several years to deacidify the soil. At that point the Brazilians have "soil" with the consistency and nutrient profile of beach sand that requires double to triple the amounts of fertilizer, pesticide, herbicide, and especially fungicide required per crop compared with such needs in the temperate-climate zones. It is the equivalent of growing crops in a petri dish. No added nutrients, no growth.*

---

* Brazil doesn't produce jack for organic output. Stuff wouldn't grow, and even then it would be devoured by critters.

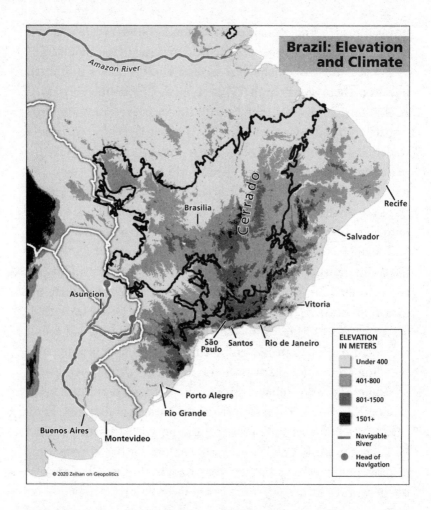

Brazil: Elevation and Climate

Amazon River

Cerrado

Brasilia

Recife

Salvador

Asuncion

Vitoria

São Paulo   Santos   Rio de Janeiro

ELEVATION IN METERS

Under 400

401-800

801-1500

1501+

Navigable River

Head of Navigation

Porto Alegre

Rio Grande

Buenos Aires   Montevideo

© 2020 Zeihan on Geopolitics

Animal agriculture doesn't get a pass either. Tropical conditions plus traditional cattle generate beef that is, well, kind of gross. Typically, ranchers address this issue by not using traditional cattle at all, instead opting for zebu, a South Asian variant, which has much lower fat marbling (and so is also kind of gross). Prime beef does *not* come from Brazil. Between higher disease rates and lower quality, Brazilian animal products are far lower in value added and far higher in cost than their competitors.

That's before transport is considered. Most products from the interior seeking export must descend the cliffs to ports on Brazil's southeastern coast. Roads aren't much easier to run down cliffs than rail lines, so all traffic must concentrate into a handful of trans-Escarpment routes—with all the traffic one would expect. Brazil is the world's only major grain producer that transports the bulk of its exported cereals by truck. During harvest time, miles-long traffic jams are not uncommon.

Considering the limited infrastructure routes, high rate of wear and tear, and the need for various input materials, per-mile Brazilian infrastructure costs are typically quadruple those of a flat, arable, temperate territory—with additional premiums for the roads that must pierce the Escarpment.

Colonial Brazil wasn't settled by pioneers who could work the land to become wealthy, but instead by people who *started* wealthy and so already had the money to lay down to build the necessary infrastructure. Rich Portuguese settlers established company towns to serve their specific corporate needs and manipulated a stilted political system and narrow economic system, enabling them to maintain control. Brazil today has a fractured economy, each piece run by different members of an entrenched oligarchic class, many of whom are those settlers' descendants.

There was no Brazilian equivalent of the American pioneer era, wherein common folk could obtain free land. Instead, the only way to acquire Brazilian land in the colonial era was to receive it as a direct gift from the Portuguese emperor. It wasn't until the first coffee boom in 1850 that land purchase by private interests was even legal. Today 1 percent of the Brazilian population owns half the country's real estate, with the *six* wealthiest Brazilians having as much wealth as the bottom 50 percent of the Brazilian population. In contrast, in America, the pioneers became freeholders became the middle class. Democracy developed in parallel, being reinforced by economic trends. In comparison, Brazilian democracy is weak and recent, truly putting down roots only in the 1990s after the military handed power to civilian authorities.

Brazil's geographic challenges have political consequences beyond the oligarchs. Because Brazilian coastal towns have so few connections among one another, most have found it easier to integrate with foreign entities on the southern Chinese model. Likewise, inland cities—where transport at least doesn't involve piercing cliffs—tend to be physically larger with lower living costs. Such lack of economic connections and the different development patterns conspire to make Brazil among the world's least unified political systems. Until Brazil's political and economic refounding in the late 1980s and early 1990s, Brazil was a confederation, with the provinces holding far more power than the central government.

One result, among many, is a deep culture of corruption—by *necessity*. The oligarchs correctly believe that most of what progress Brazil has experienced since the initial European settlement is a direct outcome of their collective investments. They see little reason why they should share their laborfruits, so they directly control all aspects of life in their fiefdoms: economic, social, cultural, and especially political. Because power in Brazil devolves to the regional and even the city levels—levels the oligarchs largely control—Brazil cannot achieve national economic goals without oligarchic buy-in, and that requires some palm-greasing. For example, in Brazil trade pacts aren't about liberalizing trade, but instead about ensuring that this or that oligarch's economic advantages are *not* exposed to international competition.

The split between those with the money to carve a future out of Brazil and everyone else has dominated the country's economics and politics since the very beginning of the colonial effort three centuries ago. Add the extreme cost of infrastructure and living expenses in Brazil's swarming-like-ants coastal cities, and Brazil suffers some of the world's most extreme economic inequality.

All the political complications of economic inequality so familiar everywhere else exist in Brazil, but with a much harder, more desperate edge. Brazil has one of the world's highest rates of urban assaults, with a murder rate five times that of gun-happy America.

The culture of crime isn't simply endemic; it's structural. The favelas do not benefit from traditional city services—not even utilities—so local mafia groups serve as the de facto providers of power and water, muscling the inhabitants out of whatever they can manage. As if that were not bad enough, these mafia groups tend to be run by the police.

Combine these factors, and Brazil faces chronic skilled-labor shortages:

- Partly the reason is geographic. Low connectivity among regions means fewer economies of scale, driving up education and relocation costs.
- Partly it's climatic. Until the late 1970s, nearly all the crops grown in Brazil were traditional tropical crops: coffee, bananas, sugar cane, pineapples, sweet potatoes, and so on. Such crops must be planted, tended, harvested, and boxed *by hand*. Tropical agriculture is the quintessential low-skilled labor industry, and it was the backbone of the Brazilian economy for three centuries.
- Partly it's structural. While an oligarch certainly has an interest in educating his children, there isn't much incentive to educate the staff—especially if the most value-added thing they do is box mangos. Besides, school might give them ideas. . . .
- Partly it's political. The oligarchs rule their roosts and don't like to contribute financially to the roosts of others. They use their local pull to minimize tax payments, reducing the resources available for any sort of national educational or infrastructure systems.
- Partly it's racial. People with poor backgrounds often have never had anyone in their extended families who have ever completed secondary school, much less received higher education. Considering the general lack of opportunities in the Brazilian economy, it is a difficult cycle to break.

This labor-quality problem effortlessly feeds into all Brazil's other issues:

- Because of the lack-of-winter insect kills, an accommodating climate for bugs, and land stewardship practices that are not exactly cutting edge, Brazilian bugs and weeds tend to develop chemical resistance two to three times faster than those in more temperate climates. Keeping up with the pests—both insectoid and herbaceous—requires adaptive technology that demands skilled labor that Brazil doesn't have in sufficient quantities. Despite the country's headline agriculture numbers, it is among the *most* dependent upon foreign chemical companies for its agricultural inputs because it cannot keep up with the chemical design at home.[*]
- A possible solution to Brazil's chronic high-cost development issues would be the indigenous development of new technologies more appropriate to the Brazilian climate and geography. Brazil has experienced some success in this endeavor via the state research firm Embrapa, which focuses on developing seed strains that can thrive in the Brazilian tropics. But the bureau faces the constant constraint of insufficient scientific staff. Anyone skilled enough to pick away at Brazil's issues can make metric gob-tons of money doing the same thing in the private sector—outside of Brazil.[†]
- This chronic skilled-labor shortage guts what forms the centrist core of most modern democracies: the middle class. Consequently, Brazil's political system oscillates wildly among the interests of regions unwilling to share power with the national government, oligarchs unwilling to share power

---

[*] Despite all the vowels in the name, Monsanto is no Brazilian firm.

[†] Embrapa's biggest victories involve its work on soil deacidification, developing nitrogen-fixing microbes to help boost plant growth, and the development of tropical strains of more traditional temperate-zone agricultural products. It's had a big impact: in 1970 Brazil was a marginal soy producer. In 2019 Brazil was the world's largest exporter.

with anybody, and a vast, barely educated mass of unskilled, low-wage laborers who are rightly indignant that they've been ruled over by people who do not have their interests at heart.*

- The skilled-labor issue is the greatest limiting factor on Brazilian advancement. Whenever international and local factors align and allow Brazil to eke out some economic growth, the increased demand for skilled labor rapidly gobbles up all that is available. Labor costs then rise stratospherically. The resulting inflation chokes off growth, and the entire system stalls. Despite "all that land" and "all those people" and "all that potential," Brazil's economy in the twentieth and twenty-first centuries has consistently been the most underperforming major developing economy in the world because any success it gleans quickly self-suffocates.

MAKING IT WORK ANYWAY

There are not many ways around such crushing geographic and climatic challenges, but there are a few. The trick is to manipulate the markets that drive the Brazilian economy enough so that what's out of whack isn't so much fixed—after all, you cannot change geography—as compensated for.

Under the imperial system, this was fairly easy. Colonial Brazil produced tropical agricultural products. Brazil's colonial master—Portugal—was in Europe, a decidedly nontropical locale. Brazilian products sold in Europe earned top escudo. The

---

* A large middle class is considered *the* goal of the developing world. A strong middle class can be economically dynamic by providing semi-skilled workers as well as *consuming* the things an economy produces. Similarly, it can be stabilizing in politics. It is a class that owns enough to want to avoid redistributive politics but not enough to allow a plutocracy. It tends to be socially on the somewhat conservative end, which stifles the kind of mob-driven change that can upend a government. In other words, a middle class keeps things decidedly . . . middle.

influx of capital helped the rich Portuguese settlers develop their plantations.

But the products in question generated consequences for Brazil.

Central was sugar—a product that requires vast tracts of land, backbreaking (and life-threatening) labor, low margins, and centralized negotiating power. In the preindustrial era, the only way to make that work was to crush labor costs as close to zero as possible. That meant slavery: first of the local indigenous populations, and later by tapping other Portuguese colonies in Africa. Sugar and tropical products like it first generated and in time locked in hierarchical economic and political norms, often based on race.

Once the slavery seal was broken, it was an easy decision to apply slaves to other less-than-fun jobs. Colonial Brazil was also known for its slave-driven gold mines, for example. Colonial Brazil hosted the highest proportion of slaves of any of the New World colonies. After Brazil broke away from the empire in 1822, independent Brazil did not simply continue the practice; Brazil expanded it. Best guess is some 4.9 million Africans were sold into Brazilian slavery—more than *half* of the *global* total. So dependent was Brazil that it did not criminalize slavery until 1888, making it the last country in the hemisphere to do so.

That legacy has folded racial bigotry into what is already one of the world's least egalitarian societies. The United States is no paragon of virtue when it comes to racially charged policing; Human Rights Watch estimates that an average of three African Americans are killed by police daily. The figure for Brazil, a country with one-third fewer people, is fourteen.*

After the end of slavery, independent Brazil struck upon a

---

* This figure includes instances when police officers were caught participating in off-duty death squads, as well as when they *open fire from helicopters into crowded neighborhoods* to make sure they get their guy. On the flip side, "their guy" tends to shoot back. Brazilian police officers are four times as likely to perish in the line of duty as their American counterparts.

different strategy: a starkly statist governing regime designed to minimize imports so domestic producers could operate behind a massive tariff wall. The oligarchs—loving the idea of a completely captive market—piled on. The result was a series of unofficial government-sponsored monopolies. Most were not nationwide, because few oligarchs held nationwide assets. Instead, the process only balkanized the Brazilian economy further, driving up costs of everything for everyone and further exacerbating the inequality issue. On the upside, limiting economic access to the wider world made it somewhat easier for the state to maintain central control.

Blinding torrents of inflation cruelly tore down occasional bursts of moderate growth. Massive economic inequality massively retarded social progress. Vicious cycles raised and viciously crushed Brazilian hopes over and over. Life . . . absolutely . . . sucked. Then all at once, all because of one power, everything changed.

BRAZIL'S DAY IN THE SUN

Brazilian-American relations have ebbed and flowed over the years. In the 1800s Brazilians got along famously with Washington. After all, with the Monroe Doctrine, the Americans were working to keep the Europeans out of the Western Hemisphere. For disparate, disunified, structurally weak Brazil, anything that kept the Europeans off the coast was grand.

But by the early years of the twentieth century, the Brazilian mind-set was evolving. The Americans had by then invaded Mexico twice, and Theodore Roosevelt modified the Monroe Doctrine to justify a more aggressive pace of intervention. Marry that imperial attitude to the paranoia of the early Cold War years, and the Americans proved willing and able to "nudge" Latin American governments out of power if they got too friendly with the

Soviet Union either geopolitically or ideologically. In 1964 it was Brazil's turn to undergo a military coup quietly supported by the Americans.

With that sort of backstory, it should not come as much of a surprise that the Brazilians were lukewarm toward the Order. The military government was perfectly willing to accept the security guarantees the Americans offered but chose not to embrace the economic aspects. With the full support of the oligarchs, who were happy being big fish in a small pond, the brass walled Brazil off from the world.

Then Brazil got *really* lucky. Two completely unrelated things progressed more or less in parallel.

First, the military government restored democracy in the early 1980s, going back to the barracks in 1985. A new constitution that partially centralized power followed in 1988, taking a bite out of the oligarchs' power and giving enough tax-levying power to the central government that it could invest meaningfully. Three years later Brazil helped found the regional trade bloc Mercosur, giving the country its first real—though still limited—experience in the modern, international economy. Brazil revamped its monetary policies, intending to break perennially high inflation, a process that included launching a new currency in 1994. In essence, Brazil reincarnated itself as a modern state.

Second, the top tier of the Soviet leadership became good enough at math to realize that the Soviet Union was doomed and that broadscale confrontation with the Americans was a recipe for disaster. Successive premiers pushed for an end to the Cold War . . . and overshot a bit. The superpower rivalry wasn't just over. The Soviet Union was no more.

The timing of these new developments—one local, one global—was coincidental but quite fortuitous. At the same time that the global system granted Brazil the greatest possibility of success, Brazil was ready to try something new.

While it would be an exaggeration to call Brazilian growth in the post–Cold War era explosive, it has certainly been stronger, more reliable, and less inflationary than the Brazilians of ages past were used to. A series of more-or-less centrist, technocratic governments bit-by-bit chipped away at many of Brazil's structural issues. These results were as impressive as the growth. Exports surged *without* an increase in government debt. Brazil was no longer some random place beyond the horizon. In nearly every agricultural and nonagricultural commodity market, Brazil had become *the* critical new source of supply. The increasingly confident Brazilians felt they had finally escaped history.

No offense intended to the men and women who reformed the Brazilian system in the 1980s, 1990s, and 2000s, but ultimately Brazil's recent success can credit the same general process that generated previous bursts of growth: someone tampered with the market. Only this time it wasn't someone who spoke Portuguese; it was the Americans with their post–Cold War incarnation of the Order.

Three market-tweaking factors drove Brazil's recent successes:

CHEAP, PLENTIFUL, RELIABLE FINANCING. Brazil needs vast supplies of capital to overcome its local topographic and climatic restraints: money for roads up the Grand Escarpment, money for transport of goods from deep in the interior to the coast, money for ports on tiny plots of land, money to clear and level the land, money for mass volumes of pesticides and fertilizers, etc. Initially, this money came from rich Portuguese settlers, who themselves often borrowed it from Dutch or British interests. Today it comes from foreign investors and foreign loans. Not surprisingly, about one-fifth of the *cerrado* territories are foreign-owned. (In the United States, the total proportion of foreign-owned farmland is only about 3 percent.) Much of this capital availability is possible because of the financial bubble from the aging global Boomer cohort.

CHEAP, SAFE TRANSPORT ON THE SEAS. Nearly all of Brazil's vast commodity production is not simply exported. Many of its co-continentals in South America produce (and export) identical products. Brazil must send its output beyond the continent. In terms of shipping, South America is the world's most remote populated landmass, and most Brazilian exports must flow to Northeast Asia, a sail of fourteen thousand miles—because of the arrangement of the continents, it is more than halfway around the world—making it one of the world's longest export runs. That leads us to Brazil's final requirement.

STRONG, CONSISTENT, EXTERNAL DEMAND. Brazil is no low-cost producer, and compared with global norms, neither are most of its exports high quality. Foreigners tend to be interested in Brazil's output only when global prices are high, global supplies are insufficient, or the insurance of having an additional supply stream is desired. The post–Cold War period has experienced the fastest global economic growth *ever*. Between general industrialization and rising living standards throughout the developing world, Brazil's farms and mines—high cost though they were—were the only ones that could expand output to service new global needs.

It hasn't proven nearly enough to enable Brazil to turn the corner.

FEAR THE FALL

Brazil's relative success required the extra layer of stability and wealth that only the *post*–Cold War Order could provide. Brazilian success required more than strong demand; it required the bottomless, price-insensitive demand of the Panda Boom. Brazilian success required more than cheap, abundant capital; it required the supercheap, hyperabundant capital of the height of the Boomer cadre's financial wealth in the moments before the Boomers retire en masse. These factors aren't simply one-offs, and they are not simply fading; they are turning inside out, heralding tragic outcomes for Brazil:

- As a high-cost producer, Brazil depends upon bottomless global demand to sell its output. Without Northeast Asia—and in particular, China—demand for Brazilian ores will collapse. Brazil's primary competitor in the iron ore and bauxite markets, Australia, is both closer to end markets and has lower capital and development costs.

- Physical risk on the seas will restrict Brazil's ore sales to markets in which they face steep competition. The European neo-empires will be able to source inputs from their African neo-colonies, while the United States will prefer supplies from domestic producers and next-door Canada and Mexico.

- Brazilian foodstuffs face even a harsher adjustment. One perk of tropically produced soy is that it has a higher protein content and so has higher demand for use as animal feed. That fact, combined with an era of global stability, has catapulted Brazil to its 2019 position as the world's largest soy exporter. In a world of broadscale economic collapse, demand for animal protein will plummet, taking with it Brazil's top export earner. In contrast, Brazil's primary soy competitor, the United States, is seeing its production costs go *down* courtesy of the shale revolution.

- Likewise, most Brazilian tropical cash crops are considered exotic luxury foods. Substantial price rises due to higher production and internal transport costs will shift them out of reach for most of the global population, reducing the oligarchs' interest in producing them in the first place.

- The rapidly aging global Boomer population combined with general global economic dislocation will vastly reduce the amount of capital available worldwide *while also* concentrating that capital in locations perceived to be safe havens (i.e., not Brazil). With capital supplies drying up, the production of most Brazilian output, as well as the maintenance of basic Brazilian infrastructure, will become difficult to impossible. That will drive up production costs while also reducing the volume and quality of subsequent production.

From the Boomer investment collapse alone, Brazil likely faces a lost decade because of starkly higher financing costs. The loss of a global market adds at least a second decade to what will be a painful adjustment period. Even *that* assumes Brazil gets the politics right.

Don't count on it.

At the time of this writing, Brazil's endemic corruption has, in effect, caused its political system to seize up. When any progress on any meaningful issue requires oligarchic buy-in, everyone who has ever achieved anything has been a participant on one side or the other of the palm-greasing industry. What started as a drug-money-laundering investigation has since ballooned into a national force that is part anti-corruption effort, part paralytic agent. Since beginning in 2013, this "Car Wash" investigation has turned juggernaut, convicting hundreds of top officials in Brazil's political and economic elite. Of Brazil's seven former presidents since democratization in 1985, two have been impeached, two have been imprisoned, and two more (as of February 2020) remain under investigation. Politically, Brazilian democracy is utterly incapable of pursuing the sort of economic, class, and constitutional reforms required to manage Carwash's outcomes, much less the broader challenges the Order's end promises.

Even now, as the lingering Order provides that cherished external and global market stability, the Brazilian system is already failing. Brazil is *already* losing control of its cities and its northwest frontier. Brazil *already* suffers from race riots. Brazil *already* has problems keeping the lights on. Brazilian manufacturing is *already* stalled. Brazil *already* faces flagging export income. Brazil *already* is contending with a populist uprising from both the Left and Right. Brazil *already* faces challenges to central control. Brazil has *already* elected a president who thinks the biggest issue with the Nazis was that they were too soft on their domestic foes. Local government control of Brazil's cities is *already* dissolving due to rampant, rising crime. All in a time of relative stability and wealth. As the global Order's failure slams into Brazil like a freight train, nothing short of the internal Brazilian "order" is at risk.

For the Brazilians, the most likely path forward is equal parts bifurcated and unenviable.

Brazil's regions disassociate from the national government as well as from one another. In times past, most of Brazil's regions managed their own affairs, both because of the intensely local nature of Brazil's disparate geography as well as the political and economic self-centeredness of the oligarchs. As input costs for Brazil's input-heavy export products rise, Brazil's exports will shrink. This, almost by definition, will hurt some provinces more than others, once again driving the country's various pieces in different economic directions.

Politics will follow economics; the old normal of decentralization will become the new normal. The national government, already a bit of a zombie on a good day, will enter full rigor mortis. Brazil will devolve from a nation-state to something far more familiar to Brazilian historians—a much poorer series of disconnected regions that look more like regional satrapies than provinces of the same country.

That makes the future of Brazil—and predicting the specifics of the future of Brazil—messy. While none of Brazil's provinces will do better in the Disorder to come than under the Order just ending, their rates of decline will be as varied as their local challenges:

- São Paulo will be the first among not-even-remotely equals. Its position atop the Escarpment grants it cheap, easy physical expansion opportunities the coastal cities lack; its size grants it the economies of scale *all* other Brazilian cities lack, *and* it commands the best physical access from the interior to the coast. In a devolved Brazilian system, it is most certainly the regional power broker.

- Rio Grande do Sul also stands apart (in a good way). It won't suffer nearly the economic disruption of its compatriots. The province already enjoys Brazil's most temperate climate, its least onerous infrastructure constraints, and its best physical access to the wider world. Unsurprisingly, it is Brazil's second-largest soy producer. Rio Grande do Sul's independence from the broader Brazilian system might even prove to be in name as well as in fact.

- Minas Geras depends heavily upon mining, but everything mined there can be produced more cheaply elsewhere in the world. In terms of relative loss of wealth, it may suffer the most.

- Rio de Janeiro is by far Brazil's best-known city, and its decline will be emblematic of several of the country's coastal cities. It's too far from the Northern Hemisphere to be involved in manufacturing supply chains, too isolated to serve as entrepôt or processing center, and too densely populated to be safe.

- Mato Grosso do Sol in the deep interior was the most recent Brazilian province to garner the technology and capital to link into broader infrastructure networks, and it has emerged as a vibrant new soy producer, accounting for roughly one-fourth of Brazil's total output. The province now faces the worst of all worlds. All its inputs—seed, fuel, fertilizer, pesticides, and so on—flow *into* it on roads from southeastern Brazil. Its *out*put either has to go allll the way back to southeastern Brazil, or instead travel six hundred miles *farther inland* on poor roads that are not completely paved to a river port such as Porto Velho for loading onto barges that then sail downstream on the Rio Madeira to the Amazon where the barges must be off-loaded onto oceangoing bulk ships. Between zero local capital supplies and an unfinished development process, if *anything* goes wrong *anywhere* in the Brazilian system, Mato Grosso do Sul's future will simply wash away along with its as-yet-unpaved road network.

- The province of Pernambuco, near Brazil's eastern tip, is too far from the more successful provinces to hitch its ride to their stars, *and* it is far closer to First World markets, *and* it grows predominantly tropical products that will be in short supply in the rich world. The likely outcome is a de facto partial colonization by American corporate interests who know an opportunity when they see it.
- Even more remote are Roraima or Amazonas in the far northwest. Bereft of subsidies from Brasília, both will get a new breed of oligarch: drug lords. Their future will likely be little more than a flimsy front for illicit production and transport operations.

Each region will follow its own logic and leadership and be challenged by its own populist insurgency. Foreigners may play on the coast, siding with this or that regional governor or rebellion as their desires dictate.

Countering such centrifugal spins on topics of national identity and regional security and national disintegration will be a single centripetal force that, depending on one's politics, isn't necessarily better: military rule. With economics and geography pulling Brazil apart, Brazil's future as a unified state cannot survive in the Disorder without some forced integration. That requires a degree of centralization that Brazilian democracy has proven incapable of providing. A flat-out coup is possible, but that's *probably* not the path Brazil is on. Something else is in the air.

Brazil's civil decay has *already* reached the point such that law enforcement in the cities cannot cope with either petty or organized crime. If anything, local police are *part* of the crime problem. Consequently, the military often patrols Brazilian cities, the city of Rio de Janeiro going so far as to place its local police command under military authority in 2018. The only real question is whether the Brazilian military becomes the de facto government of *all* of Brazil or merely the bits that can't keep the lights on and the trains running without bringing in non-hired guns.

This is not a happy forecast because this is not a happy situation. The Brazilian military walked away from power in 1985 and is totally gun-shy about returning to politics. That rehabilitation is not only a topic of discussion within Brazilian political circles but also that the Brazilian military is *already* gaining hands-on expertise running some aspects of city services at those cities' *request* is a sign of just how desperate the situation has already become. And this is *before* the Disorder rolls in.

The halcyon period of recent decades—improved political representation, economic and educational opportunity, sanitation—is unwinding, and Brazil will degrade not just politically and economically but socially. Many had hoped Brazil's 1990s and 2000s stabilization, democratization, and growth would make it a leader, a model, and the center of gravity in a new, more integrated South American system. As Brazil crumbles, the opposite is far more likely.

The implications for the wider world are less than ideal.

Throughout history, food supply has typically been the most significant limiting factor. If you couldn't feed your people reliably, you tended to lose your government (and perhaps your head), never mind any chance of national and/or imperial expansion. Famine is the ultimate continuity-ending event. Most would-be imperial territories were evaluated first and foremost for their agricultural potential.

The Order did more than break the governance link between imperial center and colony; it freed the former colonies, like Brazil, to generate output that was useful to people other than their former masters' populations. Global energy and manufacturing industries brought a millennium of technological development to some of the world's most rural and primitive zones, resulting in mass access to mechanization, fertilizer, herbicides, and pesticides.

The Order's improvement of global security and distribution capacity put famine on the run, but it was Brazil's surge in output in the post–Cold War era that started the entire world on its march to better nourishment. Remove Brazil and the Order, and fifty years of nutrition gains collapse all at once, with the worst impacts falling on countries early in their industrial modernization and at the lower end of the income scale. Central Europe, Southeast Asia, and the northern third of South America top the first list. Central America and South Asia top the second. Sub-Saharan Africa is unfortunately on both. China too is a big casualty, primarily due to its size: there aren't many places that can even hope to take the edge off China's appetite.

That doesn't mean there won't be winners.

In a higher-cost environment, the United States will be the most capable player in the hemisphere by far. It's also only the United States that will be able to either keep Brazil afloat or ensure Brazilian goods can reach end markets. However, contemporary Brazil's agricultural exports are a huge competitor to the American farm sector. Why would a more protectionist, isolationist America do anything to boost the competition? Likewise, a pair of American friends—Australia and Canada— are Brazil's primary competitors in ores. It's hard to see them lobbying Washington on Brazil's behalf.

A world with a lower Brazilian profile looks interesting to a pair of other competitors. In Asia, India often finds itself filling a similar niche: high-cost, low-quality foodstuffs and high-cost, low- to mid-grade ores, and India is *far* closer to global sources of demand—itself, for example. Closer to home, it is difficult to see the Argentines doing anything but beaming broadly. Argentina is as temperate as Brazil is tropical, flat as Brazil is rugged, low-cost as Brazil is high-cost, high-quality as Brazil is low-quality. It isn't every day one's primary regional competitor implodes.

What is great for Brazil's agricultural and industrial commodity competitors is often disastrous for consumers. The collapse of Brazil's agricultural export capacity will hand the world deep food shortages. For the Americans and Europeans that's a bit of a wash, for they live in the heart of the world's greatest producing zones. But for Africa, the Middle East, and East Asia—regions

that vacuum up massive volumes of imported foodstuffs—the future is a hungry one.

AND NOW FOR SOME GOOD NEWS

If this all sounds bleak, it is. But bad is not the same as gone, and worse is not the same as dead. Brazil faces neither the broadscale collapse of China nor the demographic implosion of Russia nor the economic and security catastrophe awaiting Germany. Brazil will look and feel and act different, and most of those differences will not be for the better, but this isn't the end of history for the Brazilians.

The sheer cost of making the Brazilian geography work means Brazil produces either at scale or not at all. All Brazil's would-be exports require foreign inputs, whether slaves or capital or fertilizer. Securing such outside support requires something that's in vogue (such as sugar or coffee), a demand of empire (gold), or an economically insane, voracious consumer market (such as soy). Brazil will always be dependent on outside forces, not only for big capital projects, but for immediate, everyday inputs. Those inputs and the exports they enable make or break Brazil. Without them, Brazil looks more like Africa—another plateau region with no coastal plain. With them, Brazil is an unequal society. That's far from perfect, but it does grant Brazil a large, demand-driven market at home. That's something the Chinese would kill for.

There *are* reasons to hope.

First, the Brazilian system *has* gotten better, particularly in agriculture. Techniques like no-till reduce the need for fertilizer. Crop rotations help break the fungal and insect cycles in a manner somewhat akin to winter in temperate climates. The Brazilians have largely perfected the liming process, both shortening the amount of time required to bring new land under till and reducing the volume of lime required to keep their lands productive. Improved access to financing at the local level—think fintech-enabled smartphone apps—has helped smaller farmers get their fertilizer via something other than barter. While the system is *still* capital intensive and inefficient despite these improvements, all this does lower costs and increase output levels.

The transport situation has arguably improved most of all. While undoubtedly a fair amount of loans from abroad and government outlays disappeared into this or that pocket, Brazil has nevertheless completed a great deal of infrastructure during the Order Without Borders era: expanded road corridors up the Grand Escarpment, canyon-spanning bridges to improve transport links in and out of São Paulo, even a few hybridized truck/canal/river systems to ship some agricultural products from the

Brazilian internal territories *north* to tributaries of the Amazon. Some of these routes have proven so successful that nearly one-third of Brazil's combined soy and corn exports now use them.

The point is that much progress *has* been made, and everything that reduces the country's ridiculous internal transport costs and agricultural input costs increases the agriculture sector's viability. The trick moving forward will be to protect such gains in the face of the tropics' normal deteriorative impact upon infrastructure of any sort. Even standing still is in and of itself a capital-intensive process.

The already "terraformed" status of much of the *cerrado* also argues for some economic dynamism. *Cerrado* soils are functionally infertile; they support the sort of plant life people can eat only if the land is fertilized. If the fertilizer disappears, it isn't as though it instantly reverts to jungle. It turns barren. For a few years at least, an untended cleared *cerrado* will look more like Tatooine than Dagobah. If the money becomes available, the land won't take too much prep work to bring it back into production. That grants at least a few of Brazil's provinces a bit of time to find their feet.

Second, Brazil isn't in the Eastern Hemisphere but instead safely inside America's Monroe Doctrine. Brazil faces foreign dabbling, but *not* foreign invasion. Its local sea lanes will remain safe from neo-empires and pirates alike. It can still access America-centric supply chains for fuel and agricultural inputs. If the stars align, it is at least possible for Brazilian farmers and miners to play a role in supplying food and minerals to a hungry, broken world. That's a far better position to be in than some of Brazil's peers, like Ukraine or Congo.

Third, while the cheap, easily securable financing that enabled Brazil to become a qualified success under the Order is gone, that is not the same thing as saying that *no* financing will be available. Brazil might not be a gold-star investment destination, but in the limited options after the Order, that hardly means it would be a stupid one.

In addition to a lower security-risk profile than many would-be production centers in the Eastern Hemisphere, most of Brazil's exports are oligarch-managed. These are not small family farms or artisanal mines but massive properties covering hundreds of square miles. Because of their scale, they can seek and secure financing from sources a continent away far more easily than, say, a small American family farm. The fact that the Brazilian currency will likely be scraping bottom in the Disorder means that the oligarchs' labor and tax costs will be a fraction of that of competitors' in more stable locations, at least compared with their financing terms, which will be denominated in US dollars.

Such factors make it likely that Brazilian interests will be able to secure at least *some* credit. Most will funnel into one of three types of schemes:

- First are the export-oriented projects that foreigners find interesting.
- Second comes energy. Brazil's oil sector is largely offshore and is among the world's most expensive and capital intensive. It will be a huge credit sink, but the alternative is importing foreign energy at top-dollar prices.
- A distant third will be for food production for the domestic market. It is worth noting that most fuels, pesticides, herbicides, and fertilizers are likely to be imported, so their costs will be fixed in US dollars.

Such a short rank-order has political consequences. Foreign investors want reliability—especially when they'll be asked for up-front investment to get Brazilian infrastructure, farms, and mines working again. Populist, democratic provinces would have a tough time attracting cash, while oligarchic, military-run neighbors might roll in it. Constant populist backlashes will be Brazil's new normal.

Fourth, while Brazilian output may not be cost-effective under normal circumstances, there is little about the Disorder that fits recent definitions of "normal." Many of the world's major agricultural producers and exporters will face drastically reduced production capacity, as well as threats on the high seas that will threaten goods available for export. Because Brazil's farmland faces no external security risks, the "only" challenges are inputs and infrastructure. As global food supplies crash, Brazil—despite its myriad geographic challenges—will still be available as a high-cost production site. The higher global food prices go, the more likely enterprising investors are to revisit Brazil.

Lifelines might even come from the most unexpected of locations. American demand for coffee first manifested in the nineteenth century. At the time, much of the world's coffee originated atop Brazil's Grand Escarpment. The resulting capital flows enabled much of the highlands of the greater São Paulo region to build their first real infrastructure. Part of that infrastructure was a rail line that quite literally *went up a cliff*. Americans decided that they wanted—nay, *needed*—coffee and so were willing to defy economics. And geography. And gravity. Hope is not lost, but for Brazil it might get a little weird.

Yes, Brazil faces a hard stall. Yes, Brazil will become a much more violent place. Yes, Brazil's much-lauded progress on so many fronts will cruelly unwind. Yes, Brazil may end up being a country in name only. But except for Brazil's western extremes and Amazon fringe, outright civilizational collapse is not in the cards. Neither is war. Nor famine. Compared with the likely futures of China and Russia and Germany and Iran, that's practically paradise.

BRAZIL'S REPORT CARD

BORDERS: Jungles and mountains largely insulate Brazil from its neighbors, few of which pose any significant military threat. Brazil's *real* barrier is a break in elevation between its interior and its own coast, which separates Brazil's production zones from its primary population centers and ports.

RESOURCES: Minerals, coal, offshore oil and natural gas, and, with enough investment, *huge* agricultural lands.

DEMOGRAPHY: Brazil's is the second-largest population in the Western Hemisphere, but it's aging several times faster than that of the United States, Europe, or even East Asia.

MILITARY MIGHT: Brazil makes its own mid-grade fighter jets, but the Brazilian army has had more experience storming drug dens and running local police forces than waging wars.

ECONOMY: With the second-largest economy in the Western Hemisphere, Brazil is a major agricultural exporter that also produces aircraft, automobiles, textiles, steel, and a swathe of low- to medium-value value-added goods.

OUTLOOK: Brazil owes its modern existence to globalization and the Order. Without the foreign capital to fuel its infrastructure and agricultural sector, without safe transport to send its beef and soy to customers around the world, Brazil will struggle to maintain its economy on its own.

IN A WORD: Nuts.

CHAPTER 13

ARGENTINA

THE POLITICS OF SELF-DESTRUCTION

Argentina's defining geographic feature is also the bedrock of its success. A series of South American rivers meet and empty into the Atlantic Ocean one-third of the way down Argentina's Atlantic coastline. This confluence of rivers, named the Rio de la Plata, is one of the largest estuaries in the world and is a massive commercial hub. The three rivers that merge into it are themselves navigable: the Uruguay, Paraguay, and Paraná Rivers snake deep into the country's northern reaches, including through most of the populated parts.

The rivers' placement is nearly ideal, directly overlying the Pampas region—the world's fourth-largest contiguous chunk of arable, temperate-zone agricultural land and one of the world's most naturally fertile. At first the Paraná and Uruguay Rivers bracket a slice of territory that is among the world's most productive rangelands. Territory to the rivers' west is particularly good for corn and soy. To the south, the land dries out and becomes perfect for wheat. The entire Pampas is far enough south of the equator that it sees actual winter, unlike most of the countries farther north in South America. The winter insect kills are sufficient to keep pests under control, drastically improving public health and agricultural productivity without the need for scads of medications and pesticides and herbicides. The Pampas is the Midwest of South America.

But that's where the similarity ends, for this region is not tacked on to other zones that are part of Argentina. This *is* Argentina.

At the very center of this entire system—in the middle of the agricultural lands and at the head of the Rio de la Plata, where all those rivers discharge—lies the gleaming city of Buenos Aires, one of the greatest metropolitan regions in the New World. Buenos Aires serves as the central node for all the country's agricultural processing and export, all its financial and import activity, most of its industrial base, and its population core, cultural heart, and political center. It is New Orleans, Chicago, Los Angeles, Houston, Detroit, Minneapolis, New York City, San Francisco, and Washington DC all in one.

The world has many primacy cities—places where countries concentrate the bulk of their economic, political, cultural, and military power to ensure the dominant culture's preeminence in all things. Paris, Moscow, Tokyo, Beijing, and London are quintessential examples, and in those cases, the choice was starkly deliberate. France, Russia, Japan, China, and the United Kingdom all have outland territories sufficiently rugged or distant to house restive minorities. And even after a millennium of consolidation (and ethnic cleansing), groups like the Basques, Catalans, Chechens, Ukrainians, Ainu, Tibetans, Uighurs, Cantonese, Scots, and Welsh regularly vex the French, Russians, Japanese, Han Chinese, and English. For France, Russia, Japan, China, and the United Kingdom, establishing their capitals as primacy cities was, and remains, a requirement of consolidation—a means of securing national identity in a heterogeneous milieu of geographies and ethnicities.

Not so for the Argentines. Buenos Aires isn't simply a primacy city. It's a naturally occurring one. Deep indentations in the country's coastline—again that Plata estuary—make it not only the country's best port but also place the city in the middle of the Pampas. All the core Argentine territories lie within five hundred miles of the capital. Cargo shipments within the core have little choice but to originate or terminate in the capital, and there are no geographic barriers whatsoever within that core zone. Everything naturally gravitates to Buenos Aires, no social engineering required. Many non–Buenos Aires Argentines may resent their capital's control over their economic, political, financial, and social lives, but it is damned hard to resist the city's centralizing pull.

The country's omnipresent plains and plentiful rivers make it easy for the state to project power, and grants would-be rebels nowhere to hide.

External challenges are barely different. Just as it has proven simple for the Argentines to consolidate rule over Argentina, once one moves away from the Argentine core, shifts in the broader region's climate, soil makeup, and geography make protecting the Argentine territories child's play.

A meaningful amphibious assault is functionally impossible. Both Miami and London are nearly a seven-thousand-mile sail. Even Africa is five thousand miles away. Nor is the Southern Cone on the way to anywhere. European sails to India and the Orient are a cool six thousand miles shorter using Suez or sailing around the Cape of Good Hope. American East Coast trade with Asia would rather transit Panama. The Southern Cone is the most remote populated location on Earth.

Back on land, to the west of the Argentine core towers the meridional section of the Southern Andes, steep-sloped granite peaks so far south of the equator and so high that most boast permanent snowcap. Few passes thread these mountains at all, and winter snows and avalanches typically seal off all but four of

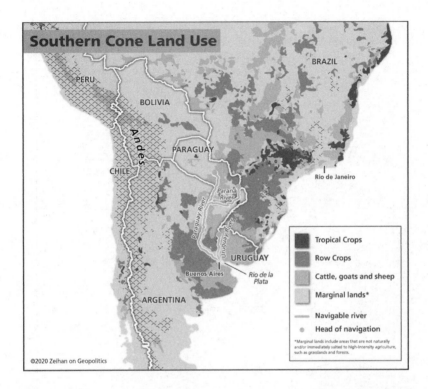

Southern Cone Land Use

Tropical Crops
Row Crops
Cattle, goats and sheep
Marginal lands*
Navigable river
Head of navigation

*Marginal lands include areas that are not naturally and/or immediately suited to high-intensity agriculture, such as grasslands and forests.

©2020 Zeihan on Geopolitics

those. The barrier is so extreme that despite sharing the continent's longest border—and the world's third-longest—Argentina and Chile might as well be on different planets.

To the northwest of the Argentine core lies the Gran Chaco, a grab bag of generally crappy geographies—blazing hot summers, low rainfall, erratic fertility, thorny scrubs, and the occasional swamp. While not technically barrens, most portions of this climatic region are far more trouble than they are worth and have witnessed development only in fits and starts. Most of the territory is so useless that even the nationalist-minded early Argentines had little issue ceding much of the Gran Chaco to the bitter, politically riven, largely powerless rump state of Bolivia. Other (slightly less useless) pieces serve as the western lobe of Paraguay. The better bits of Gran Chaco make up much of Argentina's northwest frontier.

To the north lie the only land borders that might generate a

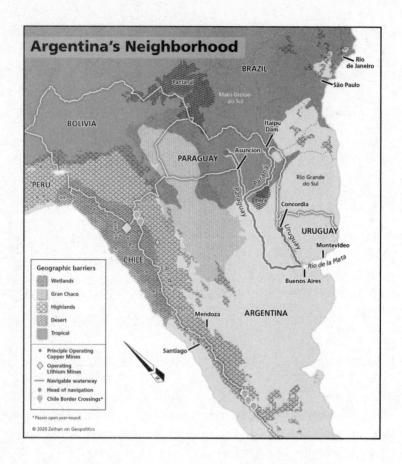

Argentina's Neighborhood

BRAZIL

Rio de Janeiro

São Paulo

Pantanal

Mato Grosso do Sul

BOLIVIA

Itaipu Dam

Asuncion

PARAGUAY

Paraná

Rio Grande do Sul

PERU

Paraguay

Iberá

Concordia

URUGUAY

Uruguay

Montevideo

CHILE

Rio de la Plata

Buenos Aires

Geographic barriers
- Wetlands
- Gran Chaco
- Highlands
- Desert
- Tropical

- Principle Operating Copper Mines
- Operating Lithium Mines
- Navigable waterway
- Head of navigation
- Chile Border Crossings*

Mendoza

ARGENTINA

Santiago

*Passes open year-round.

© 2020 Zeihan on Geopolitics

bit of strategic heartburn, yet even these aren't exactly invasion routes.

Directly north of Buenos Aires, Argentina's core territories sink suddenly into the Iberá Wetlands, a marshy zone that blocks the most direct overland route between the Argentine core and the arable, eastern half of Paraguay. To the northeast, the Uruguay River is navigable only as far north as Concordia. Beyond that point, it is packed with rapids that complicate any meaningful crossing and end any possible routes for shipping. The mighty Paraná is traversable for a considerably longer distance, but even that abruptly stops at Guaíra, the world's largest waterfall by volume—roughly nine times the flow of Niagara. Or at least it did. In the twentieth century, Guaíra was drowned by the creation of the Itaipu Dam, creating the world's largest hydropower scheme. In essence, Argentina's northern border has shifted from being damn-near impossible to penetrate to completely impassable.

That leaves the possibility, however remote, of a land invasion that is capable of crossing Argentina's two remaining—and fairly narrow—frontiers: land approaches that follow the shores of the Paraná and Paraguay Rivers from the country of Paraguay, and across the Uruguay River from Uruguay.

The struggle over these approaches is the story of modern Argentina.

INDEPENDENCE AND THE SHAPE OF THINGS TO COME

The British colonial experience in the New World was starkly different from that of the other imperial powers, most notably from Spain's.

The eastern coast of North America boasts barrier island chains and the Chesapeake, features that provide a plethora of coastal purchase and shielding for communities while still enabling direct sea access for all. Even inland urban zones like Philadelphia could reach the ocean via the Delaware River, while even the poorest farmer could access the maritime network to sell grain downriver. Infrastructure needs were minimal, so the need for a government was minimal. As such the American colonies were characterized by lots of smallish cities and great swathes of rural development that incrementally pushed back the frontier line.

In contrast, most of the Mexican and South American coasts were not temperate while quickly giving way to uplifts, denying both easy coastal access and internal transport options. Artificial infrastructure was an absolute necessity, and locations without it tended to be empty. Economic activity and political power concentrated in a handful of zones centered upon major cities. Such concentrations were easy to control. While the Americans right from the start became used to living and working and thriving in a world separate from—or, at most, parallel to—the government, Spanish settlers found themselves under the thumb of individuals dispatched by the imperial center in Madrid, the viceroys. These agents of the crown expressly enjoyed maximum latitude to manage the New World as they saw fit so long as they maximized revenues for Madrid. Trade with foreign countries was expressly criminalized.

Different geographic and economic features necessitated a different sort of immigrant. In proto- and early America, the limiting factor wasn't transport or capital, but labor. A horde of hale and hearty immigrants with can-do attitudes (plus a sprinkling of criminals who were not allowed to return to Britain, backed by a far from insignificant number of slaves) fit the bill perfectly. In contrast, the Spanish New World desperately needed capital, and so attracted a cut above: the wealthy, especially those willing to relocate for a shiny new business opportunity. They'd make the trip in style, bring a bevy of peasants with them, enslave the local natives with the viceroys' blessings, set up company towns, and get on with the business of getting richer. They were also (much) pickier about the specific spots they would settle in. While American smallholders came predominantly from poor stock and would take whatever tiny parcel of land they might be able to call their own, Spanish settlers only wanted the *really* good lands. They would set up the equivalent of company towns encircled by mines and farms adrift in large swathes of wilderness. Smallholders of the American style were largely unheard of; there is no Spanish version of *Little House on the Prairie*.

The difference made Spanish America nearly empty. By the time the newly independent American government was firmly established in 1790, the young nation held four million people to the Spanish system's seventeen million, but those Spanish colonies occupied *ten* times the land area and had been founded two *hundred* years earlier.

The Rio de la Plata region was more different still, with much of the variation boiling down to isolation.

The range from the southern reaches of the Spanish New World colonies near Buenos Aires to the northern reaches at Santa Fe is a cool six thousand miles as the crow flies—over twice the distance from Madrid to Moscow. The functional distance was much farther than it sounds. Columbus's first landing in the Americas was in the Caribbean, and the Spanish crown went with what it knew. Contemporary Cuba, Mexico, and eastern Colombia were the original Spanish colonies. Then Spanish penetration stalled. South of contemporary Panama and Venezuela there just weren't many opportunities. On the Pacific, the mountainous jungle coast gave way to mountainous desert coast. On the Atlantic, the Portuguese snagged the Brazilian territories. Chewing through or skipping over all the hostile lands took a couple centuries. The rest of the colonized Americas literally had generations of development under their collective belts before the Rio de la Plata region received its first formal viceroy in 1776.

The result was a mix. In a way, the top-down, viceroy-commands-all political system of the rest of Spanish America concentrated power even *more* in the Rio de la Plata. Most of the best lands were in the interior, in particular the Pampas and lands extending north through what is today eastern Paraguay and edging into parts of southern Brazil and southeastern Bolivia, a zone slightly larger than the US states of North Dakota, South Dakota, Nebraska, Kansas, Minnesota, and Iowa combined. In controlling Buenos Aires, the viceroy commanded it all.

And in a way the boundless horizons and labor dearth of the American model was even more applicable. The region's singular access point combined with its navigable rivers combined with its late opening meant the entire region was thrown open to settlement at once. Add in the fact that by the time the Argentine lands started settlement the Spanish crown had outlawed native enslavement and chronic labor shortages were the norm. Robust capital plus free (high-quality) land plus massive labor shortages meant that none of the rich settlers wanted to farm—ranching was the way to go, with the wealthy settlers finding themselves toiling right alongside their gauchos.

All the output had no choice but to sail down to Buenos Aires to be processed for export. Merchants there *loved* that a third of a continent's worth of bounty was a captive flow, while simultaneously viewing the viceroy's dictates that *all* output was to be shipped *exclusively* to Spain as profoundly irritating. Yet even here the labor shortage shaped outcomes. As late as the American Revolution, Buenos Aires had but twenty-five thousand residents, making it damnably difficult for the viceroy to provide sufficient customs staff. Smuggling ran rampant.

Both in town and in the country a hybrid mind-set evolved that combined the arrogance, entitlement, and connections of old money with the sweat, pride, and ferocity of new money. It was this rural-urban split-class of old-new rich that started rebelling against the Spanish crown in the early 1800s.

Whether the Spanish crown could have maintained control of its Western Hemisphere colonies is an open question, but in the end, the future of Spanish America was decided by military developments in Old Spain. In 1803 the Napoleonic Wars began in Europe. In those wars' many twists and turns, Spain shifted from neutral to ally of France to occupied by France. Before it was over, Spain itself imploded politically, consigning the former imperial center to political instability, economic breakdowns, and foreign invasion.

Without a functional government at the imperial center and with uprisings throughout Spanish America, the New World viceroy system shattered. Rebellions flared everywhere based on the timing and fervor of local issues. The Buenos Aires uprising took shape in May 1810, ending the city's role as the regional seat of Spanish power. The settlers and especially second-generation merchants who were born in the New World began overturning what remained of Spanish control in favor of their own local interests. Formal separation was declared in 1816, and save a couple islands, all of Spanish America had shrugged off Spanish control by 1825.

While technically this separation meant independence, it wasn't "Argentina" that jettisoned Imperial Spain in 1816, but rather the United Provinces of the Rio de la Plata—less a country than an ad hoc council of local bigwigs. It disintegrated quickly. In part due to infighting. In part due to opposition from the wealthy settlers in the interior. In part because of the rise of a new class of elite: men who could rally the region's scarce labor with populist calls, men who could then press guns into hands and overnight raise a militia stronger than that of a viceroy, men who could seize control of large swathes of land and set themselves up as kings. Once stirred into the mix, these caudillos ensured Argentina's opening decades would be brutal and bloody. Some refer to this period of Argentine history as a civil war (and this period of South American history as a series of civil wars), but doing so implies that there were conflicting visions for Argentina at play. It was a great deal messier than that, because there *wasn't* an Argentina.

Each caudillo and each of the remaining crop of original landholders ran his own government in his own town with his own populist patronage network based on his own personality and economic interests, and saw no reason to cede any of the political or economic authority he had created to others—most certainly not to some cocky dudes in Buenos Aires who thought they could simply step into the viceroy's shoes. The result was a region of fractured to nonexistent loyalties, generating a system that was more or less all against all.

A mere month after the independence declaration, the fighting began. Imperial Portugal tried to mop up the formerly Spanish lands, particularly the Cisplatine, a territory that includes the contemporary country of Uruguay as well as the Brazilian province of Rio Grande do Sul. Loose coalitions of caudillos tried to fight them off, but each caudillo was interested only in his own lands. Earnestness was hard to come by; betrayals, less so.

Brazilian independence arrived in 1822. Now the Portuguese were not only trying to supplant the Spanish, but also fighting their own colonials. In the chaos of *that* transition, many Spanish caudillos saw an opportunity to take it *all*. In 1825 a broad caudillo alliance launched the Cisplatine War, crossing the Uruguay River and wreaking so much havoc that Brazil was forced to abandon the southern Cisplatine. At the war's conclusion, in 1828, Uruguay was carved out of Brazilian territory to become an independent nation.

Then everything went to hell. The weak coalitions that held together the newly independent Argentines, Brazilians, and Uruguayans all collapsed in one way or another.

Considering the Darwinian melee of the region's politics at the time, it shouldn't come as a surprise that the resulting Uruguayan civil war quickly sucked in various caudillos of all flavors, first indirectly, and in time with troop commitments, leading to the messy Platine War in 1851. It should further come as no surprise that the war wasn't simply "Argentina" against "Brazil," but Argentine caudillos against their Brazilian equivalents as well as internecine fighting and strongmen from either side courting and fighting with peers from the country opposite. National authorities not only couldn't rein in their regions; they couldn't even field forces. The war ended as inconclusively as it had been fought. Such developments repeatedly and starkly showcased Argentina's (and Brazil's and Uruguay's) general political, military, and strategic disarray. Simply achieving meaningful armistices to the various Argentine, Brazilian, and Argentine-Brazilian conflicts generated a booming industry in conflict mediation.

Sinking in such quicksand of internecine conflict was more

than self-destructive; it nearly ended Argentina *and* Brazil *and* Uruguay before they could establish themselves as countries. For there was one caudillo whose "company town" was a country, and as the big fish in the pond, he nearly ended it all.

SCARED STRAIGHT

Francisco Solano López of Paraguay was a strange cat in that he hadn't lived his entire life in South America. In López's younger days, his family dispatched him to Europe as an ambassador, where he had a front row seat for one of the greatest military conflicts of his generation: the Crimean War of 1853–56. In that conflict the French and British used the first generation of industrial weapons for the first time. Breech-loading rifles and ironclads and naval shells combined with railroads and the telegraph. It made for an awe-inspiring example of the world as it was about to be . . . and it couldn't have been more different from what López had experienced back home. By the time of Spain's fall, the Spanish were Europe's technological laggards by a century. Spain's colonies were a century behind *that*.

No more. Upon López's return to Paraguay in 1855, he immediately launched a modernization campaign. In a mere decade "tiny, poor, backwards, landlocked" Paraguay wasn't simply the most technically advanced country on the continent; it boasted an army over double that of Argentina, Brazil, and Uruguay *combined* despite having a population that was less than one-twentieth its rivals' size. A hot peace was as stable as the Plata region had been for the past half century. Conflict was inevitable. The Paraguayan War began in October 1864.

It was all about transport. Paraguay needed to deny its foes their river transport capacity, and that meant capturing Buenos Aires. Unfortunately for Paraguay, the Brazilians sank Paraguay's only fleet at the Battle of the Riachuelo, stalling Paraguay's initial assault while freeing the Triple Alliance of Argentina, Brazil, and Uruguay to use the region's rivers as a troop shuttle system. Five grueling years later, Paraguay wasn't so much overrun as destroyed. Over half the country's prewar population lay dead.

History is written by the winners, and most Latin American historians record López as unstable, maniacal, narcissistic, and vengeful. This (and more) is almost certainly true, but it misses the real points of this little side trip through history.

First, Paraguay and López didn't just die fighting; they died fighting *well*. Despite the fact that the Triple Alliance picked the time and place of nearly every battle after Riachuelo, the Paraguayans gave far more than they took; in one particularly nasty battle the casualty ratio was in excess of twenty to one in Paraguay's favor. López's forces delivered painful lessons as to how deadly a unified, professionalized, industrialized military boasting a telegraph communication network and a railed transport system could be versus a fractured, ad hoc, musket force.

Second, the conflict bankrupted Brazil while empowering the Brazilian army at the expense of the civilian government, setting the stage for a long progression of coups, countercoups, military takeovers, debt binges, and economic collapses that has defined and hobbled the country ever since.

Third, power was also centralized in Argentina but with different implications. A few months into the war, the Argentine elite realized just how narrowly they had escaped mutual eradication. They threw *everything* they had at Paraguay. The war necessitated the centralization of the country's military, economic, and political life under Buenos Aires. It took the death-scare of upstart Paraguay to forge Argentines into a nation. And in a textbook example of geography's influence, in Argentina it took.

Within months of victory over Paraguay, the Argentines turned their newly formed military south, culminating in the "Conquest of the Desert" to eliminate the indigenous tribes living between the Negro River and the mythic mountains of Patagonia. The Conquest featured the same genocidal features of America's Indian Wars: one-sided battles, the killing of unarmed civilians, land rushes, ethnic cleansing, forced marches, and the intentional spread of diseases. But here it happened with the best military technologies the early industrial era had to offer, and so it was all over in a *decade*. Repeating rifles versus clubs tends to compress the time frame. The result was as final as it was predictable: the writ of newly empowered Buenos Aires became firmly and permanently entrenched throughout the territories we now know as Argentina.

SUPERPOWER, INTERRUPTED

With the wars over, the Argentine government got busy doing the things governments do: establishing a continuity, forging a mass educational system, threading infrastructure throughout the territories. All along the way, mighty Buenos Aires easily absorbed all the commerce and dynamism the young country could produce, yet a mixture of past and future ultimately sent the place off the rails.

First, the past. While achieving economies of scale in Argentina was *soooo eeeeasy*, things still needed to be built, and that took money. A torrent of primarily late-Imperial British money helped bring the country's natural agricultural advantages into full bloom. But that foreign money wasn't lent to just anyone, only to men who had economic and political track records worth trusting—only to the regional oligarchs who had inherited their mantles from the original (rich) settlers and merchants and, more recently, the caudillos. By the headline figures, Argentina may have been doing well, but wealth inequality ran as wild as wealth, and the debt overhang was truly massive. When Britain found itself in need of cash to fight the First World War, the Brits called for repayment, triggering Argentina's first financial collapse.

Second, the future. Argentina wasn't the Western Hemisphere's only big chunk of arable land shot through with navigable waterways. For the first 125 years of the United States' history, the Americans mostly kept their nose in their home continent. But part and parcel of the post–Civil War Reconstruction effort was to knit the country back together with the latest in industrialized infrastructure. When the Americans had their coming-out party in the 1890s, their historically unprecedented economies of scale took global agricultural markets by storm. Argentina may have had great lands and great rivers, but America's were even better. And unlike the United States of the Order era, pre–World Wars America was downright mercantilist and not about to cede market share to anyone.

To compensate, the Argentine state waded into agricultural markets with borrowed funds, subsidizing the oligarchs' enterprises. Social inequality and economic desperation of the lower classes boiled over just *before* the Great Depression hit. To prevent anarchy, the military launched a coup in 1930, setting off decades of juntas, coups, and populist uprisings.

Crystalizing the patterns of chaos and dysfunction among plenty, which defines Argentina to the current day, was the leadership of one Juan Domingo Perón, a populist who ascended to the presidency in 1946.

Perón's cult of personality is a study in contradictions. Peronism views the state as the ultimate guarantor of worker's rights, but also dissolves unions and criminalizes workers' right to protest while lobbying on their behalf . . . to factory managers the government commands. It makes the government the primary mover of the economy but uses a communist-*cum*-fascist approach to public and private ownership of the means of production.

Under Peronism and oftentimes violent bouts of anti-Peronism, Argentina lost almost everything. Once a leading power in wheat, corn, pork, beef, oil, and natural gas markets, Argentina nearly became an importer of *all* of them. Once the undisputed second power of the Western Hemisphere, Argentina is barely even relevant within its own region. All in all, pretty much everything that made the Argentines rich, sophisticated, and powerful has bled away.

Yet even suffering from a history of politicians who were, shall we say, unique in their deliberate incompetence, Argentina is set up phenomenally well for the Disorder to come.

First, Argentina is beyond the back of beyond. While in the contemporary era of nuclear-powered aircraft carriers the Southern Cone isn't nearly as isolated as it was in the sail era, it is still a long trip to Buenos Aires from the Eastern Hemisphere. America's de facto enforcement of the Monroe Doctrine will keep any would-be predators out of South America. Add Brazil's general discombobulation, and there are no countries on the planet except New Zealand with as good a physical security position as Argentina.

Second and somewhat ironically, despite Argentine safety, the country remains packed with everything the wider world needs.

Part of this is obvious: despite the recent ravages of Peronist governments in the 2000s and 2010s, the country's export potential for any number of temperate-zone agricultural products remains massive. It wouldn't take a very forward-looking government policy to upgrade Argentine output to the point the country could storm into the world's top ten list of producers or exporters of corn, soy, wheat, poultry, and beef.

Part of the Argentine bounty is less obvious. Argentina sports significant shale petroleum reserves. While shale deposits are fairly common globally, most of the world's are not all that useful. Most are thin and petroleum-poor, and located inconveniently far from population centers. Not so in Argentina.

Here they rival those in North America: petroleum-dense and proximate to preexisting oil transport infrastructure already linked to major metropolitan regions.

Bringing such fields online would take a few years instead of a few decades. Argentina's coming shale boom (probably) won't be as transformative as the United States', but it will still be large enough to make Argentina a significant exporter of natural gas and oil—just as it was in the mid-twentieth century. Argentina *already* boasts the world's third-largest shale oil and natural gas production, behind only the United States and Canada. As an additional kicker, Argentina is coolly in the top five countries globally for solar *and* wind potential. Every electron green power can generate locally is another bit of petroleum available for export.

Even better—and rarer—Argentina's population structure is sublime for a country at its stage of development.

In part, this is an inadvertent legacy of Peronism. One of Peronism's most pernicious impacts is its subsidization of basic goods. Lower prices for food and electricity encourages the population to use more than they would normally. At first, the government has to pay for the difference, but once the government runs out of money and maxes out its ability to borrow, the cost is passed on to producers. Faced with government price caps that are *below* the cost of production, most producers do what comes naturally: they stop producing. The result is massive state debt, huge distortions in supply *and* demand, endemic corruption, resource mismanagement on an epic scale, a collapse of the productive sector, and goods shortages in pretty much everything. Argentina has lived through this cycle several times already.

However, one perk of the Peronist style of subsidization is that cheap food, housing, electricity, and water mitigate the cost of raising children in an urban environment. Argentina boasts a *preindustrial* population structure with the skilled labor set of an *industrial* society. Such a huge proportion of skilled labor compared with the overall population puts Argentina at the opposite end of the scale from Brazil. Even in *worst*-case circumstances, the Argentines are both young enough and aging slowly enough that they will not face an American-style—much less a European-style—retiree budget pinch until 2070 at the *earliest*. For comparison, the Brazilians—crammed as they are into tiny urban plots—will hit that point by 2045 at the *latest*.

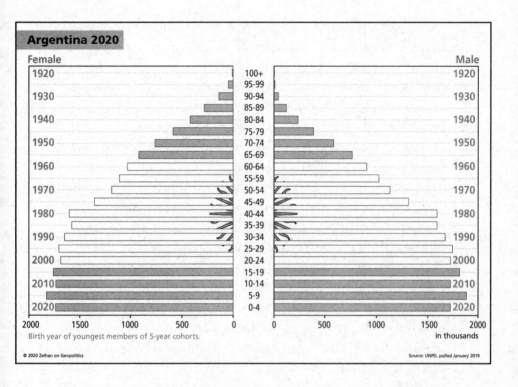

Argentina 2020

Female / Male population pyramid, ages 0-4 through 100+, in thousands. Birth year of youngest members of 5-year cohorts.

© 2020 Zeihan on Geopolitics

Source: UNPD, pulled January 2019

In a world degrading into energy shortages, physical insecurity, and demographic collapses, Argentina boasts the resources, land, rivers, geographic position, *and* demographic structure to make the most of an age of Disorder.

There is even a tantalizing possibility that Argentina could develop a South American echo of East Asian manufacturing. It has a similar labor-market structure and the kind of physical and energy security that underpinned the 1970–90 Asia boom. Even better, with the sole exception of the United States, no country has enjoyed such domestically powered consumption-led growth since before the global baby bust of the 1980s. "All" Argentina needs is some raw materials access (which it can source from Brazil), some imported capital (likely from the United States), and a better legal and regulatory environment (which is entirely up to the Argentines).

It all adds up to make Argentina a bit like Turkey, but without the external security complications. With the Peronist rise, Argentines, for the most part, have alternatively wallowed in narcissistic populism or domestic reactions to Peronism. They've gone from being globally significant in the 1880s through the 1920s to barely being even regionally relevant today.

But with the world falling into a deglobalization death spiral, one of two things is going to happen.

In option one, the Argentines elect governments that unwind many of the crippling impacts of Peronism. In such a circumstance, Argentina's underlying features shine through, and it becomes a continental—a *global*—success story in a decade or two.

In option two, Peronism persists, and the Argentine economy remains statist and inefficient and self-destructive. Investment would lag, and general degradation would be the order of the day. But even then Argentina has a near-perfect demographic profile, which is rare. It has resource wealth, which is rare. It has dreamy geography, *exceedingly* rare. Most of all, the Argentines simply have a *lot* more hands-on experience operating in a world ruled by dysfunction and populism and conflict than anyone else. Argentina might not shine, but it will *still* look better than nearly every other country in the world.

Regardless of whether the Argentines move to the middle or the middle moves to Argentina, Argentina's superior geography and demographic structure ensure its success moving forward. Argentina's impending golden age is only an issue of degree.

DEFINING SUCCESS

Organizationally, economically, politically, and diplomatically, Argentina is starting from century lows. The ravages of Peronism and anti-Peronism have wrecked much of the country's capacity for even *thinking* about the outside world. Even if Buenos Aires came up with a deviously brilliant master plan today and executed it competently and faithfully, the cultural hole the country is in is so deep that Argentina would not become a force in global affairs beyond its product sales within two decades.

Instead, regardless of intent, Argentina will become the natural centerweight of its region. Geographic patterns make it obvious how it will all progress.

Step one: just as the rivers gave the Triple Alliance the ability to smack down López in the Paraguayan War, they remain the key to controlling Argentina's neighbors. During the Peronist implosion of the 2000s, the Argentine system weakened so much that local authorities stopped dredging the shipping channels of the Uruguay, Paraguay, and Paraná Rivers. During the same period, a perfectly timed burst of infrastructure policy in Brazil enabled Paraguayan agricultural producers to send their output through Brazil via truck to the wider world. Brazilian roads are wildly more expensive than Argentine river ports, but Argentina proved itself so unreliable, the exporters had no choice. Consequently, Brazil is now far and away the dominant economic power in both countries.

All Argentina needs to do to ensure that these countries fall back into its orbit is keep its rivers' shipping channels functioning, something already in the Argentines' domestic self-interest. Basic economics will see to the rest.

But more than trade-related income is the strategic value of Uruguay. Its capital of Montevideo sits almost directly opposite Buenos Aires on the Rio de la Plata. If anyone ever were to wish Argentina ill, Montevideo would be *the* launching point. Control of the city threatens outbound Argentine shipping, giving Montevideo effective command of both the Argentine capital and the entire Argentine core. The Argentines are not used to thinking strategically, but as the world shifts and an older sort of thinking creeps back, they will recall their battles with the Spanish, the British, the Portuguese, *and* the Brazilians over just this scrap of land.

Step two involves Argentina's other pair of border states—Bolivia and Chile—and requires slightly more thinking.

Chile is a schizophrenic place, split between a single valley system packed with farms, orchards, and vineyards in the country's middle, and tenuously linked to a series of copper and lithium mines in the country's northern deserts. Such an odd setup has only persevered for three reasons.

First, the Andes prevents significant interaction with Argentina. Weak Argentina or strong Argentina, the Andes will always limit contact, helping preserve this Chilean geographic oddity.

Second, the only country with sufficient exposure and interest to challenge Chile is Bolivia, and Bolivia is just as split as Chile. Two-thirds of the population (largely natives) lives in the highlands, while two-thirds of the economy (largely agricultural) is generated from farms in the lowlands that abut Brazil.

Way back in the War of the Pacific of 1879–84 the Bolivians and Chileans crossed swords. As part of the postwar settlements, the Bolivians had to surrender to Chile all their ocean frontage as well as much of the Atacama Desert—the same desert that is home to most of contemporary Chile's copper and lithium production. Bolivia remains just as pissed off at Chile now as it was when it lost the war over a century ago.

The feud gives Argentina a wealth of opportunities. Argentina used to provide Chile with natural gas, but Argentina's 2000s collapse pretty much ended that. More recently, natural gas production and exports are again trending upward. Recovery in other sectors will naturally make Argentina Chile's primary supplier for most everything food- and energy-related. For its part, Bolivia refuses to ship its exports to the wider world via Chile, so when Argentina was in its 2000s Peronist funk, Bolivian mine output from the highlands made the long, arduous trip through the Chaco and Brazilian territory to the Atlantic Ocean. Very light tweaks to Argentine policy would make transport *much* easier. The Chilean capital of Santiago isn't even all that far from Argentina's world-class solar and wind potential. Argentina can expand its economic footprint in both countries while also becoming essential for both countries' expanded economic interaction with each other.

That just leaves Brazil.

Much like Argentine relations with Uruguay and Paraguay, the shifting balance with Brazil does not require much planning or a particularly heavy hand. It doesn't take a policy of infiltration out of Buenos Aires to translate economic growth in Argentina to greater influence over the Brazilian border regions. The Paraná River is navigable all the way up into central Paraguay. It would be far easier and cheaper for Brazilian farmers in the province of Mato Grosso do Sul to ship their output to Asunción for loading onto Buenos Aires–bound barges than to truck it up and over Brazil's Grand Escarpment.

To put it simply, Argentina's geographic features condemn it to being a successful country, while Brazil's features restrict it to being successful only in very specific circumstances that are both beyond its control and no longer extant. Brazil's fall enables Argentina to *passively* leach into becoming something more. It adds up to a different sort of Southern Cone, with Argentina not simply eclipsing Brazil as the dominant regional power but putting the best territories of the South American continent all within a single sphere of influence.

For those who find all things South American a bit of a yawner, consider this: if Argentina succeeds in reconsolidating its own nation and achieving primacy in Chile, Paraguay, and Uruguay, then Argentina will command the world's fourth-largest chunk of temperate-zone arable land via the world's second-largest naturally connected, naturally navigable waterway network in a regional geography where it has no competition. As Argentina recovers, it will quickly become a sandbox superpower, but in facing no local threats whatsoever, in time the Argentines will find it really easy to project *out*. It is precisely this combination of factors that in time created the *American* superpower.

ARGENTINA'S REPORT CARD

BORDERS: Argentina enjoys some of the most secure geography on the planet. The Andes to the west, and the Atlantic to the east. The only thing coming for Argentina are the Argentines themselves.

RESOURCES: Argentina boasts a near-dizzying array of mineral and agricultural wealth. It is a leading producer of beef, grain, soy, silver, copper, wine, oil, and natural gas.

DEMOGRAPHY: Unlike neighboring Brazil, Argentina enjoys a relatively young and healthy demographic profile. (Mass subsidies make babies.)

MILITARY MIGHT: Heh. No. A history of coups, mismanagement, and deliberate demilitarization makes the Argentine military a bit of a joke. But Argentina's safe frontiers and lackluster local competitors make military might unnecessary.

ECONOMY: Argentina's economy is a textbook example, literally, of how political decisions can stymie even near-perfect geographic advantages. A unique, indigenous mix of nationalist-socialist-fascist policies have resulted in waves of massive inflation, capital flight, sovereign debt defaults . . . and yet, Argentina's future remains bright.

OUTLOOK: Once a political ideology more conducive to . . . sanity takes hold, Argentina has everything it needs to dominate its neighborhood.

IN A WORD: Mulligan.

CHAPTER 14

THE MISSHAPE OF THINGS TO COME

THE FUTURE OF AMERICAN FOREIGN POLICY

The Order has been unique not just from the view of global management, but in American history as well. In the pre-Order days, US foreign policy was *not* bipartisan. It would flip back and forth between various flavors of engagement and disengagement. An early version of engagement tinged with anti-British feeling committed the United States to the War of 1812. An early iteration of isolationism tinged with local imperialism generated the Monroe Doctrine. Reconstruction and the Great Depression generated new isolationist impulses, while the Spanish-American War and World War I thrust the United States into the limelight.

But despite these oscillations, in no version of American pre-Order strategy were the Americans fans of international bodies, multilateralism, or restrictions on their room to maneuver. For most countries, such fluctuations aren't common. In most cases, a country's geography hardwires certain strategic needs into its

DNA: Russia must expand to more defensible boundaries; Japan must secure its supply lines; Germany must act first or risk being overwhelmed; Brazil must seek external financing. Such perennial demands limit options, focus minds, and make it less relevant which political ideology dominates on any given day.

Not so for the United States. American power is both huge and insulated. Until the rise of the Soviet threat forced the bipartisan unity of the Order, American policy varied with the political passions of the day. Now, with the Order's end, the United States returns to a more "normal" strategically unfettered state. Predicting French or Turkish or Argentine actions is child's play compared with guessing what the Americans will do now that they are freed from concerns of context and consequence.

That forces us to do something most students of geography loathe: dive into what actually *will* guide American foreign policy for the next decade or so. It is time to dissect American domestic politics.

The United States is a democratic republic whose Constitution dictates a first-past-the-post election scheme based on single-member districts. Translation: political candidates run for a specific seat representing a specific constituency; to win, a candidate does *not* need a majority of the votes, but only one more vote than whoever comes in second.

This constitutional feature heavily shapes the makeup of the country's political parties. Any party that throws a narrow net concentrated on local or ideological rather than national issues is never going to gain traction outside of its immediate area or idea set, and so gets swallowed up by larger, more national groups. Similarly, single-issue parties don't generate a wide enough appeal to be competitive because getting a few percentage points— even nationwide—won't gain them any seats in Congress.

Such features make America's political parties both stable and weak. They're stable in that, once a coalition of factions is formed under the umbrella of a party, that party tends to stick around for decades. Each faction's power can ebb and flow within the party coalitions without disrupting the overall political system. They're weak in that each faction has its own ideas about what the party should be about, and corralling the factions behind any specific goal requires a great deal of legwork.

America's two established parties in 2020 are in a state of flux. Factions are not simply rising and falling within the existing parties, but even jumping party affiliation. Previously disguised tendencies in portions of the electorate are emerging into the daylight, while previously prominent tendencies are falling into shadow. This is a moment of renegotiation, of scrambling to build a party with as many of your faction's characteristics as possible while also building a large enough base to, you know, *win*. What does that mean in practice? Apparent chaos, and a feeling that you've been cursed to live in interesting times.

Let's begin with America's center-right party, the Republicans, who in many ways have been victims of circumstance since 2000. Events as large as September 11, the subsequent involvement in the Iraq and Afghan Wars, the aftermath of hurricanes Katrina and Rita, the Baby Boomers' aging into a very ungraceful retirement, would put stress on any political alliance. That the Republicans were in power for these developments and didn't exactly acquit themselves professionally didn't help.

The core problem, however, is that the Republican coalition was too perfect. The Republican Party is broadly an alliance of six groups: the business community, national-security conservatives, fiscal-primacy supporters, evangelicals, pro-life voters, and populists. Each has a core issue they are passionate about, and none particularly care about anything else. It's child's play to build a platform that takes every group's interests into account while avoiding clashes among the factions. Such a narrowcast structure—while it can generate relatively few votes—tends to win the big elections because infighting is minimal.

From 2000 on, however, the coalition began to fall out. The

costs of the Global War on Terror horrified the fiscal faction and the business community. When those groups succeeded in wrestling the executive into winding down the wars (without meaningful settlements), national-security voters felt deep unease. Government incompetence over the hurricanes didn't please anyone, but the storms hit areas so teeming with evangelicals that even the rightmost of the right got pissed off. The Republican coalition had so little experience in managing internal disputes over policy that its leaders handled these new clashes badly. Bad feelings spread and rooted throughout the party.

The deepest alienation, and from it the biggest cracks in the coalition, was felt among the populists. America's right-leaning populists have always been a bit of a grab bag—especially when it comes to motivation: local-authority seekers, conspiracy theorists, poor whites, racists, and in general anyone who thinks that someone, somewhere is hurting America, whether that be a bomb-toting Arab, a cheese-eating Frenchman, or someone sitting in a posh office on Wall Street.

The populists have always been the crazy uncle of the Republican alliance, and the other factions have tolerated them only because the populists were wildly disorganized and so provided a vote bank without ever really being able to shape the Republican agenda. Based on whose data you're using and which national election you're looking at, such populists made up 5 to 15 percent of the national voter base. On the low end, that's bigger than the gay bloc, while on the high end, it's bigger than either the black or Hispanic vote.

Times change. In a world of social media, the obstacles to organizing, raising funds, and making your views known have dwindled to zero. The same general progression that enabled the rise of Howard Dean, Barack Obama, and Bernie Sanders—the technologies that enabled them to bypass the Democratic Party machinery and go straight to potential voters—also created the Tea Party and the alt-Right.

While every national candidate is to a degree a compromise among the factions, most candidates favor one faction over the others. Ronald Reagan understood the business community. George HW Bush was tight with the national security–minded. His son—W—elevated the evangelicals to prominence. In all cases, the populists were in the background. But courtesy of social media, the populists were able to storm the Republican primary system in 2015 and put their man—Donald Trump—in the party's driver's seat. The populists' previous disorganization meant the rest of the Republican coalition didn't really have a grip on what the populists were after in terms of policy. Now with the populists calling the shots, the rest of the old Republican coalition is somewhat gobsmacked. The populists' position on social programs alienates the fiscal conservatives. Their views on national-security policy infuriate the military and intelligence communities. Their immigration goals split the evangelical community down the center. And their thinking about finance and regulations has banished the business community into the wilderness. With the exception of pro-life voters, the old Republican coalition is completely shattered. Beyond repair.

Beginning with the Republican Party convention in 2016, Donald Trump has broadly succeeded in purging the executive branch of "traditional" Republicans of the fiscal, national-security, and business factions. This effort culminated in the 2018 midterms with the broadscale ejection of those factions from Congress as well, and their replacement with Trump-supported populists. Entire constituencies were put out to pasture. "Republican" doesn't mean what it used to.

It's no better among the Democrats. The problems of America's center-left party are linked more to faulty strategic planning than any particular governing foible. Beginning in the mid-1980s, the Democrats became fixated on the hot topic of demographics. Reading the tea leaves, the Democratic leadership concluded that due to immigration, differences in birthrates, and the rising women's and gay rights movements, if a party could lead a coalition of blacks, Hispanics, gays, and single women, it inevitably would become the country's "natural" governing party.

The demographics of California suggest that this is a winning strategy. Nonwhites have outnumbered whites in Cali since the 1960s; since 2010, new Hispanic births have outnumbered white births two to one.

And yet, the Democrats have found gaining power to be *more* challenging, rather than less. From 1969 through 1993, Democrats held both the Senate and the House for ten out of thirteen sessions. From the mid-1990s on, supposedly when the Democrats' "rainbow coalition" strategy should have started some serious fruit-bearing, it was the Republicans who held power in the majority of the Congresses. The Democrats held a majority in both houses of Congress only *once*, in the immediate aftermath of the hottest part of the unpopular Iraq War.

Part of the failure is an incomplete reading of demographic trends. The racial breakdown of the Californian demography exists only in Cali. In the American South, Midwest, and Northeast, the racial breakdown is about as opposite of California's as is possible. America is becoming more ethnically diverse, but outside of California it is proceeding at such a slow pace the Democrats' strategy will not have a chance to prove correct for decades to come. And losing election after election is not a great way to keep all the various minority factions bound within the Democratic party.

It is worse than it sounds (for the Democrats).

The country's most politically liberal region, the Northeast, is strongly majority-white and will remain so for the rest of the twenty-first century. It is also the country's most rapidly aging area. Retirees tend to be far more strongly socially conservative than liberal, retirement being the milepost in life when the "get off my lawn" mentality tends to kick in. As the Northeast eases into mass retirement in the 2020s and 2030s, a growing slice of

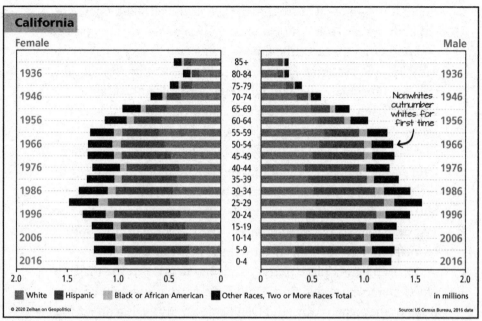

California

Female / Male

1936, 1946, 1956, 1966, 1976, 1986, 1996, 2006, 2016

Age groups: 85+, 80-84, 75-79, 70-74, 65-69, 60-64, 55-59, 50-54, 45-49, 40-44, 35-39, 30-34, 25-29, 20-24, 15-19, 10-14, 5-9, 0-4

Nonwhites outnumber whites for first time

2.0 1.5 1.0 0.5 0 | 0 0.5 1.0 1.5 2.0

White   Hispanic   Black or African American   Other Races, Two or More Races Total   in millions

© 2020 Zeihan on Geopolitics

Source: US Census Bureau, 2016 data

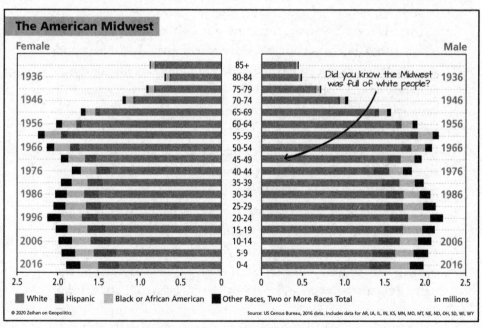

The American Midwest

Female / Male

1936, 1946, 1956, 1966, 1976, 1986, 1996, 2006, 2016

Age groups: 85+, 80-84, 75-79, 70-74, 65-69, 60-64, 55-59, 50-54, 45-49, 40-44, 35-39, 30-34, 25-29, 20-24, 15-19, 10-14, 5-9, 0-4

Did you know the Midwest was full of white people?

2.5 2.0 1.5 1.0 0.5 0 | 0 0.5 1.0 1.5 2.0 2.5

White   Hispanic   Black or African American   Other Races, Two or More Races Total   in millions

© 2020 Zeihan on Geopolitics

Source: US Census Bureau, 2016 data. Includes data for AR, IA, IL, IN, KS, MN, MO, MT, NE, ND, OH, SD, WI, WY

the country's "liberal" base that the Democrats depend upon will naturally age into something more populist and more conservative.

Another big piece of the failed Democratic strategy has to do with the way the Americans industrialized—or more specifically, the *pace* at which the United States urbanized. Because the United States has so much cheap, high-quality land, it is the least urbanized of the world's modern states relative to the land that can be used for something. Rural voters and voters in towns with fewer than fifty thousand people—roughly 30 percent of the American population—tend to be more socially conservative, but *not* politically mobilized. In the age of social media, for the first time they truly *were* mobilized, and the surge in their voting participation most certainly moved the political needle.

An even bigger piece of the puzzle has to do with the makeup of the Democratic coalition. Unlike the narrow and focused coalition of the Republicans, the Democratic coalition is a broad and squabbling one.

Most blacks, Hispanics, and union members lean left on economic issues, but are staunch social conservatives. Most gays and single women may lean left as regards individual political rights, but they tend to be middle-of-the-road to center-right when it comes to economic policy. Even in places where it seems the Democrats' diversity alliance strategy should work (outside of California), it doesn't. Texas is perhaps the best example. It is the state with the demography most like California's, but there are precious few souls who would describe Texas as the land of the blue, in large part because once the state's Mexican Americans become second generation, they reliably vote Republican.*

The coalition's fractured nature combined with a faulty grand strategy has proven consistent in delivering deep defeats at the ballot box at the state and national levels even as national de-

---

* A fun, if awkward for the Democrats, fact: American Hispanics are the most reliably anti-immigration party. They want a liberal system for family reunification, but only for *their own* family.

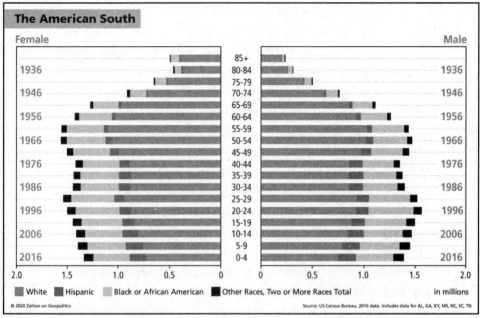

The American South

Female / Male — population pyramid by age group (85+ down to 0–4), with birth-year labels 1936, 1946, 1956, 1966, 1976, 1986, 1996, 2006, 2016.

Legend: White | Hispanic | Black or African American | Other Races, Two or More Races Total — in millions

© 2020 Zeihan on Geopolitics

Source: US Census Bureau, 2016 data. Includes data for AL, GA, KY, MS, NC, SC, TN

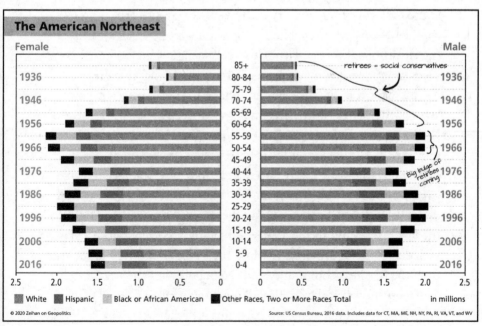

The American Northeast

Female / Male — population pyramid by age group (85+ down to 0–4), with birth-year labels 1936, 1946, 1956, 1966, 1976, 1986, 1996, 2006, 2016.

retirees = social conservatives

Big bulge of retirees coming

Legend: White | Hispanic | Black or African American | Other Races, Two or More Races Total — in millions

© 2020 Zeihan on Geopolitics

Source: US Census Bureau, 2016 data. Includes data for CT, MA, ME, NH, NY, PA, RI, VA, VT, and WV

mographics steadily shifted in the strategy's supposed favor. Between 2000 and 2016, the Democrats lost over a thousand elected positions, along with control of some three-quarters of the nation's governorships and state legislatures. Barack Obama is an obvious exception to the trend, but it is worth noting that the Obama campaign* operated *outside* of the Democratic Party's institutions rather than within it. It is equally notable that the subsequent alienation within the American left contributed both to the election of Trump—whose rise to power cost the Democrats control of all three branches of the national government—and to the arrival of deep progressives like Bernie Sanders, who proceeded to eviscerate the Democratic Party from within.

The United States' 2016 federal election season was a watershed in that it shattered the coalitions that composed *both* parties. A founding member of the modern Democratic coalition—the unions—switched sides. The populists of the right expanded their reach, but at the cost of driving away two-thirds of the groups that normally consider themselves Republican. The populists of the left weakened the institutional forces within the Democratic Party and, as the Democrats' defeat sunk in, proceeded to shatter what remained.

This isn't the end. Economics change. The balance of power among the states shifts. Technology proceeds. Society evolves. Factions form. Factions merge. Factions split. Factions die. Politics moves with all this. But it takes time for the swirls and eddies that color the American polity to congeal into electable, sustainable party structures. The Americans have been in this sort of internal discombobulation before. This is the seventh reshuffling. Before the last reshuffling in the 1930s and 1940s, the Republican Party included African Americans while the Democrats were home to both the business community and the populists of the *right*.

America's political factions will reorganize into a new set of alliances, but it will take time. In part, this is because the parties have

---

* And arguably the Obama presidency.

not had experience crafting new alliances in living memory. In part, it's because, until new parties emerge from the detritus of the old Republican and Democratic parties, they cannot even begin internal discussions as to what American foreign policy *should* look like. In part, it's because the people who lived through the World Wars and developed an instinctive understanding of just how insane and dangerous and genocidal a world without an American-led Order can be are almost gone. But mostly it's because the bipartisan agreement that the Order served American strategic needs is gone not only because the Order's time has passed, but because a bipartisan agreement requires, well, partisans.

The inability of the Clinton, W Bush, and Obama administrations to even *begin* the conversation as to what might replace the Order, complemented by the breakdown in the Democratic and Republican parties, means Americans cannot even begin to define what they want out of the world until at least 2030. In the meantime, what passes as foreign policy in the United States will be erratic, guided by a weird mix of the passions of the moment and the lingering inertia of a system that is decades out of date that becomes more discredited by the day. Getting past this won't be pleasant or easy; structural realignments rarely are. Ten to fifteen years for it to all shake out seems a reasonable stab.

Given time, the United States will settle into a new groove, adopting an outcomes-based foreign policy based on a mix of concerns strategic, economic, and moral.

But as Churchill famously said, "You can always count on the Americans to do the right thing, after exhausting all other options." What follows are precisely those "options." The Americans, lacking any unifying strategic vision, will execute *all* of them in a mishmash of colliding interests and inertias that shouldn't be confused with a coherent policy. Don't think of them as themes, as that word implies an organization that does not exist, but instead as threads that will come and go on their own independent politicized logic.

THREAD 1: UNWINDING THE GLOBAL WAR ON TERROR

The September 11, 2001, attacks sent the United States smashing sideways into the Middle East, and what started as a straightforward, focused effort rapidly devolved into something much ganglier.

The Americans invaded Afghanistan with the express goal of eliminating the attacks' perpetrators, the militant Islamic group known as al Qaeda. American forces quickly discovered that al Qaeda was only a piece of the problem. It was only a very small part of a constellation of militant groups that were either from Afghanistan or that were using the chaotic country as a base of operations.

Both Pakistan and Saudi Arabia viewed such militant Islamist groups as tools for manipulating events in other countries. Along with Syria, the pair of countries exported their disaffected youth to such groups so they would cause trouble somewhere other than at home. That led the intelligence authorities of these states to tip off al Qaeda and its Islamist allies, making it impossible for the Americans to hunt them effectively.

A manhunt is not like a war. Military operations in the broad, arid lands of the Middle East do not go down in the same way as they do in the relatively tight confines of Northern Europe. American military strategy had spent decades preparing for integrated land/air/sea battles with the high likelihood of the odd tactical nuclear weapons strike, informed by a cadre of intelligence personnel steeped in Russian language and history. It was all but useless. Retooling and retraining and re-equipping for infantry patrol and Arabic and Pashtun and Dari was an intensive, expensive, decade-long labor.

Needing more eyes and ears (and boots on the ground), the Americans activated their Order-era alliance to secure assistance.

Of the handful who responded, most either had no experience in this sort of manhunt (Poland, Georgia), or had rules that prevented their troops from shooting at anyone (Japan, Italy), or wielded tools of limited usefulness (Denmark, Canada). The result was erosion of the Cold War alliance network, even among those states that honestly tried to help.

Needing to pressure more countries into cooperation, the Americans invaded Iraq, not because Iraq was affiliated with al Qaeda (it wasn't), but because US forces based in Iraq could threaten states that were in a position to threaten al Qaeda. It worked, and al Qaeda's regional networks fell before the security apparatuses of Syria, Saudi Arabia, and Iran.

However, the American footprint in Iraq generated its own outcomes. Saudi Arabia saw the Americans degrade from guarantor to bully and started crafting a foreign and security policy that made al Qaeda look peaceful. Part of that policy involved encouraging the suicide-bombing of American forces in Iraq. The Iranians, incensed to have US forces on their border, upgraded from anti-American diplomacy to anti-American paramilitary action. The Americans found themselves enmeshed in Iranian-instigated civil uprisings throughout western Afghanistan and southern Iraq. The Syrian regime (among others) cracked under this and unrelated pressures, and the nation devolved into a civil war that rapidly spilled over into western and central Iraq.

These three theaters—Saudi Arabia, Iran, and Syria—began intermingling and clashing. Iran worked to seize control of the new Iraqi state, Saudi Arabia mobilized Sunni groups to attack their foes, and Syria began to lose control of its borders and territory. The Americans were caught in the middle. Understandably the American citizenry has lost interest in the region, and any sane strategist can see there isn't anything the Americans can do to put Humpty Dumpty back together again. The only options seem to be to fight a war against guerillas until the end of time, or leave and let the locals duke it out.

Americans ultimately did achieve their goal of destroying the people responsible for the 9/11 attacks, but the way they did so critically degraded their global alliance structure along with the functionality of many countries in the Islamic world. They guaranteed that the region would be fertile ground for groups that thrive in areas of weak central control—groups like al Qaeda. America's obsession with the region has faded to disenchantment, and the country's ongoing disengagement will only deepen such impacts.

This takes us four places.

First, at least in the beginning, the Americans are certain to see all of this as a *good* thing. A broadscale Iranian-Saudi competition is certain to occupy the attention of both—win, lose, or stalemate—until at least 2040. The degradation of Syria, combined with the ongoing struggle between Iran and Saudi Arabia, is certain to hobble any attempt by Turkey to focus on any of its other horizons. From the American point of view, this ties the whole region up with a nice bloody bow. A generation from now, when someone emerges victorious, that might be a problem,[*] but it's a problem for another day.

Second, such multivector competition in such a messy region keeps most of the state sponsors of terrorism focused on attacking one another. This is likely to make the Americans reconsider the threat of terrorism. There are going to be hundreds of examples of militant groups attacking in-region countries that the Americans do not care for. That is likely to break the American mind-set developed during the Global War on Terror that all terrorism is bad.[†]

---

[*]   OK, that absolutely *will* be a problem.

[†]   If this sounds ridiculous, please consider that in the 1980s the Americans supported an Afghan-based Islamist group—the mujahideen—that gathered recruits from all over the Middle East, provided them with intelligence and equipment from Saudi Arabia and Pakistan, and battled against the Soviets. One of the middlemen between the Americans and the mujahedeen was one Osama bin Laden. Whether you call them dastardly terrorists or courageous freedom-fighters largely depends on whom they're shooting at.

Third, the more spontaneous, unsponsored terror groups are also likely to find themselves sucked into the maelstrom. With no American troops in the broader Middle East, some of the more traditional triggers for regional violence will dominate: Shiites v. Sunnis, Sunnis v. Alawites, Christians v. Muslims, Jews v. Muslims, Kurds v. everyone. All are highly motivational for folks who are already on the trigger-happy side of life, and the Americans soon will not be directly involved in any of them. For those Middle Eastern groups that still choose to ignore the omnipresent and emotionally charged targets and causes in their front yard and strike farther abroad, it is far easier to walk to Europe than it is to fly to North America.

To be blunt, most Americans will see all the explosions and suffering as someone else's problem. That's not the sort of attitude that ensures a loyal, functional alliance in *any* part of the world, much less in one so defined by grievance as the Middle East.

Finally, another potential "positive" outcome is that, with the Americans no longer being occupied by all things Middle Eastern, they can at least begin to think about other topics. The greatest beneficiaries of the Global War on Terror were not the American people (who are arguably less safe now than they were on September 10, 2001), but instead a pair of countries who see themselves as rivals to the throne: Russia and China. While American policy toward both was hardly warm and fuzzy in mid-2001, events were unfolding even then that would take relations in a sharply colder direction. Sophomore Russian president Vladimir Putin was already taking his country in an obviously statist, authoritarian direction. The Chinese government was beginning to lean on all its neighbors to rewire East Asian security relationships.

Had the 9/11 attacks *not* occurred, American strategic policy versus *both* rising powers would have been quite different.

Versus Russia, the goal was to ensure that the Russians could never possibly regenerate into something even remotely resembling the geopolitical power of the Soviet Union. Considering Russian abject weakness in the aftermath of the 1990s collapse, this was an eminently achievable goal. All it required was reasonable American support for the new tranche of NATO members that joined the alliance between 1998 and 2004—a list that included all the former Soviet satellite states in addition to the Baltic trio—and the negotiation of a sort of demilitarized zone in the lands between NATO's eastern edge and Russia's western edge. With Russia weak and the United States strategically unfettered, establishing such a cordon sanitaire east of the Carpathians and the Baltic would have broken any hope of Russian reconstitution and condemned Russia to an unanchored demographic collapse.

Instead, America's obsession with the Middle East denied its new NATO allies any meaningful support. The Americans *needed* active Russian assistance to access Afghanistan to begin their manhunt, leading to a series of strategic compromises. Even worse, distracted attempts to box in or pressure Russia convinced the Kremlin that the Americans were going for the Russian throat. America's split attention granted the Putin government the room it needed to rebuild, while Russia's fear-fueled mobilization induced them to abandon any semblance of restraint.

Two decades later, Russia is the strongest it has been since before the Soviet collapse. It has established deep energy dependencies in countries as far west as Germany and Italy, it has troops occupying pieces of Ukraine and Georgia, and it regularly interferes in the internal political affairs of the entire NATO alliance (up to and including the United States) without fear of retribution.

Versus China, a resetting of the relationship was in progress. In the early 2000s, the legitimacy of the Chinese Communist Party was based on a combination of subsidized exports made possible by a bottomless supply of near-slave labor and exorbitant levels of technology theft. Job losses in the United States manufacturing sector, already in flux as the NAFTA accords were finishing up their first decade, had gotten large enough to trigger a political backlash. Simultaneously, the Americans were beginning to tinker with the insertion of environmental and labor-protection clauses into their trade deals. There was broad agreement among both the Democrats and Republicans that a come-to-Jesus meeting with Beijing was required, and an equally broad (and accurate) expectation that China could be forced into a newer relationship more reflective of power balances. (Keep in mind that at this time the Chinese economy was but one-sixth the size of the US economy, and China's naval buildup had not yet begun.)

Instead, broad plans to pivot the American security footprint from Europe to Asia never materialized, either under W Bush or Obama. Afghanistan turned into Iraq turned into Syria and Yemen, and the sharp end of American power remained a continent away from Chinese shores. Lacking the political and strategic bandwidth to both fight the Global War on Terror and confront a rising China, the Americans left the China relationship on cruise control. The Chinese seized the opportunity with both hands and gradually but ever-more forcefully pushed their economic reach and in time their military reach to every corner of East Asia. Between 2000 and 2020, the Chinese economy nearly quintupled in size.

These missed American opportunities are personified in one Donald Rumsfeld, secretary of defense for George W Bush from 2001 to 2006. History remembers Rumsfeld as the man who botched the American occupation of Iraq, resulting in far more American and Iraqi casualties than were necessary.

What history forgets is that Rumsfeld was not a military man, but instead an over-the-horizon strategist and logistics whiz. W didn't bring Rumsfeld in to do battle with the Middle East, but to do battle with the *Pentagon* over force structure and procurement.

The intent was for Rumsfeld to comb through the military's many weapons systems—both in the field and on the drawing board—culling anything that was foe-specific to the Soviets and focusing resources on enhancing America's technological edge, the idea being that when the next peer foe arose, US capabilities would be so overwhelming, there wouldn't even be a fight.

The 9/11 attacks largely doomed the Rumsfeld plan. America marched to war with the military it had rather than the military it needed. Instead of sprinting ahead technologically, the Americans found themselves bogged down in a conflict that required *less* advanced equipment. Russia and China seized their window of opportunity technologically as well as strategically and worked diligently to narrow the gap.

The time of imbalance is nearly over. Without the strategic distraction of the Middle East, the Americans will *at a minimum* put more effort into weapons systems that emphasize power projection over distance rather than those that help with manhunts.

Digitization and materials science and energy storage have come a long way since 2000. Applying the technologies of the 2020s to the military will generate stealthy drones that can loiter above a region for upwards of a month, rocket-powered *artillery* shells with ranges in the hundreds of miles that cost less than 1 percent of the price of a cruise missile, and finger-size drones that can do reconnaissance or targeted strikes. It is going to be awesome and terrifying in roughly equal measure.

No matter what the Americans think about contemporary Russia and China—friend, foe, or something in between—America's return to focusing on the sorts of weapons systems that they do best cannot help but limit Russian and Chinese options in the long run.

THREAD 2: THE ORDER HANGOVER

The Soviet Union and the risk of mutual civilizational destruction assured by the existence of tens of *thousands* of nuclear warheads made the Cold War more than the usual existential battle for survival; it was one that focused minds. For a country as physically large, economically diverse and ethnically varied as the United States, such a unifying and motivating factor played no small role in guiding and enforcing policy.

It's an issue of time constraints. The United States has a presidential system in which all the cabinet secretaries serve at the president's pleasure and can be dismissed at any time for any reason. In addition, while Congress can vet would-be cabinet secretaries, the only way *it* can dismiss them is via the impeachment-and-conviction process, which is so laborious that that particular power has never been used successfully. This effectively concentrates all power over strategic and foreign policies into the hands of a single person: the president. And the president has a *lot* of things on his plate at any given time.*

When the Cold War ended, the nuclear boulder hanging over the American foreign policy and security and intelligence institutions dissolved—and took with it that unifying focus. During the Cold War, American global primacy was a requirement to fight the Soviets. Without the Soviets, simply maintaining American global primacy has become the de facto goal of US strategic policy.

Having power for the sake of having power is sustainable neither strategically nor politically. At the point of the spear, it isn't clear who the spear should be pointing at, so it tends to point

---

* Technically, there are two other people in the United States who are responsible for forming foreign policy: the national security advisor and the Federal Reserve chairman. However, the national security advisor also serves at the president's pleasure, while the Fed chairman is primarily concerned with the effects the world has on the American economy rather than the other way around.

itself at anything and everything that might theoretically become a problem. Issues as diverse as the Fulda Gap, the Taiwan Strait, al Qaeda in Afghanistan, the Korean Demilitarized Zone, South African Apartheid, the Rwandan Genocide, the Mexican drug cartels, the Yemen War, the Haiti refugee crisis, the independence of Kosovo, the Troubles in Northern Ireland, Central American migrants, and Venezuelan oil subsidies to London all end up in the same bucket. The American people understandably don't understand why America's national-security institutions want to hit everything with a hammer, nor is funding for that sort of scattershot strategic policy as easy a sell as preventing Soviet-initiated nuclear annihilation.

It isn't that America's military and intelligence professionals are obstinate or inflexible but that they're bereft of leadership. Seven decades is a long time to do the same thing. Plenty of time to develop expectations that your current courses of action are the norm—the right, the good. Without a unifying strategic vision appropriate for the global environment, the entire American security bureaucracy relies upon firm guidance from the very busy man at the top, and if that is lacking, policy becomes driven by what the United States *can* do rather than what the United States *needs* to do. Worse, the actions required to counter many of the new threats inadvertently bolster others. Without a clear, defining goal, success in one part of American policy ensures that other aspects will fail.

At the other end of the spear, without a grand strategy things get . . . muddy. It's easy to motivate both sides of the political aisle as well as the broader population when the issue is clear. "Contain and beat back the Soviets so we don't all die in an atomic holocaust" is about as clear as it gets. But when that threat ended and no new vision replaced it, Americans bit by bit became less interested in an increasingly amorphous world. The recent political orphaning of America's national-security conservatives removes the sliver of the population from the American political system that even thinks about these things.*

The disintegration of purpose isn't something limited to America. It has percolated out to the Americans' alliance structure as well. Everyone allied with the United States during the Cold War understood who the bad guy was and what was at stake, and if anyone wavered, the American president, ambassadorial core, and intelligence and military officials were not shy about reminding people.

After the Cold War, however, the allies are confused because Americans are confused. Is it American foreign policy to oppose terrorism or to oppose Iran? Because the countries that do the most to generate transnational Islamic terrorism are the countries that do the most to oppose Iran.† Is it America's goal to pursue human rights or expand trade? Because many of the countries who most excel at dehumanizing humans are among the most lucrative trading partners.‡

Without an overarching goal, America's priorities change not year by year, but often hour by hour, with diplomatic, military, and intelligence efforts often working at cross-purposes. Even if allies *tried* to please the Americans in all things, in doing so they would automatically find themselves inadvertently opposing some aspect of American foreign policy because American foreign policy is now inconsistent.

However, the allies don't always try to do what the Americans want:

---

* Bereft of a unifying vision, even among those within the national bureaucracy, there isn't much strategic thinking going on among the professionals whose job it is to think strategically. One of the downsides of how the W and Obama administrations fought the Global War on Terror is that most arms of the intelligence services dedicated to over-the-horizon threat detection were refocused to operational support—less, who will the next war be with? And more, what side of the door are the hinges on? Such taskings are critical, but when you only look at hinges you tend to miss the army marching toward you.

† I'm looking at you, Saudi Arabia.

‡ I'm looking at you, China.

- China obviously wants trade expansion, but Beijing certainly wants a big chunk of the world for its own and often uses its influence in North Korea expressly to stymie this or that aspect of American policy.
- Turkey wants the European Union's wings clipped, but Ankara sees no reason why it can't pursue a closer relationship with Iran.
- France and Germany like America's occasional emphasis on human rights, but Paris and Berlin don't see why that should limit their options when making deals with Russia. They certainly don't like the Americans mucking around in the Middle East in a way that roils energy markets.

Both within and beyond the United States, this policy schizophrenia is simultaneously maddening and entirely understandable. Consider Russia:

- Since the Soviet collapse, some aspects of American foreign policy have kept up the mantra of containment, and rather than seeking a rapprochement with the defeated Russians instead have sought to deconstruct the entirety of Russian influence everywhere. Part and parcel of this strategic hangover was the uprooting of a Russian ally in the Balkans: Yugoslavia. From that came American bombing campaigns in Bosnia, Serbia, and Kosovo.
- As the Russians have become more aggressive (cyberattacks in the Baltics, invasion of Ukraine and Georgia, election meddling, etc.) there is a strong case to be made that the United States should prosecute a neo-containment effort against the Russian government with the intent of crushing Russia the way the Order defeated the Soviets.
- Strategic distraction in the Middle East prevented the Americans from putting their back into the effort, however. The result was a muddling through, with the United States

instituting an echo of some of the old Cold War–era containment policies, including a sanctions regime and the rotation of a trip-wire contingent of a couple thousand American and allied troops at a time through several of NATO's border states in order to increase the credibility of American security commitments.

* In response, the Russians have unofficially abrogated some of the key treaties that closed out the Cold War. The first—the Treaty on Conventional Armed Forces in Europe—regulates how many troops can be in what states and near what borders in what concentrations. The second—the Intermediate-Range Nuclear Forces Treaty—bans the development or deployment of missile systems with a range of roughly 500 to 5,500 kilometers.

During the Cold War, the Soviets' fielding of new intermediate-range weapons would generate a direct and harsh American counteraction at the highest level, which would immediately trigger a direct and harsh counteraction that would make the Godfather blush. A command from the American president would quickly fan out through the American bureaucracy to slam American influence into points of contact throughout the world. The Soviets would quickly discover that their one violation would have consequences in Iceland, Britain, Denmark, Germany, Turkey, Pakistan, Japan, in grain markets, in oil markets, and so on.

Lately, however, the United States—under three different presidents—has done . . . nothing. Washington largely stood aside when the Germans went out of their way to establish new energy links to Russian natural gas fields—along routes specifically designed to cut out countries like Ukraine (whose independence is central to keeping the Russians unanchored) and Poland and Slovakia (which are NATO allies). France even (almost) sold the Russians amphibious assault ships that really could have had no use *except* against NATO allies in the Baltic Sea littoral.

These are not the sort of actions allies take. They certainly are not the sort of actions allies put up with.

If the goal is to constrain Russia, the United States is left with the worst of all worlds. Halfhearted actions to counter Russia not only were insufficient, but their poor execution split what's left of the alliance *while also* encouraging Russia to be more aggressive in dismantling the institutions that hold the West together and constrain Russian actions.

It isn't so much that the Americans will get this all sorted out in time, but instead that sliding away from the Order means first and foremost the Americans' sliding away from the burdens of European security. By commission or omission, that means the end of NATO as a functional alliance. The question becomes whether *any* of the Americans' relationships with *any* of their Order-era allies in *any* region can segue into something that can outlive the Order:

- It is difficult to see how NATO's unraveling will enable the Americans to maintain a constructive relationship with the Central European nations. Without the more capable Western European nations anchoring the alliance on the Continent, the defense of Latvia and Slovakia and Romania is simply impossible. Furthermore, it seems there cannot be *any* meaningful alliance-based security relationship with Germany any longer. For some, this might be a good-riddance-to-bad-rubbish moment, but consider what happened the last two times the Americans and Germans didn't have a strong security understanding. The unentangled Americans smugly watched European events in 1914 and 1939 . . . right up to the point they found themselves with boots on the ground in Europe. The best way to preclude such a historical repeat would be to retain a constructive relationship with the swathe of countries from Estonia to Bulgaria, but just because those states would be useful in precluding the war of tomorrow does not mean they are useful to the America of today.

- It is safe to say most Americans would *love* to disengage from the Middle East completely, and economically the shale revolution helps the United States to do exactly that. Many in the United States would enthusiastically down bushels of popcorn as the Saudis and Iranians rip into each other, and as stratospherically high oil prices crush the Asian rim economically. But fast-forward to the next day. What happens when either the Saudis or Iranians *win*? What happens if* a more capable and consolidated country like Turkey inherits the entire region, and along with it absolute control over the bulk of the world's easily produced crude oil? The best way to prevent the rise of such a Middle Eastern hegemon with global reach would be to keep several of the small states of the Persian Gulf independent. That means American forces on the ground in *all* of them, but an *increased* American commitment to the security of oil-rich statelets is the furthest thing from most Americans' minds.

- America's paranoia of the day is China, and most Americans will applaud *anything* that takes the Chinese down a notch or seven. But China isn't likely to be the power that dominates the East Asian rim a generation from now. The future of Asia is one where a remilitarized and almost certainly nuclear-capable Japan rules over an associated network of allies throughout the Asian rim that it keeps in economic thrall. The last time the Japanese *attempted* something like that, it ended in the Pacific theater of World War II, and this time the Americans will instead *encourage* the Japanese to achieve it? The best way to prevent such an outcome would be a powerful American presence not only in Japan and Korea, but also the Philippines,

---

* When?

Vietnam, and Taiwan, both to crush the Chinese system and then to manage whatever regional alignments come next. But the United States has no appetite for a containment regimen on the scale of the Cold War, and even the most stalwart anti-China hawks at the Pentagon are sure to pale at the suggestion of US troops in Vietnam.

In all three theaters, the seeds are being sown for much bigger competitions with much more formidable rivals two or more decades from now. In all three regions, American unconcern now leads to the broad-scale abandonment of the sorts of allies that could nip those competitions in the bud. America's inability to voice what it wants from the world is condemning it to a series of future conflicts that are both eminently preventable *and* wouldn't have to be fought alone.

THREAD 3: STRATEGIC RETRENCHMENT

Americans aren't just *tired* of global engagement; they're confused by it. They don't understand how everything fits together— because it doesn't any longer. American policy thrashes about from crisis to crisis, and such spasming generates a huge disconnect between American internal politics and the broader world.

Confusion is not an unrealistic point of view, nor is disengagement an unintelligent conclusion. Between America's huge standoff distance and its relatively slim economic interest in the global system, the United States *can* wash its hands of everything outside of North America and go home. And since the world has become increasingly hard to understand even for someone who has made its study his life's work,* it is *really* easy for someone who sees the world as just one of many issues at hand to say the world just isn't that important.

For most Americans, the conclusion that the United States should be less involved in managing the world is as reasonable as it is worrying. Such a view is now dominant on the Right and Left in American politics, across both the centrist and populist factions.

Somewhat belatedly, America's strategic drift since the Cold War's end is becoming entrenched in policy as well. The United States now has fewer troops stationed abroad than at any time since the Great Depression, and all current indicators point toward slimming down what remains. Under the past four presidents, American relations have degraded with *every* country on the planet. Within the next few years, American forces will almost certainly be removed from operational bases in Afghanistan and Iraq, but also politically sensitive locales such as Turkey, Qatar, and Germany. US troop commitments in legacy

* That would be me.

locations—South Korea, Japan, and Italy come to mind—won't be far behind.

Such drawdowns are easy to argue for. Most relate to wars long past (Germany), wars recently ended (Iraq), or wars the Americans never want to fight (Korea). And whenever one of the strategic hangovers of the Order or the Global War on Terror runs headlong into the wall of Americans' preference for general retrenchment, retrenchment tends to win—with lasting mal-effects on all sides.

For example, there's the issue of alliance involving the Trump administration and the Persian Gulf statelet of Qatar.

If the United States is going to do anything in the Middle East that involves real power, America must have a military footprint in the Gulf more reliable than a carrier or other mobile base.* Instead, America needs a base on land that can support large ground forces. That requires safe harbor and hefty logistical support, which in turn requires a very willing partner. Options are limited. American policy says Iran is a problem, so no base there. The United Arab Emirates is too close to Iran politically. Oman tries to be neutral. Even if Iraq were not full of bad memories, it is too unstable. Saudi Arabia is too bossy. Israel has bad optics. That leaves Bahrain, Kuwait, and Qatar, and so that's where the United States maintains its military facilities.

Tiny Bahrain is a naval base and lacks the land area for anything grander. The governments of Kuwait and Qatar are critical to the webs of political, intelligence, and financial support that make extremist militant groups like al Qaeda and ISIS possible. If a leading reason to be involved in the Middle East is to fight Islamic-themed terror, placing bases in countries that are leading

---

* Carriers are great, but they can carry only so many jets, and they cannot carry troops. Plus, if your goal is to exercise power in a specific theater *all* the time, a carrier isn't the right tool for the job, because it must return to its home port from time to time for reprovisioning, maintenance, shore leave, and crew swaps. Carriers are tools of *mobile* power projection, not *constant* power projection.

supporters of such activities is among the most asinine actions possible.

In 2017 Trump called a duck a duck and publicly condemned the government of Qatar for being a de facto sponsor of terrorism. His statements were entirely accurate. The American defense and intelligence community, however, had a herd of cows. They can't carry out their jobs in the region without the base in Qatar to serve as the operational headquarters for the Army and Special Operations branches of United States Central Command, the pieces of the Department of Defense that fight the wars in the Middle East. *Their* concerns on the topic were also entirely accurate. Both were completely right, *and* both were completely wrong.

Trump in effect peeled away a critical US ally, whose basing enables the American military to operationalize strategy against factions in Saudi Arabia, Syria, Iraq, Iran, and Afghanistan, and replaced it . . . with absolutely nothing. Constructive policies in any of these countries are next to powerless without Qatar, and yet Trump was 100 percent correct: Qatar *is* a state sponsor of terrorism—often against American interests—and calling a duck a duck publicly (in a region flocking with ducks) means Qatari cooperation with the United States on anything has plummeted.[*]

Precisely these sorts of fallings-out with allies are being repeated around the world, and they are hardly limited to Donald Trump. They have occurred with Japan over legal troubles in Okinawa (Clinton, W), in Korea over the placement of bases in the Seoul metropolitan region (W), in Germany over the degree to which Berlin has oversight of US military activities that might use base facilities (W, Obama), in the United Kingdom over the fate of American forces as relates to issues of Brexit and Scottish

---

[*] There are plenty of signs of deployment shifts addressing this reality. Since peak deployment levels, American forces stationed in Qatar have plunged from 15,000 to under 1,000, in Kuwait from 58,000 to under 2,000, and in Bahrain from 9,000 to under 4,500. Again, the Americans are simply leaving without resolving anything.

independence (Obama, Trump), and in Turkey over America's preference for working with the Kurds (W, Obama, Trump). There just are not going to be many places where both (a) the locals want US forces, and (b) the United States has a strong strategic interest in having forces there in the first place. In less than a generation, the United States will have gone from the greatest, broadest, deepest list of allies in history to a mere handful.

The second issue is one of capabilities. The sort of military structure, equipment, training systems, and posture required to maintain a global Order is fairly specific. It consists of three pieces:

1. Large numbers of static troops deployed at friction points to serve both as strategic trip wires and credible deterrents. Large-scale Cold War–era permanent deployments in Germany and South Korea are the obvious examples. The entire Vietnam War also fits the bill.

2. A moderate-size ground force that can deploy in force quickly to meet challenges. The Marines fit this requirement to a T, as do some of the United States' more celebrated forces, such as the 101st Airborne and 10th Mountain Division. As a rule, these forces are meant to be kept well out of harm's way until the day they are unleashed.

3. A broadly deployable and deployed navy that can enforce freedom of the seas on a global scale. Throughout the bulk of the Cold War, the US Navy had in excess of 550 ships, with an emphasis on long-range destroyers.

This is no longer the force structure the United States maintains. Downsizing in the 1990s reduced the personnel roster of the American armed forces by over half, with the biggest drawdowns within the army. Even at the height of the infantry-heavy Afghan and Iraqi wars, overall American force structure did not appreciably shift back. Deployment patterns changed as well. The United States' large static forces were used for occupations rather than deterrence, while the overall reductions in manpower brought about by the Cold War's end meant that those rapid-reaction forces were no longer held in reserve, but were instead deployed day-in, day-out for fifteen *years*.

Even more notable is what has happened with the US Navy. Relentless modernization and drawdowns reduced the number of vessels by half even as the number of carriers increased. That forced a concentration of the remaining naval vessels around the all-important carriers. (Even when not expecting combat, each carrier typically has a dozen escorts.) The Cold War–era Navy of 550-plus ships could fairly reasonably disperse to provide global coverage. Not so in 2020, when the Americans' naval vessels total fewer than 300—nearly one-third of which compose the concentrated forces of the carrier battle groups. What they *can* do is kick in any door on the planet.

Combined, these changes reflect the sort of strategic power the United States can be and wants to be. Maintaining the global Order isn't simply something the Americans don't *want* to do anymore. The shift in alliance priorities and the change in force structure means the Americans *could not* maintain the Order even if they wanted to.

Between inability and lack of interest, change is cascading through the entire alliance system. The de facto abandonment of traditional allies combined with the military's reorientation from an Order-oriented global deployment to something more appropriate for a narrower definition of national interest makes US policy less predictable, more kinetic, and potentially much more disruptive.

THREAD 4:
PROFITS WITHOUT BORDERS

In many ways, America's "problem" is an issue of strength—
specifically America's wealth and productive capacity.

First, wealth. The United States is by far the largest economy
in the world and will remain so for at least the rest of this century. While inequality is a big issue in the country, even people
laboring full-time at minimum wage pull down about $15,000
annually. That might not sound like much, but it is *five* times the
median annual *household* income globally. By global standards,
Americans are sick rich, and sick-rich people don't do certain
types of things. They don't sew a lot of T-shirts or cobble out
footwear or snap together computer boards or assemble most
cars. Instead, they art-up T-shirt graphics, carve out would-be
shoe designs, fabricate deep-memory chips, and design the next
generation of automobiles.

Americans do much of the value-added work of the world,
then outsource the manufacturing process to countries where labor is cheaper and less skilled and less educated and less creative.
Compared with the value of high-end design, the value of low-
end manufacturing is tiny.

Second, productivity. Most successful countries have one of
two things going for them. Some have a decent geography that
allows for significant production of this or that product: South
Africa has the agricultural bounty of the Johannesburg plateau,
Australia the mineral wealth of the Outback, Russia the oil fields
of Siberia, Canada the timber of the Rockies. Others have a population that due to educational quality and/or spatial layout can
add some serious value to products: Korea in smartphones, Thailand in semiconductors, Mexico in automotive manufacturing,
Switzerland in biotech.

Few countries have both, and the marriage of a high-value-added workforce to a high-value commodity system generates scads of economic opportunity. The United States sits at the very top of this very short list of countries, and the scale of American production nudges two other factors into the economic math. First, because there are 331 million rich, relatively high-value-added Americans in play, American businesses often need to—and can—differentiate their products and services to make their mark. Spend a weekend on the Kentucky Bourbon Trail to get an idea of the spread of possibilities.*

Combine differentiation with the reach of modern digital technologies and shipping, and much of this specialization can sell globally. American finance doesn't manage just American money, but global money. American agriculture doesn't feed just Americans, but the world. American *high-end* manufacturing is what makes Chinese *low-end* manufacturing possible. A big chunk of American *exports* are of products Americans produce too much of for their own market. You can drink only so much of any specific bourbon before you have to push stuff out to the global system.

While the United States' need of and tolerance for the rest of the world is thin, what interest there is is heavily concentrated into a very slim slice of the American population: business leaders—particularly those in the sectors in which the product in question is eminently transportable, due to either technology or geography, and in which the imbalance between domestic supply and domestic demand is most lopsided: agriculture, finance, and tech.

This takes us some interesting places.

First, because this external orientation is more an issue of specific products, like soybeans or processing chips, rather than broader categories, such as commodities or manufacturing, even among the business community those looking beyond America's

---

* For God's sake get a driver and *don't try to do it in a day!*

borders are in the minority. This is not a politically powerful group, and it cannot shape American policy on the front end.

Second, because these business leaders *are* used to maneuvering globally and because they are in *business,* there is still a lot of money at stake.* While money is far from a nonissue in American politics, it is *the* issue in the politics of most countries where American businesspeople operate. In a world in which the Americans have retrenched, that makes these outwardly oriented, globally savvy, cash-rich American businesspeople the dominant American presence in much of the world.

Third, because the United States government will have few strategic interests in the broader world, these outwardly oriented American businesspeople will become the wavers of the flag. Their corporate policies will be perceived around the world as de facto American foreign policy. Deals with them will be perceived as deals with Washington, while attacks upon them will be perceived as attacks on Washington.

Put simply, it is this thin slice of the American business community—in and of itself a thin slice of the American electorate—that will largely determine what American interests *are.*

We have seen this before. During America's Reconstruction period, most American economic interests focused on internal issues: rebuilding the South, settling the West, knitting the country together with railroads. By the turn of the twentieth century, most of that work was completed, and American business leaders began sniffing around the world for opportunities.

Simultaneous to the end of Reconstruction, the Americans had a "Hey, we're back!" party—the Spanish-American War of 1898. A decade later, President Teddy Roosevelt sent the country's

---

* While the US is the least involved country of size in the global economy as a percent of GDP, America's GDP is so huge that even that thin sliver of involvement still makes it the world's largest importer, second-largest exporter, and one of the top five players on at least one end of every trade imaginable.

new battleships—the famed Great White Fleet—around the world on a meet-the-neighbors tour.

The disconnect is telling. The Americans sent the bulk of their navy away from US shores for fifteen *months* just after the Japanese had wrecked the Russian fleets and *while* the Europeans were gearing up for World War I. This may seem strategically unfathomable to us today, but at the time, the United States saw itself as sufficiently physically secure to send its best ships on an extended "Hi! How are ya?" trip in a different hemisphere. It is hard to come up with an example of a military in any era engaged in more thumb-twiddling than the Great White Fleet.

Enter the business leaders. In the early twentieth century there weren't any broadly successful countries in Latin America or East Asia save Argentina and Japan. National governments were weak, infrastructure was fractured, local manufacturing was thin to nonexistent, and labor was cheap. Outward-oriented American businesspeople—many with experience in carpetbagging during Reconstruction—would flood into foreign countries and set up a mix of mining, basic manufacturing, and import-export businesses. As most of these places had horrid logistics, anyone who could provide financing, technology, jobs, services, or access to external markets or goods could make some serious bank.

That's the positive side. The less positive side is that these were still *countries*. Many aspects of American law don't necessarily translate well to foreign lands or vice versa, and not all American businesspeople were angels. Exploitation wasn't exactly rare. Excessive debt was issued, and these being countries, sometimes sovereign defaults occurred. Sometimes American businesspeople got shot.

Back in Washington, an unfettered strategic picture, combined with poorly or undefined strategic goals and a military with nothing to do, enabled Washington to come to the rescue. Naval vessels would impress upon locals the importance of allowing American merchant ships to come and go. Troops might go ashore to ensure that American contracts were upheld. Perhaps a brief national occupation was required to convince local authorities that those bonds they issued to American investors needed to be paid back in full after all.

This fusion of corporate interests with otherwise listless American power is called dollar diplomacy, the idea that *business* interests will by lack of competition define *American* interests, and the US diplomatic and military community will use the tools of statecraft to reinforce business decisions. It isn't the prettiest chapter of American history, and it—along with the Monroe Doctrine—set the tone in Washington that Americans can more or less do what they want in Latin America. For their part, most Latin Americans see Monroe, dollar diplomacy, and the Americans' Cold War–era penchant for supporting coups against Latin American leaderships they disliked as variations on the same policy theme.

Dollar diplomacy version 2.0 is just around the corner. This time, however, the American corporate focus will look quite a bit different. In the early 1900s version, the imperial era was still large and in charge, and huge swathes of the world were under this or that empire's control. This time around not only are the empires not present, but the dislocations of the Order will be raging like wildfire, and only the Americans will have access to the sort of resources and foodstuffs and financial wherewithal and credit availability and maritime strength required to ensure the commerce that so much of the world will so desperately need:

- Disorder-era China will lack reliable supplies of almost everything. Many Chinese coastal cities—especially those in the south—will eagerly embrace anyone who can provide them with imported goods, especially in the fuels and foods categories. *Especially* after China's central government proves unable to continue the policies of bottomless credit that have defined Chinese economic growth since the death of Mao.

- It will take a decade for the United Kingdom to recover from its hard Brexit. Part of the Brits' fall will include the debasing of the pound, a factor which will provide a legion of opportunities for American businesspeople and US financial markets to almost take over the UK economy.

- Most of the former imperial colonies that achieved independence only under the Order cannot see to their own agricultural, manufacturing, or credit needs. Many American businesspeople will thrive in just such an environment.

- Whether Germany or Russia or Iran or Saudi Arabia or China or Japan, whoever *loses* the wars of the future will be a target for the sort of carpetbagging activities that so seared the American South under Reconstruction and Latin America under dollar diplomacy.

- The breakdown of East Asian manufacturing supply chains will spawn huge interest in the United States in re-forming those chains with more of an American emphasis. American businesspeople will be particularly interested in integrating with pieces of the old system that can be salvaged. That suggests a heavy American hand in Thailand, Singapore, Malaysia, and Vietnam.

- Americans won't ignore Latin America. While most of these states will do reasonably well as Order decays into Disorder, there are exceptions, as well as those that cannot manage their resources without foreign money, foreign technology, or both. Brazil, Cuba, and Mexico in particular are likely to see an influx of American corporate interest.

If this sounds a bit mercantilist, that's because it is. If it sounds *too* mercantilist for the American ethos, consider that mercantilism was the norm before 1939. If it seems the United States has evolved in a more ethical direction since 1945, consider that even if the United States wanted to keep global commerce safe, it is still about to have to make some choices.

Under the Order, the global merchant marine expanded by an order of magnitude. In a world of Disorder, the United States Navy would, conservatively, need to triple in size to provide global shipping security. Such an expansion is not possible on anything other than a generational time frame, and so the Americans will be forced to choose which shipping to protect. They will start with their own. The process of choosing whom to leave hanging already starts the *government* on the path of picking winners and losers. Outward-oriented American businesspeople will helpfully and happily assist with the selection process.

THREAD 5:
DESPERATELY SEEKING INSTABILITY

The first four threads of US foreign policy are less notable for what they are than for what they are not. Getting past a strategic hangover or disengaging from an outdated commitment is a process, not a policy. Dollar diplomacy can exist only because there is a hole where policy normally would be.

The hole isn't limited to American foreign policy. Since 1946 American foreign policy has determined the shape of the world. When the Americans stop holding up the roof, they will note that the resulting power vacuum is bad, exceedingly bad, for most of the world. But they will also notice that in many cases the chaos works *for* America.

It's a shift in mind-set that ongoing structural evolutions will support.

America's recent wars have shaped what Americans consider to be "normal" military activity. Of late very little of America's war fighting has involved, well, fighting a war. Instead of clashes with other organized militaries, American forces have alternatively been doing a lot of patrolling and sweeping and hunting and drone warfare. When the goal is to pacify and stabilize large swathes of territory, large numbers of troops are required, and they are subject to high levels of irregular attacks. This is noisy and bloody; the American public notices it, and they don't like it very much.

As the Iraq and Afghan wars ground on, however, it became obvious to the W and Obama administrations that these places could never be "fixed" as Americans defined the term. Goals narrowed from reconstructing Iraq and Afghanistan in America's image to setting up local forces that could carry on the fight to setting up a government structure that could survive with minimal American support to simply ensuring that, when the bulk of

American forces left the countries, some bases would remain in American hands to insert Special Operations forces and drones as needed.

It's an issue of numbers and information. At their heights, the Iraq War had 148,000 American troops in theater and the Afghan War had 99,000 (not including contractors or allied forces), out of a total personnel roster (including reservists) of 2.3 million. Patrolling central Iraq requires a dozen major bases *in* central Iraq and tens of thousands of troops.

In contrast, drone and Special Operations require only a single small footprint that doesn't even need to be in-country. America's *entire* Special Operations forces—even with their tenfold increase in budget during the Global War on Terror—number fewer than seventy thousand. Add that Special Operations by their very nature are less public than conventional military operations, and that drones don't have families back home, and most Americans are quite comfortable in (or at least resigned to) their ignorance about the way the US now fights.

What America finds ideal, however, the rest of the world finds problematic. The United States has become comfortable with a war-fighting method that exposes few Americans—civilian or military—to risk, but enables US power to reach anywhere in the world on short notice in an often deniable manner. Marry the new techniques to a broadscale and deepening American indifference to global stability, and the results are disruptive, bordering on explosive.

It does not take much of a leap in logic to see the next step here. The United States will begin to view disruption in and of itself as a tool, perhaps even a *goal*. It sounds dangerous (and there certainly is a risk), and it sounds irresponsible (and by some measures, it is) but that doesn't mean it's a bad strategy. Consider the following:

- The United States is not a trading nation, but all its current perceived and likely future competitors *are*. Fully 100 percent of trade in Northeast Asia uses oceanic shipping, as does over 95 percent of Europe–Asia trade and 70 percent of global oil trade. In contrast, less than half of American trade (comprising less than 8 percent of total American economic activity) uses the ocean, and much of that is within the Western Hemisphere and so is immune to chaos beyond it. Targeted disruptions, even *broad* disruptions, do not only make a great deal of mercantile sense, but they also make would-be trading nations strategically dependent upon American goodwill even if they are not US allies.
- As of early 2020, the United States has just become a net crude-oil and products exporter, and it had already been a net exporter of every other finished energy and petrochemical product since early 2018.* Disruption in energy or chemical flows would be economically crippling to competitors, whether importers or exporters, while a policy tweak at the presidential level would shield US consumers from global price rises. For decades, Moscow has had a delightful time throwing wrenches into the Middle East to complicate global energy flows and thereby American strategic policy. Now it's America's turn. Considering American naval power, achieving something an order of magnitude worse than the 1970s and 1980s oil shocks would be child's play.
- Modern manufacturing rests upon the idea that any particular location can thrive by mastering individual steps of the

---

* For those of you who love to nitpick, to consider the United States a net oil exporter you must include the shale sector's four million barrels of daily production of natural gas liquids. NGLs, as the name suggests, occupy a gray area between natural gas and liquid oil, and are heavily used in refining and especially petrochemicals.

manufacturing supply-chain process. Global safety enables global supply-chain integration. That requires absolute freedom of the seas as well as access to a near-bottomless supply of capital to fund constant industrial plant overhauls as technologies evolve. The United States is not only the provider of the security and commander of the financial access, but the NAFTA network is the only manufacturing system on the planet that does *not* require global maritime access. A small degradation in maritime safety not only destroys most manufacturing supply chains, but also forces many of them to relocate to the only place where inputs and production and consumption are co-located: North America. Almost anything that reduces supply-chain security globally *increases* the case for positioning industrial plant within the NAFTA system.

- Agricultural supply chains are not as fixed and vulnerable as those in the manufacturing sector, but the impact of their disruption is far more terrifying. Oil is processed into fertilizer that helps grow foodstuffs, which are transported by ocean the world over. Any interruption along that chain of events means food does not make it to mouths. Few countries have their entire supply system internally; fully 75 percent of global oil is wrapped up in transnational trade, and the United States' *internal* supply-chain system for all the various inputs is by far the world's largest. American farmers will cringe at the carnage of continent-spanning famine . . . and then giggle all the way to the bank.

- The single largest source of stable, productive financial resources comes from a population cohort that is nearing retirement. Workers in their thirties save a small sliver of a small income for the future. Workers in their fifties save a larger wedge of a larger income. Workers just shy of retirement save as much as humanly possible. With global aging, there are a *lot* of these near-retirement populations, so global capital is plentiful and therefore cheap. But very few countries don't face demographic hollowing-out, and on average the global Baby Boomer population flips into mass retirement in 2022. Upon reaching retirement, one lacks income and so has nothing new to save. The result? Very few countries will have a stable, indigenous capital supply in the future. The American Boomers, however, had kids, so America's demographic decline will be far slower and far less painful. This not only grants the Americans cheaper capital than everyone else, but also makes the United States the primary global destination for capital flight. Chaos, war, and depression—*disruption*—encourage capital flows to the American system. From 2015 through 2019—a time of strong global growth and relative stability—that capital flight probably topped $8 trillion. Imagine what's coming. Imagine what's coming if the United States puts its thumb on the scale.

- Most raw materials worldwide—I'm referring here to non-energy, non-food products such as iron ore or copper—are processed in regions the Americans consider to be economic competitors: Germany and China consistently rank high. Yet in most cases the raw materials themselves don't originate in Germany and China, but in the Southern Hemisphere. Disruption either in or near Germany or China or in the waters between them and the source material forces metals processing and refining to shift to other locations—locations that are more stable with cheaper and more reliable electricity. Places like the United States.

- Today most global trade is denominated in US dollars, and any significant global degradation will reduce the stability of nearly every currency to the point that nearly *all* future trade will be USD-denominated. The Americans will be able to selectively grant or withdraw access to global finance on a whim, so even countries with only limited security exposure must consider the consequences of impinging upon American interests or souring the mood of the American president and Congress. Failure to do so could easily reduce them to barter.

- By doing some minor tinkering in the financial markets, it would be easy for the United States to crush systems that are financially overexposed. The obvious candidate here is China, where any financial interruption quickly generates political chaos—which would be *great* for US manufacturers. There's also Brazil, where a timely interruption in agricultural finance first generates local economic breakdown and, a season later, seizures in global food supply—which would be *great* for US farmers. Don't forget Turkey, where the political alliance that supports the current government is built upon tens of billions of dollars of cheap construction loans—the disruption of which would be *great* for US energy interests. Moreover, the entire eurozone faces risk, and not only basket cases, like Greece, or those with rotted banking systems, like Italy. Germany's economic model is based on deep, reliable financing to keep the industrial base updated with the latest technology, while state intervention in the economy is a hallmark of French policy. In an era of more mercantilist and populist policies, American *non*intervention in the world of finance does not come for free.
- One side effect of trading systems that are less global and more regional is that instability in this or that region has less of an impact upon others. The primary reason the global recessions of 2007–9 were not worse was that the US Federal Reserve stepped in to provide unlimited volumes of dollar-denominated bridge loans to *any* peer institution that needed the liquidity. At a minimum, that prevented a much deeper financial crisis in (from mildest to greatest danger) Australia, Canada, Brazil, New Zealand, Norway, Denmark, the United Kingdom, Japan, Mexico, Sweden, South Korea, Singapore, Switzerland, and the eurozone. If the Fed no longer fears contagion, similar assistance in the future might not be forthcoming—or at least not without a significant price tag. And even *that* assumes the United States wouldn't deliberately nudge financial systems that were unstable or overexposed. It wouldn't

take much prodding to upend Argentina, Brazil, Venezuela, Peru, Saudi Arabia, Egypt, Turkey, Russia, Ukraine, Kazakhstan, Japan, South Korea, China, Taiwan, Thailand, Pakistan, South Africa, Canada, or the eurozone (and within it, Italy, Greece, Cyprus, Belgium, Portugal, Ireland, Spain, the Netherlands, and Germany are particularly vulnerable). It isn't that the United States expressly views any of these countries in particular as targets, but rather that in a changed world the US dollar's status as the only truly viable currency gives Washington incredible reach into and influence over foreign economic systems.

Marry an American strategic, willing disregard for global security to a public that has become more comfortable with low-level, disavowable military activity to a military with global reach to a more mercantile approach to the world, and it is Anakin-falling-to-the-Dark-Side eeeeasy to envision a United States that *seeks* disruption rather than stability as both a tool and a desired end of foreign policy.

CHAPTER 15

THE UNITED STATES

THE DISTANT SUPERPOWER

An American strategic emphasis on unenlightened self-interest will be positively imperial, almost Russian, and it will hugely shape America's list of relations. From an American point of view, all relationships will fall into one of three categories.

GROUP 1: THE COALITION OF THE WILLING (AKA THE ALLIES)

First, consider where there is a degree of common cause. A disconnected and disengaged America in a disintegrating world won't have any enemies in the traditional sense. In a world without China, the links to Japan and Korea fade. In a world in which the Americans don't care about Russia, relations with Germany and Poland don't matter much. In a world of shale, Iran doesn't generate the same degree of heartburn it once did. As soon as Americans allow some emotional distance to reframe relations with Iran, Saudi Arabia looks a *lot* less attractive. There won't be any serious strategic foes out there until at least 2030—maybe even 2045. It will take time for countries like Japan and France and Turkey and Argentina to reshape their regions into something the Americans might find threatening. And the word there is *might*. The likely paths of Turkey, France, and Argentina, in particular, are not tracks likely to generate any strategic anxiety for at least a generation.

That doesn't mean there won't be issues and locations of concern. The vicissitudes of American domestic politics can always tilt the Americans toward this or that cause. A bit more militancy in the American environmental movement could crack the seal on the use of *military* tools against global coal, putting countries as varied as Germany and China and Spain and Australia front and center. A reinvigoration in respect for human rights would have no end of targets, especially as global wealth levels plummet.

The War on Drugs already has a military angle, but it could easily be amped up and applied not only to countries that *want* American involvement, but also to those that *don't*. Think Mexican cartels, Brazilian and Venezuelan smugglers, Ecuadorian and Bolivian growers. And why not a "war" on sex trafficking? Slavery? Migration? Several of the 1990s interventions claimed similar ethical and moral justifications.

What about the fights that others pick? Without American forces broadly engaged throughout the world, there are not as many Americans to target. And the *last* thing a sane aggressive power with regional ambitions would want to do is involve the Americans—distant, uninterested, but *strategically unfettered and armed to the teeth*. While it is exceedingly unlikely, it is *possible* that some breathtakingly stupid foreign leader would go well out of his (or her) way to target the United States. God help them.

A far more likely stimulus for meaningful American foreign military action would be a nonstate actor whose actions triggers a new manhunt. However, after America's less-than-rewarding experience in the Global War on Terror, any new such countereffort on the American side will be started from a more experienced point of view, with a more jaundiced eye, and using more surgical tools. Fewer armored divisions, more Special Operations forces.

Any allies in this "common cause" category, therefore, are not going to be standing shoulder to shoulder in the trenches with the Americans in any sort of mass warfare. If the Americans are mildly interested in the topic at hand, they'll let their ally do the heavy lifting and relegate themselves to providing technical, logistical, and matériel support. And if the Americans are fired up and decide to act directly, all they need are some temporary basing rights.

At the top of the short list of allies-in-common is a country that has found itself in a bit of a pickle: the **United Kingdom**.

At the time of this writing, the Brexit process is well into its third long year, with the British government sacrificing all its economic, financial, and political capital to maintain a never-ending Brextension. That has made the divorce harder than it needed to be and delayed British efforts to figure out how they fit into a world of Disorder, also known as a world without the EU. Instead of getting a head start, the British divorce is now happening at the same time the global system is breaking down. That timing limits options.

London's post-EU plan was to shift its trade from the euro-zone to the rest of the world with a heavy emphasis on the former empire. That strategy has a severe problem: extracontinental trade requires global stability that will soon no longer exist, and the United Kingdom's post–Cold War military downsizing means the UK lacks the ships to protect its own long-haul shipping.

That unfortunate restriction will do two things. First, it will condemn the Brits to a multiyear depression. With the end of the Order and the EU, this outcome was unavoidable. If anything, the timing here is a plus. Because the Brits are beginning their Disorder-decline earlier, they will be able to hit bottom faster and start on whatever is next.

Second, the Brits' trade will largely be deals with countries that will be safe regardless of how bad the Disorder gets, or that can protect their own shipping (so the Brits don't have to). The world has only three large trading and supply-chain systems. Whether due to Brexit or the EU's own end, the European market will be largely off-limits. Distance rules out East Asia. That limits meaningful interaction with the NAFTA system.

The weird makeup of the early-Disorder British navy—a pair of supercarriers lacking escort rings—makes the British dependent upon the Americans for deployment. This will last until the British can expand their navy to compensate. That will probably take until 2040.

The weird makeup of the early-Disorder British economy— minimal European access, but large-scale North American access—makes the British dependent upon the Americans for growth. With the American population aging far more gracefully than the British, this dependency is for the long haul and will become more intense with every passing year.

Many Brits wanted to leave the EU because they didn't like having bureaucrats in Brussels set their policies. In the early Disorder, the Brits will have handed *more* control to the bureaucrats in Washington. Simply put, the Brits are supplicants with no other options, so the terms of American-British interactions will be wholly American-determined. Regardless of how the specifics unfold, this is what full strategic capture looks like.

It adds up to an incredibly lopsided relationship the British won't be all that culturally comfortable with no matter how much strategic and economic sense it makes. For all practical purposes, in the Disorder the United Kingdom will no longer be an independent country.

With the Americans distracted—especially once British strategic policy is lashed to the American will—the future tenor of the relationship is largely up to the United Kingdom. Whispers that increase in volume to conversations will increase to a public debate about just how close a relationship with the Yanks is appropriate.

NAFTA inclusion? Certainly. Commonwealth? Possibly. State-hood? It might not seem all that likely due to issues of physical and cultural distance, but it *is* the fate of most aging parents to move in with the kids.

The states of the North Sea littoral—**Norway, the Nether-lands,** and **Denmark**—will find themselves experiencing an echo of the British situation. All broadly share America's stra-tegic concerns about the Continent and Russia and so will work to stay as close to US defense policy as the Americans will allow. All will look for ways to piggyback on Britain's economic inte-gration with North America (rather than throwing themselves at Washington's mercy in a bilateral deal as the Brits will be forced to do). They'll undoubtedly have more luck with the former than the latter.

Germany will find itself in a fluid, multivector competition with Russia, and that will require it to militarize and develop an independent deployment and intelligence capacity. Everyone in the neighborhood will be understandably concerned. The Amer-icans? Not so much.

Rather than trying to contain or dissuade the Germans, the American goal will be to keep Germany as integrated into Amer-ican military systems as humanly possible. Part of the rationale is German and economic. It isn't so much that the Germans lack the Reich stuff, but more that an economy the size of Germa-ny's preparing for a conflict with a country as big as Russia is going to need a *lot* of equipment, and it's an open question just how quickly the Germans can retool Volkswagen, Siemens, and Bosch to make tanks, jets, and drones. The Americans can help fill the gaps . . . and make a *lot* of money on the side.

The bigger question is strategic. Reliance on the American security blanket has let German strategic planning atrophy much like American strategic drift in the late-Order decades. But while the American drift is on a near-bottomless budget, so nearly *everything* gets funded, the Germans are on a near-zero budget, so *nothing* does. With the Germans about to face real needs and threats, the opportunity for the Americans is sublime. The more dependent the Germans are on American systems in the conflicts to come—especially on intelligence systems—the less a Disorder-era Germany (or even a post-Disorder Germany) might pose a threat to the United States. Enabling an independent Germany is the best way to constrain it. For the Americans, that is. Everyone else will think this is a horrible idea.

The goal here—very unofficially—is the end of **Russia** as a state. In the history of the American republic, only three countries have ever endangered any of the continental American territories. The first, the United Kingdom, has already been strategically neutered and will soon be lashed to the American will. The second, Mexico, ceased being a strategic threat after the Mexican-American war ended, in 1848, and is now a partner. That leaves just Russia.

In the many conflicts of the Disorder, the Americans are unlikely to play mastermind, much less assume decisive roles on the ground, but that hardly means they won't play favorites. And pretty much *anyone* who finds themselves in a struggle with Moscow can count on weapons, fuel, and intelligence support. No one more than Germany.

It could all backfire. After all, the last time the Western nations hoped for a German-Russian smackdown, it instead generated the Molotov-Ribbentrop Pact, a truce between Stalin's Russia and Hitler's Germany. Shortly thereafter the Nazis dominated most of Europe. But just like the West's last great strategic miscalculation, a failure this time would be Europe's problem, not America's.[*]

On the opposite side of Eurasia lies **Japan**, with which America has one of its most complicated relationships. A big piece of the complication comes down to the difference in development patterns between the Japanese and the Europeans.

---

[*]   At least at first.

The Europeans within the early-Order alliance were technological peers. Most had interacted with one another for centuries before the modern age, most rose through the Industrial Revolution, and most experienced life in the Order under American hegemony. The commonality of experience made things like the partially pooled sovereignty of the EU possible. Some (many?) intra-European conflicts are inevitable, but they are neither sought nor yearned for, nor are they likely to devolve into a war of all against all.

Northeast Asia has no such common bonds. There was never a European-style balance of power in Northeast Asia because the ruggedness of the Korean Peninsula, the disunity of the Chinese core, and the island nature of Taiwan and Japan heavily restricted interaction. Even industrialization wasn't a common experience, but instead, the Japanese preying on the others. All except North Korea were Order members, but none (again, except the territory we know now as North Korea) shared land borders, limiting integration opportunities. Even at the Order's height, the Northeast Asians regularly fumed at one another. The Koreas, along with China and Taiwan, still cling to a military standoff so entrenched they remain technically at war. Japan's relations with China are beyond adversarial, while even Japan and South Korea—supposedly core American allies—regularly have deeply personal fallouts over history, trade, tech, textbooks, and uninhabited islets.

Unlike the Europeans, the Asian powers never even tried to put their pasts behind them. They lacked the unifying impact of a common American occupation force after World War II. Economic integration started later and never penetrated as deeply, leaving the Northeast Asians with fewer and less recent common bonds. They never attempted to meld into something new. Driven by feelings of technological, economic, racial, and historical superiority and grievance, there are powerful factions within *all* these countries itching for a fight. And this time the Americans will not be there to stand in the way.

The way the Americans are departing *increases* the scope and depth of the coming conflict, for they have unofficially anointed Japan to be their successor as regional hegemon. They have officially urged the Japanese to reinterpret the constitutional clauses that limit military action—clauses the Americans wrote—so the Japanese can take on the Chinese directly. And the Americans have used a mix of intelligence sharing and military-technology transfer to make the Japanese as formidable as possible.

One final difference between Europe and Northeast Asia will shape American-Japanese relations for the foreseeable future. In Europe, no matter who emerges victorious between Germany and Russia, both states will remain constrained by other powers. Conflict's end in Europe begins the next chapter of European stalemates and struggles. Conflict's end in Asia heralds the dawn of a fundamentally new age of Japanese primacy, not just in Northeast Asia, but in Southeast Asia as well, with tendrils of economic and military influence reaching to the Persian Gulf.

At some point, the Americans will behold what Japan hath wrought and have some very serious second thoughts. Considering the time it will take the Japanese to consolidate their gains in the face of their demographic decline and the time it will take the Americans to shake themselves out of their internal political narcissism, that beholding is unlikely to happen until late in the 2030s. There are *plenty* of ways that the next chapter of American-Japanese relations could *not* devolve into conflict, but that doesn't change the fact that Japan is still the most likely candidate to top future America's list of concerns.

A surprising entry into America's "allies of the future" list are a number of states in **Africa**. Economically, there isn't all that much to be concerned with. The bulk of the African continent is a series of stacked plateaus with such steep slopes that running even basic infrastructure into the interior is all but impossible except in a precious few corridors. The continent may have reasonable quantities of crude oil, but the United States' ample local supplies limit interest in even that.

A far more likely lure will be a lingering American distaste for Islamic militancy. In the Muslim expansions of the tenth century, traders either sailed around or pushed south of the Sahara. The rugged topographies of sub-Saharan Africa stopped them cold, but not before they put down deep cultural and religious roots in the transition zone between the Saharan barrens to the north and the uplands and tropics to the south.

This region, the Sahel, offers little for human development: brutally hot summers, heartbreakingly brief winters, tragically fickle precipitation. Life is precarious. The Sahara ebbs north and flows south with the decades, making long-term human settlement tenuous at best, and that's before considering most climate-change models, which suggest that the Sahara is going to flow south pretty damn aggressively for the next half century.

Even during the height of the Order, the Sahel was a dangerous place, where regional governments could barely provide the basics of water, food, and security. With the Disorder reducing global financial and food availability as it guts the broader security environment, the Sahel is set up for the worst of all worlds: weak government, low security, and desperate people becoming more desperate. That's the trifecta for militant groups looking to put down roots. The Sahel is about to become Africa's version of southern Afghanistan.

In part, the militants in this region will be those squeezed out from other conflicts. Algeria's civil war against Islamic fundamentalism has generated just as many militants with no home as Egypt's return to insular dictatorship. The winding down of civil wars in Libya and Syria—or just as likely, those countries' complete collapse—will leave glory-seeking militants looking for a new place to jihad themselves upon. As Saudi Arabia loses its former security guarantor, its willingness to visit violent delights upon places near and far will increase.

Local governments already face thin and thinning resources. Such "trade" in terrorists, combined with an increase in home-

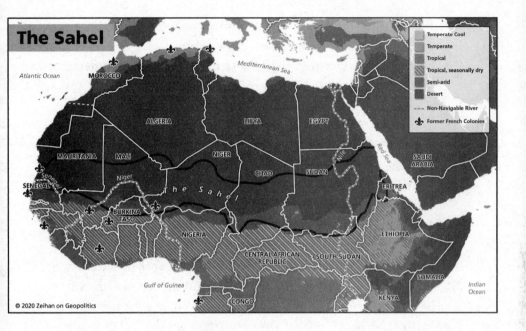

The Sahel

grown militant forces, will leave them horribly and inescapably in the lurch.

Enter the Americans. Most Americans citizens—hell, most Americans in the armed forces—had no idea their military was engaged in active measures in the Sahel country of Niger until a botched mission in 2017 resulted in the deaths of four American soldiers. The operation was hardly a one-off. In the military lexicon, Niger is something called a "cooperative security location," party to a sort of mini–basing rights agreement, which enables the Americans to surge forces in to deal with this or that security threat as needs evolve. Similar, highly active agreements also exist with Senegal, Mauritania, Burkina Faso, Mali, Nigeria, Chad, Ethiopia, and Eritrea.

It is here, in the Sahel, that the Americans will fight their final battles in the War on Terror, not because the Sahel is or is adjacent to an area of core American interest, but instead out of a semi-happy intersection of several unrelated facts and forces:

- First is the Americans' steadily diminishing habit of defining anti-Muslim militancy as their cause célèbre.
- Second is the fact that the Sahel's overstressed governments *invite* the Americans in; charges of Crusading imperialism are not nearly as sticky south of the Sahara.
- Third, the African governments are not bystanders, and it matters hugely to the American public that their hosts also bleed in antiterror fights.
- Fourth, the hosts are eager learners. The American presence helps them develop local capacity to carry on the fight after the Americans leave.
- Fifth, the relationship building that comes with training grants the Americans some of the best intelligence penetration into the internal workings of the militant groups, which will be all too useful near and far, now and deep into the future.
- Sixth, the heavy use of drone warfare in the Sahel turns the entire region into a live-fire technology proving ground. As Americans shift farther away from massed forces in their warfare, action in the Sahel helps them sharpen their technological blades.
- Seventh, the heavy emphasis on drones, on empowering local forces, and on classified, manpower-light Special Ops keeps America's Sahel activities out of the public eye. In an era of war-weariness, not needing to fight a public-relations battle back home makes Sahel activity a relatively easy call.

But the eighth reason is perhaps the most intriguing: the Americans won't be fighting alone, and not only because the Sahel governments are eagerly on board. Much of the Sahel is part of the Francophone world, and **France** is at least as involved in the fight against the region's militant groups as the Americans are.

The longest, most successful, and least bad-blood-ridden foreign relationship the Americans have *ever* had with anyone has been with the French. Making matters more certain for the future, the converse is true as well. The pair spit and glare at each other like estranged siblings, but they have never been on opposite sides of an honest-to-God *war*.

The intersections of American and French interests are many, though not necessarily identical in justification or magnitude. The Sahel is part of the old French Empire's sphere of influence, and the French have long felt a strong feeling of responsibility and possessiveness toward the region. France and the United States retain strong concerns about the future of Germany and Russia, the French especially about the former and the Americans, the latter. If Turkey defies the odds and achieves runaway success in its efforts to remold its region, the Americans will likely register some concern, and the French—whose physical location means they can't consider abandoning the Mediterranean, as the Americans will—will be forced to act.

These common interests will manifest as a newfound warmth in bilateral relations, with much of said warming occurring in Paris. Any version of the United States that defines its interests more narrowly is one that almost by definition gives France more room to maneuver. The end of the global Order and the collapse of the EU and NATO remove the hyperpower nature of the United States, which the French have always found overbearing, while also freeing France to be a pure national power pursuing purely national goals. The Americans will notice and appreciate and embrace the change in tenor without fully grasping the reasons behind it. The Americans will make up for their lack of comprehension by providing moderate-size flows of shale crude that should help both the French and British achieve their ambitions in a non-zero-sum manner.

The opportunities for the rekindled relationship are nearly endless. Throughout Europe, the Middle East, and Africa, France's intelligence, military, and paramilitary capability ranks, at worst, in third place. France is no weakling power that will seek American help every other Tuesday. When the French *do* make a call, they will do so with good reason, a plan of action, assets in place, and a treasure trove of actionable intelligence. Put simply, France will displace the United Kingdom as the Americans' most useful ally, becoming the perfect spotter for the American hammer.

GROUP 2: WITH FRIENDS LIKE THESE . . .

The second group of countries the United States will have broadly productive relationships with are not so much allies as neighbors, fortunate (or unfortunate) enough to be located within the Americans' sphere of primacy, places where the Americans will automatically perceive foreign intrusion as hostile and will meet it in kind. Technically, the entire Western Hemisphere falls into this expansive claim, but like all big groupings, some members are more equal than others. Motivations and stakes will vary.

The first motivation is all about the guns. The United States has always held proprietary views about the Americas, first laid down in the Monroe Doctrine: Eastern Hemisphere powers are not welcome in the Western Hemisphere, and the United States will not only resist their intrusions, but will take action to dissuade them from intruding in the first place. Under Teddy Roosevelt, that policy advanced to the point of sponsoring or stopping coups in Latin American countries to deny European powers potential footholds.

During the Disorder, the American interpretation of Monroe will be even more aggressive. American naval power will keep *anyone* from sailing in force to the Western Hemisphere, and considering how blammo the Eastern Hemisphere is likely to be, at times the American cordon will feel a bit more like a hemispheric blockade of the East than a defensive ring around the West. Western Hemisphere countries that seek to use outsiders to balance American power will find themselves in America's crosshairs.

The second motivation is all about the money. The Eastern Hemisphere is about to become a nasty place. Security concerns will dissuade most Americans from doing business there. Conversely, the upgrading of Monroe means most Eastern Hemisphere powers—whether neo-imperial or not—won't be allowed to swim in the Western Hemisphere pond. While only about 13 percent of the US economy is dependent upon imports and 8 percent on exports, the total trade flow in and out of the United States still amounts to about $4.3 *trillion* in commercial involvement. Total Latin American *non*-commodity trade is another $2.3 trillion. Most of both will by necessity become concentrated in the Western Hemisphere.

The Latin Americans will be of mixed minds on this outcome.

No Latin American states look back on the last period of American dollar diplomacy with fondness. All will seethe when the Americans take most security decisions out of their hands completely, and their loss of nearly all leverage vis-à-vis the Americans is not something that will be welcomed.

Yet all must admit that this time, things are different. The last time around, the Latin American states were relatively new, and the European empires were predatory. Incessant European actions encouraged the Americans to take a much more aggressive role in managing the Western Hemisphere's countries, including direct occupation when the mood struck. This time the Americans, courtesy of their recent wars in the Islamic world, are a bit gun-shy and so are likely to limit any military incursions. American actions will take the form of aggressive exclusion of the East more than direct country-by-country manipulation of the West.

But far more important, there aren't many players in the game. Latin America's rugged geography means it needs capital for infrastructure and development. Between demographic and capital-flight patterns, *only* the United States will have the money the Latin American states need to thrive. The vitriol most

Latin Americans feel toward the United States, rooted in American actions dating back to before World War I, will gather fresh impetus for all the same reasons, but considering the dearth of alternatives, most of Latin America will suck it up and carry on.

Yet history and geography still enable a wide degree of variation in just how aggressive the Americans can be with each country . . . and how aggressive each country can be with the Americans.

Americans rarely consider it, but the countries of the Caribbean Basin are far closer to American population centers in Houston, New Orleans, and Miami than they are to the major population centers of the Southern Cone. While much of the world must adapt to a world *without* America, these states must instead learn to get by in a world with *only* America:

- The microstates of the **Lesser Antilles** chain highlight much of the challenge the rest of the world will soon face. A combination of internal squabbling and imperial legacies has prevented them from negotiating as a whole with the Americans, reducing them to little more than destinations for American tourists. As globe-spanning passenger travel craters in a time of Disorder, the tourist influx from the rest of the world will slow to a trickle. The Caribbean islands' dependence upon the United States will become absolute, as will American penetration into them to combat the trade in illegal drugs.

- The **Dominican Republic** has struck the best deal possible, going out of its way to become included in the Central America Free Trade Agreement, itself an extension of the better known North American Free Trade Agreement. NAFTA is already the biggest, deepest multilateral trade deal of the Order era, and it will be the only one of size to survive into the Disorder.

- There is a reason that **Cuba** has loomed so large in American foreign policy, from the Spanish-American War to the CIA's obsession with Castro. The island's position less than ninety miles from Miami would enable a suitably capable and hostile power based in Cuba to disrupt *internal* American commerce. Cuba has done very well for itself serving as a platform for whatever anti-American power seeks a footprint on the American doorstep. In a world in which the Americans take a more hands-off approach to the Eastern Hemisphere but a more hands-*on* approach to the Western Hemisphere, that sort of cheekiness will not be allowed. If the Cubans handle that transition gracefully, they could become the next . . .

- **Colombia.** Even before the Colombians realized they needed American military and economic assistance to first fight and then resolve their civil war, they learned to read a map. The only country worth trading with en masse that is also within reach of their ports is the United States, and so the Colombians approached the Americans about a free-trade deal . . . in the *1990s*. Now, with the world's other trade linkages collapsing, the Colombians look like fortune-tellers. This path is open to Cuba as well, but so is the option of . . .

- **Venezuela.** The governments of first Hugo Chávez and later Nicolás Maduro allowed their anti-American sentiment to blind them to a simple fact: the bulk of the world's refining capacity that can process Venezuela's thick and dirty crude oil resides in the United States. Under their rule, Venezuela has sold its crudes at steep discounts to Russia and China . . . who turn around and sell it at full market price to the United States, pocketing the difference.* Add a degree of economic mismanagement, kleptomania, and willful ideological blindness that is well past criminal, and Venezuela has degraded from one of the most advanced countries in the developed world, in terms of education, infrastructure, health care, and diet,

---

* Global processing capacity for Venezuelan crude outside of the greater Caribbean totals about 1.4 million barrels of crude a day, all of which is inconveniently located in a different hemisphere. Under normal circumstances, Venezuela exports about two million barrels daily.

to state failure, complete with famine. In the Disorder, the Venezuelans will lose all capacity to leverage Eastern Hemisphere powers against Washington. Venezuela's future is now entirely up to the Americans. The question is whether the Americans care.

Beyond the Caribbean, things get more complicated, not because the rest of the countries of the Western Hemisphere have more options—most don't—but rather because we are used to thinking of them as countries that do.

The American attitude toward **Canada** is in the process of shifting. Under the Order, Canada held a special position in global affairs. It sits astride the ICBM flight paths from the Soviet Union to the United States. That would have enabled the Canadians to be security free-riders during the Cold War. To their credit, they were not. They openly and willingly participated in all aspects of North America security policy. But that doesn't mean they didn't use that leverage at all; they deflected it into other realms. Economically, the Canadians proved adept at eking out concessions from the Americans in trade relations, most notably in agriculture, finance, manufacturing, and services. Strategically, the Canadians played middleman between the Americans and pretty much everyone else. With most US foreign policy made by a single person, few aside from the Canadians had a reliable in. Such top-level access meant the Canadians wielded amazingly outsize influence in a world of global interconnections.

No more. The Americans' Cold War mentality is gone. Washington no longer sees value in engaging the world's many countries multilaterally, so while the need for an "American whisperer" might be higher, the Americans have become mildly offended that anyone might play that role at all. More to the point, the Americans no longer fear a Soviet nuclear assault, so there is no longer leverage to be gained from being on the missile flight paths. Any special regard the Canadians once commanded, any leverage they might have wielded, is dust in the wind.

In a world of Disorder, the power balance in the Canadians' disfavor is extreme. Working purely on population and economic strength and completely leaving aside issues of military capacity and reach, the United States outweighs Canada by a ratio of roughly ten to one. Other similarly lopsided power balances around the world include India v. Bangladesh, Japan v. Taiwan, China v. Vietnam, Brazil v. Bolivia, and Germany v. Denmark.

Even such comparisons assume a fair fight. It would not be. Nearly all of the Canadian population lives in the line of widely separated urban regions near the American border: Vancouver, Calgary, Regina, Winnipeg, Toronto, Ottawa, Montreal, Quebec City. It's ironic that the one farthest from the US border—at about two hundred miles, Calgary is more than twice as far away as the others—is the one culturally closest to America. *Every* mainland Canadian province trades more with the United States than the rest of its own country and also is constantly exposed to the full array of American power.

Playing the Canadian provinces off one another would be child's play. Ontario and Quebec compete furiously for industrial access to the Canadian and American markets. The Maritimes are poor and utterly dependent upon trade access to the American Northeast. The Prairies are cultural and economic extensions of the American Midwest. British Colombia is linked to the rest of Canada tenuously at best, via a single road-and-rail transport corridor.

Even Canada's ocean ports are less independent than they seem. The country's internal maritime arteries—the Great Lakes, St. Lawrence River, and St. Lawrence Seaway—are all shared with the United States. Both Canada's Pacific and Atlantic megaports—Vancouver and St. John's—access the wider world via waters the Americans functionally control. Only Montreal grants meaningful autonomy.

It's not that the Americans have it in for the Canadians, but what has made Canada special in the international system no longer applies. Even worse, the Canadians completely escaped the Americans' first iteration of dollar diplomacy because they were still part of the British Empire. With the United Kingdom both weaker than it was in the early twentieth century *and* itself being reduced to an American adjunct, everything that once shielded Canada from American power has evaporated.

Almost overnight, Canada has shifted from the country most likely to get what it wants out of Washington to one of the least. Canada is being reduced to little more than a passive-aggressive American satellite. Even *that* assumes full Canadian cooperation. If Ottawa strikes too strident a tone with Washington, Canada will be reminded that both Alberta and Quebec have active secessionist movements. As the Canadian provinces are laid out east to west in a straight line, one leaving would break the remainder of Canada into two completely disconnected pieces—ending the existence of Canada as a unified state and leaving it to the Americans to pick and choose which pieces they desire.

Brazil is another Western Hemisphere state where the center might not hold.

The geopolitical conditions that enabled Brazil's rise since 1990 are gone, while the disadvantageous geographic features that have constrained Brazil's development since colonial times have not. There is no version of the future where Brazil performs or coheres better than it has in the past few decades, while there are *many* versions where Brazil falls apart as a country. And that's before American business interests start partnering with Brazil's central-authority-averse oligarchs.

Brazil's choice is a direct, brutal one. Encourage those American business interests—the sole source of the capital that Brazil so desperately needs—and risk the country spinning apart as local oligarchs seize operational control. Or bar the Americans, and descend into such poverty that the country becomes a riotous mass. Such options feel almost . . . Chinese.

On the flip side are a pair of countries that will not only weather the Disorder's storms in far better shape, but will also be able to deal with the Americans from a stronger footing.

The first is **Mexico**.

Existing integration between Mexico and the United States *already* makes for the world's busiest manufacturing, energy, financial, and passenger transfer border—even before illicit transfers of drugs and migrants are considered. The difficult work of integration is already completed. Moreover, of *all* the world's other major relationships, the US-Mexico link will suffer the *least* degradation as the Order ends. In a world where trade, movement, and finance are hard, maintaining the American-Mexican affiliation is in and of itself a cause for celebration.

But the Americans will see a less-than-ideal side to all this:

- In 2019 Mexico became the United States' largest trading partner. Unlike the countries in second and third place—Canada and China—Mexico's demographic, economic, and geopolitical runway is robust and long.
- Latinos are the largest minority in the United States, with Mexican Americans making up the vast majority of that community.
- Politically, Mexican Americans are the most recent aspirants to the American Dream, and thus are very politically active. They have influential representatives in Congress, as well as strong representation in powerful states, such as California and Texas, *and* in the electoral swing states of Colorado, Florida, Illinois, New Mexico, and Nevada.
- Strategically, the Mexican border is the only one the Americans have that presents any meaningful concerns, whether migration, drugs, or more traditional trade issues.

There are no easy solutions to any of the issues in the bilateral relationship. If there were, they would have been implemented decades ago. But put another way, America's Mexican association is America's only "normal" relationship. It is the only one in which the Americans cannot either dictate the result or walk away from the table and not care about what happens the next day. It adds up to make Mexico the *only* country on the planet that the Americans *must* deal with to address issues both internal and external. That gives the Mexicans something no one else has. Something the Americans truly despise: leverage over Washington.

If Mexico has leverage, **Argentina** has flexibility. In part, it's the simple issue of distance. Argentina is as far from the United States as a country in the Western Hemisphere can be, and it's also as far from nearly every other country in the Eastern Hemisphere as is physically possible. The only country in contemporary times that might even theoretically be able to complicate Argentina's existence is Brazil, and Brazil's days in the sun are over. Such insulation and Argentina's geographic strengths make for a pocket power that can deal with issues local and regional more or less at its own pace.

Strategically, the Americans are unlikely to spare Argentina many stray thoughts. It is simply too far removed to pose any threat to American interests, whether by extreme success or extreme failure. Most interaction will be via the business community, and *that* relationship could well prove phenomenally productive. Argentina's economic bones are excellent. All it needs are an infusion of capital and technology and a government committed to looking after the country's best interests.

Americans will not enjoy the sorts of advantages in Argentina they will become used to in the wider world. It all has to do with the centrality of Buenos Aires to all things Argentina. The political capital is the financial hub is the chief port is the manufacturing center is the agricultural entrepôt. Any American who attempts anything in Argentina is immediately exposed to the full cultural, economic, and political strength of the Argentine nation. That's a tough nut to crack, and it gives the Argentines the ability to shape the investment coming their way rather than vice versa. Add American strategic indifference, and the Argentines' interactions with America will occur more or less like their interactions with their neighbors: on Argentina's timetable.

The coming decades are an interesting window of opportunity for the Argentines. They have had a couple of decades to re-consolidate internally *while also* extending a creeping neo-imperial control over the countries to their north *while also* benefiting from a collapse in many of their global resource-exporting competitors *while also* absorbing and shaping vast swathes of external investment. By the time the world gets to whatever comes after the Disorder, Argentina will be a very rich, very powerful place that has *already* established itself as the undeniable hegemon of the Southern Cone, all in a geography second only to the American lands in its capacity to support a big nation projecting major power outward.

Southeast Asia, Australia, and **New Zealand** are a bit of a hybrid of the common-cause and friends-like-these categories. Relationships with all will be somewhat awkward.

Nearly every state in the region has a vested interest in ensuring that the Chinese rise is halted, but likewise nearly every state in the region has a vested interest in ensuring that the Japanese do not become all-powerful. These are both goals the Americans broadly share, but neither are goals the Americans feel the need to bleed for. All have some of Argentina's flexibility vis-à-vis the United States born of distance, while being situated in a strategically useful area gives them some very Mexican-like leverage.

But that's where the clarity ends. Sharing the same geopolitical space with Japan, China, and India can be a precarious existence. Countries in this region *love* the idea of American security guarantees, and as all of them until the early 2000s counted the United

States as one of their top two trading partners, they equally loved the economic possibilities American partnership provided.

In the late-Order era, everything got even better. The US guaranteed their security *and* enabled the Chinese boom, which generated huge demand for their agricultural, industrial, and manufacturing outputs *while also* discouraging either the Japanese or Chinese from getting any cute ideas about becoming regional hegemons. Guns *and* butter. *Lots* of butter. For the Southeast Asians, it has been the best of all worlds.

With the Order over, these states will still do well. They all produce things that people need, whether it be iron ore, wheat, beef, dairy products, rice, fuels, natural gas, engine blocks, or semiconductors. They all can maintain their own internal security, and none of them have meaningful territorial spats with one another. These states will all still do well in the Disorder, but that doesn't change the fact that the gravy train is over.

Rebalancing is their word of the era. Rebalancing their local economies away from servicing the Chinese behemoth and toward internal and inter-regional activity. Rebalancing in a new security environment defined by Chinese weakness instead of Chinese strength. And above all, finding a way to rebalance their relations with an increasingly distant America and an increasingly present Japan. That it is the Americans who are encouraging Japanese boldness just before they tune out makes this region's diplomatic acrobatics all the more difficult.

GROUP 3: ALLIES FOR HIRE

The third group of relations is more transactional. Such transactions will look quite a bit different from what came before.

In the Order, the United States would sacrifice its trade interests to achieve security cooperation. In the Disorder, this trade-off will likely be flipped. If a country wants American assistance on a security issue, then it must grant the United States deference on an economic issue. Because the Americans' external trade portfolio is so small, any inducements will need to be particularly shiny. This limits the potential relationships in this third category to countries neither unduly hampered by the Americans newfound penchant for trade disruption nor so desperate that they have no choice but to seek alliance anyway. Such conditions make for an exceedingly *short* list of potential partners.

While the United States and **Iran** have found themselves on the opposite side of many issues, they have yet to find themselves on opposite sides of a *war.*

Even assuming that every tangentially Iran-aligned militia were fully under Iran's control and only targeted US troops under Tehran's explicit orders, even if the maximum estimate of American casualties is credited to such groups, after two decades of wars in the Middle East, Iran is responsible for fewer than eight hundred American combat deaths. For comparison, the Americans suffered some thirty-four thousand combat deaths at the hands of the Chinese and North Koreans in the Korean War, and another forty-seven thousand inflicted by the North Vietnamese.

It makes for America's most curious bilateral relationship, based on some of the hottest emotions but generating some of the fewest casualties. Much of the bad blood is based upon a series of early miscalculations on both sides. Early in the Cold War, the Americans supported the coup that brought Shah Mohammed Reza Pahlavi to power, and in 1979 elements of the Iranian Revolution targeted American diplomats. It is far from accurate to call such events water under the bridge, but those events *are* over four decades old.

Even more curious is the United States' legacy relationship with **Saudi Arabia**. Few in the American intelligence community would even pretend to claim that Saudi actions in and beyond the Middle East are responsible for fewer American deaths than Iran. Al Qaeda is an indirect Saudi creation, and the 9/11 attacks alone killed 3,000 Americans. Al Qaeda in Iraq is a similar outcome of Saudi policy, and its actions are responsible for at least another 1,500 American deaths during the Iraqi occupation. The Islamic State (ISIS) is the most recent incarnation of Saudi actions, and while few Americans fell prey to the Syrian Civil War, ISIS is credited with killing some 170,000 people.

During the Order and before the shale revolution, the American-Saudi alliance was a necessary evil. No Saudi oil, no global economy, no global trade, no American alliance, no American security. But *after* the Order and *with* America's shale-induced energy divorce from the world, the Americans now have the opportunity to recognize the horrors that make up day-to-day Saudi policy for what they are.

In the shale era, the United States doesn't care about global oil for itself. In the Disorder, the United States doesn't care about global oil for its allies. With the American withdrawal from the Middle East largely completed, there are no more American deaths at Iranian hands. That leaves legacy anti-Iranian feeling as the core rationale for the relationship with Saudi Arabia. But if there is *any* place in US foreign policy for things like ethics or morals, aligning with the Persian Gulf's *less* ethical, *less* moral power doesn't make a great deal of sense.

At some point in the next decade or two, the combination of the economic distance provided by the shale revolution, the physical distance provided by American disengagement from the world, and the emotional distance of time will enable the Americans to reevaluate both relationships. Iran is unlikely to be seen as a friend, but a sort of cool disinterest will establish itself. Saudi Arabia is unlikely to be seen as an enemy but instead as the distasteful family who runs the next town over.

This process could occur gradually over many years, but it is more likely to happen suddenly.

At some point after the American disengagement, some event in East Asia or the former Soviet Union or the Persian Gulf itself will trigger a global energy crisis. The economic and strategic significance of the Persian Gulf's energy deposits are too tantalizing, and the indigenous powers' control over them so nebulous, that any external power willing to sink time, money, and troops into the region could see itself well rewarded. Only the Americans have enough energy and money and insulation to reliably resist the call. Someone—the Brits? French? Turks? Chinese? Japanese?—will feel obliged to move military forces to the Persian Gulf either to secure their interests or seize an advantage. Iran and especially Saudi Arabia will then turn their tender mercies on a new player.

All at once, things will crystallize for the Americans: the Persian Gulf is *not* our problem, and life will move on. The transactional nature of American foreign policy will kick in, and American relations with the broader region will transform.

Americans will discover that Iran—now with a vested interest in regional *stability*—is the far more pragmatic partner and one that can be leveraged to achieve midterm goals, such as propping up or smothering this or that regional state. Because Iran is more capable, has more varied tools, and is more constrained by its local geography, Tehran is likely to emerge as America's partner of choice.

Conversely, Saudi Arabia's burn-it-down mantra, while distasteful, will prove useful at times. American-Saudi relations will not be the normal state-to-state type, but instead more like the relationship of a rich tycoon with the goon squad he occasionally hires. If you want to build something or manage a relationship, you call Tehran. If you want to break some legs or burn down a building, you dial up Riyadh.

It's a system of mixing and matching that will feel almost . . . French.

That leaves a pair of powers on the edges of the Persian Gulf region.

The first, **Turkey,** is both extremely complicated and brutally simple. Complicated in that Turkey will have its fingers in a lot of pots and will prove relevant—perhaps even central—to any American goals in the Balkans, Central Europe, the Hordelands, the Caucasus, Persia, Mesopotamia, the Levant, and the Eastern Mediterranean. Managing ongoing relations with such a diverse, neo-imperial power will be a delicate affair, as American actions in one zone could easily endanger Turkish cooperation in another.

Yet bilateral relations will also be fairly simple, because in a world of Disorder the Americans *don't* have any sustained interests in the Balkans, Central Europe, the Hordelands, the Caucasus, Persia, Mesopotamia, the Levant, or the Eastern Mediterranean. The complete lack of overlap means American-Turkish relations will be akin to relations between two empires that have no bordering territories: coolly wary due to the nature of both powers, yet professional, polite, distant, and *heavily* quid pro quo.

The second, **Israel**, is more straightforward. The nature of the Israeli state will determine future American-Israeli relations. Ever since occupying the Palestinian territories after wars in 1967 and 1973, the Israelis have wrestled with an internal, existential debate: Is Israel a democracy, a Jewish state, or should it directly control all occupied lands? So far, Israel has mostly managed all three, but demographics are swinging the argument. Higher birthrates among both the occupied Palestinians *and* deeply conservative Orthodox Jews means that by 2030 an enlarged Israel will be a Jewish-minority state that will not even want to converse with the Palestinians, much less grant political or territorial concessions.

The only possible outcome is a social and security management system that makes Apartheid look progressive. Apartheid South Africa enabled the country's black population to hold jobs in white-run zones but refused integration or political rights. In contrast, in Israel the Palestinians are kept permanently in what have become massive open-air prisons.

For Americans with a sense of history or ethics or national security, this is, at best, problematic for bilateral relations. Add that Americans under age fifty have no memory of Israel battling for its life in the wars of the 1950s to 1970s, and it's easy for the American polity to view Israel as aggressor and occupier rather than plucky upstart.

Even before economics, distance, and time suck some of the venom out of American-Iranian relations, Israel's coming evolution means there is no way forward for Israel in which US-Israeli relations *strengthen*. The increasingly racial and religious diversification of the American political scene will only hasten this evolution in perceptions. The rise of American populism, particularly on the political left, has brought the questioning of the ethics of Israeli security policy into the American political mainstream, and there it will stay.

Such shifts in American perceptions do *not* present Israel with an imminent threat. The Israeli military is among the world's most competent, and the threats to Israel are manageable. Jordan already is a satellite state. Egypt and Saudi Arabia already are de facto allies. Lebanon and Syria already are teetering on the edge of failed-state status. The Palestinians are already boxed up. Post-Order Turkey already has a cool, businesslike partnership with Israel that soon will likely include joint management of the Eastern Mediterranean.

The only country with the capacity to take military action against a worse-than-Apartheid Israel for *moral* reasons would be the United States, and the Americans need to work from their current attachment to Israel to neutrality before they might even theoretically flip to hostility. That mental evolution would take *at least* a decade or two. Israel may become defined by attributes that are normally associated with pariah states, but without American leadership, international institutions like the United Nations are not likely to continue anyway. Being a pariah doesn't mean what it used to.

For the duration of the Disorder, Israel will be left alone— even if supporters of liberal democracy don't necessarily like what they see.

THE UNITED STATES' REPORT CARD

BORDERS: Lakes, mountains, forests, deserts, and vast ocean moats surrounding the best agricultural lands and largest waterway network on the planet. Nowhere else on Earth does a territory have such a beneficial balance of good lands with great stand-off distance. Americans spend little on territorial defense, freeing their military to project *out*.

RESOURCES: Nearly two centuries of industrialization have heavily tapped out a continent of bounty, but new technological break-throughs continue to surprise. The most recent surprise—the shale revolution—has made the country a net oil and natural gas exporter.

DEMOGRAPHY: The American Baby Boomers—the country's largest generation ever—are nearing mass retirement, generating a painful financial crunch. But American Boomers had kids. Lots of them. America's Millennials may be a pain, but their numbers may just save us all. *gulp*

MILITARY MIGHT: The most powerful projection-based military in world history. With the Order ending, it has . . . nothing to do.

ECONOMY: The American economy isn't simply the world's largest and most diversified economic system; it is the least dependent upon the outside world for its health. The world needs the American economy to survive, not vice versa.

OUTLOOK: The Americans excel at missing opportunities due to domestic squabbling, but there is *nothing* in what's left of the international system that will threaten the American heartland either militarily or economically before 2050.

IN A WORD: Detached.

CHAPTER 16

PRESENT AT THE DESTRUCTION

THE DAWNING OF THE FOURTH AGE

If I had to select a single word that will define the ongoing historical turning, it would be *overwhelming*.

The wolf at the door already prevents us from looking forward and thinking sensibly about what changes geopolitical and technological will bring, whether the issue at hand be environmental concerns, manufacturing supply chains, or global food and energy flows. Considering the omnipresent consequences of American withdrawal, widespread demographic collapse, and runaway technological evolution, there's a veritable *pack* of wolves at the door. And in many places, the door wasn't particularly sturdy in the first place.

A few closing thoughts.

First, few recognize just how beneficial and transformative the global Order has been to the world writ large, much less their personal lives. Globalized food supplies and manufacturing supply chains often seem too complicated or distant to feel like part of day-to-day life. Here's something a bit more tangible, and probably more than a bit more terrifying. The fact that most people now consider the Germans and the Japanese to be nothing more problematic than rude tourists as opposed to crazy-eyed armies committed to eating the horizon is a testament to the Order's profound success. Wars of expansion haven't been fully eliminated, but since 1946 they have been rare.

Second, the level of development most countries enjoyed *before* World War II will be firmly out of reach for most. The Order's impact has been deep and pervasive, so singularly effective that it suffused itself into every aspect of human existence, nearly erasing every previous structure. Its absence will not be merely non-Order, but instead a new kind of chaos. In the Imperial Age when people were miserable, they were just the same kind of miserable they had always been. But in the Disorder the sense of achievements lost will be palpable. People will *remember* a degree of security and wealth that they will *never* be able to achieve on their own.

The collapse of economic norms will shatter political norms. Among democracies this will, at a minimum, require an overhaul of the social contracts that underpin modern society. Calling it "disruption" is to undersell the scope of what's happening, much less what's coming. The changes on deck are not evolutionary but revolutionary. Even if leaders guide their countries perfectly, this sort of adaptation takes time. The European imperial centers required *centuries* of political, social, economic, and military development to reach their pre–World War II heights. Nearly all the institutional advantages they once had are now seven decades gone, either destroyed outright in the war or forcibly dismantled by the Americans in the aftermath. It will take a decade (or more) for most to find their feet. In the meantime, popular uprisings will be the norm. Even *that* prediction assumes that tensions are contained within individual countries, a painfully unlikely development. Centuries of deep-seated rivalries and hatreds are not wiped away by something as simple as a few decades of prosperity.

Third, even among the soon-to-be neo-empires of the Disorder's winners' circle, none will be as powerful as the pre-Order empires before them. In part, again, it's a simple issue of time. Even rapid geopolitical expansions, such as the westward march of the American pioneers or the vast sweeps of the Mongolian hordes, were measured not in years but decades.

But more important is that in the 2020s and 2030s there will *not* be vast technological gaps in transport and warfare. There will be no twenty-first-century equivalent of an industrialized power exploiting vast nonindustrialized territories. Countries can and will fall, and their remnants can and will be preyed upon, but even where civilization itself is collapsing, like Syria and Venezuela, firearms aren't hard to find.

The neo-imperial expansions will be more incremental, looser, and more dependent upon collaboration. From countries with no experience in running an imperial system (Argentina), those whose expertise is limited to history books (Turkey), or those whose instincts are for hands-on management (France), we can expect lots of mistakes, too heavy a hand, or both. The Americans are not so much passing the torch as dropping it. It will start quite a few fires before someone picks it up.

Fourth, for the four new regional powers—Turkey, Iran, Japan, and Argentina—allaying American concerns and courting American goodwill will be essential to long-term success.

Retrenchment, indifference, and isolationism are not the same as sequestration. The United States' military evolution since 1992 has been toward tools with greater range and precision, while its political evolution has been toward viewing military tools as more reliable than economic or diplomatic tools. Argentina can probably pull off mollifying the Americans with a nice steak dinner; France, with a few cases of wine and some intelligence sharing. Turkey will need to be a bit more forward-looking to forestall American intervention in Turkish plans. Japan, as both the most capable new power *and* the one most likely to step on American toes, will need to tread *very* lightly. After all, above all the United States is a *naval* power. No matter how much America withdraws from the world, it can steam through any part of Japan's new empire in days if relations sour, and as tends to happen when the Americans get involved, Japan's reality convulses.

For the short list of countries likely to remain in the Americans' inner circle of allies, the whole situation is actually pretty good. For the short list of countries likely to seek American alliance as a hedge against the new crop of regional powers, life isn't so clear-cut. For everyone else, waking the eagle is something to be avoided. Even if the US military wasn't more lethal and unfettered than ever, Americans' views of the world have shifted.

The "America First" of the hard right is reflexively hostile to the world. The "America First" of the hard left is reflexively hostile to American involvement in the world. The "America First" of the middle just finds the world exhausting.

In all three versions, however, Americans believe that the world is not their problem and that America's military strength will keep the world from hurting them. While a deep dive into context and history and capabilities and geopolitics can finds lots to quibble with in such a simplistic assessment, for at least the next couple of decades it *will be mostly true*. Nearly all evolutions of Americans' views on the world lead to a United States that is no longer holding up civilization's ceiling, and if poked in the wrong place, the Americans could well knock down what pillars remain.

Finally for the energy-rich, demographically stable Americans in their splendid isolation, this is all rather academic. Even in the worst-case scenario, the United States' trade disruptions will be a shadow of those of suffered by Ireland or Germany or Korea or China, and they won't affect the supply of energy or food. Instead, America's economic basis will widen and deepen as it becomes one of the few locales with markets and security and a population and financial stability, while other states must grapple with the streams of radicals and refugees that failing states generate.

Even on that last point, the Americans face more opportunities than challenges. Living as Americans do in an age of gun violence and backlash against mass migration from Central America, it is difficult to envision them reclaiming their "Shining City upon a Hill" aura, yet that's precisely what will happen. For example, think of the last time challenges on the scale of what's coming struck Germany: in the 1930s the state became so hateful that Albert Einstein felt compelled to defect and seek American citizenship. A generation of wildly educated people desiring escape from dark futures will crave new homes somewhere you can't reach by walking or swimming. If only by process of elimination, most will come to America. Just *imagine* the volume of Chinese skilled labor that will attempt refuge when the Chinese Dream proves ephemeral.

The stories of their journeys and the dying of their hopes will close the chapter on the third age of history, the age of wealth untold, of security unparalleled—the age when all seemed possible. That age was never much more than a moment, historically speaking, and its demise was always inevitable. But the end of an age is not the same as the end of history. It simply marks a new beginning.

ACKNOWLEDGMENTS

Honestly, the acknowledgments section is the best part of writing a book because I finally get to give credit fully to where it is due. On the off chance you've made it this far, you're probably the kind of person who has noted there are very few direct callouts in this book and what footnotes I have are more about add-on thoughts rather than attributions.

That's not an oversight, but instead the nature of the particular herd of beasts I work with. I am *not* a journalist. Journalism is about reporting facts and statements and events, so the criticality of citations is burned into the souls of all good reporters.

In contrast, my background is in private intelligence work. Back in the day it was less my job to uncover facts or tell someone's story, or even to pore over and make sense of a specific region or sector, but instead to weave disparate information flows from every part of the world into a tapestry that demonstrates linkages across systems. I build thought models. I interpret. I bring together topics that on the surface share no connections. I combine and compare and, of course, forecast. None of which is particularly easy, or even appropriate, to source.

Take the example of Dr. Michael Pettis, a professor of finance at the Guanghua School of Management at Peking University in Beijing. Undoubtedly one of the most intelligent people alive today, Pettis regularly uncovers, dissects, and comments upon the intricate inner workings of Chinese finance, both private and public. (He has also ran a punk-rock club, which is delightfully at odds with his denser-than-granite writing style.) I follow literally hundreds of people like Pettis who are masters of their craft; each informs many aspects of many of the topics I examine, often in completely unrelated topics. In the specific instance of Pettis, bottomless Chinese finance contributes to both broadscale Chinese military incompetence as well as higher fertilizer demand in Africa. My job is to distill the world. If I cited every obliquely contributing thought, each page would have a book's worth of citations.

What this all means is that literally *every* conversation I have and *every* question I'm asked goes into a sort of seething cauldron of thoughts and information from which *everything* I do is forged. If you've interacted with me in person, on the phone, at a conference, or via email, you've had an impact on my thought process, research goals, and writing. If I'm tapping something into my phone with a concerned look on my face, you've given me something to explore—and you've just ruined my weekend. Thank you *all*. You are wonderful! Please keep it up!

But, as they say, some animals are more equal than others.

In no particular order, these extra-wonderful people who have had an outsize impact include Nancy Szigethy of NMS Capital; Milo Hamilton of FirstGrain; Danny Klienfielder and Mark Welch of Texas A&M; Julie Hammond of the CFA Society; Bob Grabill of the Chief Executive Network; Anne Mathias of Guggenheim Partners; Thierry Krier of BUCHER Industries; Rya Hazelwood of the Industrial Asset Management Council; Adam Jones Kelly of Conway Incorporated; Vic Hayslip of Burr Forman; Vince Shiely of Lubar & Company; Carl Sohn of Northwest Farm Credit Services; John Ruffalo of Omers Ventures; Tom Fanning of Southern Company; Rebecca Keller of Stratfor; Lance Lahourcade of South Texas Money Management; Ian Bremmer of Eurasia Group; Liam Denning of Bloomberg; Kris Kimmel of the Idea Festival; Guillermo Diaz of Cisco; Don Kuykendall of Commerce Street Capital; pretty much everyone at Rimrock Capital, Knowledge Leaders Capital, and the Society for International Business Fellows; as well as Jen Richmond, Marc Lanthemann, Joe Ricketts, and Mitt Romney, who live in worlds all their own.

Many of you are friends. Many of you are clients. Many of you ask great questions. Many of you are wicked smart. Some of you are insane. Pretty much all of you fall into more than one category, and in my opinion are better people for it.

In this world of hyperpartisanship and manufactured facts, rest assured that there are still American institutions dedicated to providing the sort of unbiased information and data that has been used so rarely in the last four American presidential administrations. Not just this book, but very little that I do could even be attempted without the dedication of the hardworking men and women of the United States Geological Service, and the Bureaus of Labor, Census and Economic Analysis. While their jobs involve *lots* that will always be classified, I couldn't have begun to parse global naval forces without assistance from the good folks at the Department of Defense, and there is no better way to understand the backstory of everything than to take a deep dive into the Library of Congress. Particular thanks to the people of the Energy Information Agency and the Department of Agriculture, who are consistently the most friendly, helpful, and underloved folks within the American federal government.

Internationally, the United Nations system continues to provide stellar assistance, with the Development Programme, Comtrade, Conference on Trade and Development, and the Food and Agriculture Organization being particularly valuable. All hail the World Bank's Data Bank and the independent Our World in Data, both of which are shockingly user-friendly and collectively enable the easy access of pretty much any numbers one might find useful. The International Monetary Fund continues to provide top-notch information on national accounts. Much of my energy work would be impossible without the Joint Oil Data Initiative, nor would parsing global debt levels be doable without the work of the Bureau of International Settlements, nor could I make sense of trade patterns without the International Trade Centre's Trademap.

RAND remains the best source for tactical details on the Chinese navy. Pew Research for its background and population data. While OPEC definitely has an agenda, their data remains peerless. And the BP Statistical Review of energy is such a regular part of my staff's day-to-day lives that they commemorate its release every year in song (yes, really).

At the more national level, contrary to popular belief, the various national ministries of Japan, France, Mexico, Brazil, and even China (on most days) are absolutely solid.

There are some exceptions of course—counter-acknowledgments, if you will: Eurostat and Statistics Canada. Such great data. Such horrible interfaces. Why do you make data mining so painful? And, of course, Russia's Rosstat. I am hugely thankful for your work. Not for your data, of course. Most of it is absolute crap. Fabricated crap to support the Kremlin's renewed propaganda campaigns. But, wow, you are always good for a laugh!

In the creation of *Disunited Nations*, my team and I read oh so many books as well. Listing them all here would make your eyes bleed, but there were a few that proved particularly useful:

William Bernstein's *A Splendid Exchange: How Trade Shaped the World*.
Barry Naughton's *The Chinese Economy: Adaptation and Growth*.
Lilia Moritz Schwarcz's *Brazil: A Biography*.
Jonathan Brown's *A Brief History of Argentina*.
Max Brooks's *World War Z*. (Don't knock it—it's the best geopolitical book I've ever read and has nothing to do with that awful movie.)

Finally, there is the team itself that did all the heavy lifting.

To everyone at HarperBusiness who chose to invest their time and reputation on my work and mood swings: Thank you, and I hope you're enjoying the ride.

To Adam Smith of 713Creative: If you keep making graphics that make me look·like I know what I'm talking about, you will continue to have my absolute gratitude.

To Melissa Taylor of Taylor Geopolitical Research: Melissa doesn't simply regularly dig up bits of information that I find necessary, she doesn't simply find critical bits of information that I didn't even realize were critical, sometimes I make up crazy requests just to see if she can pull it off. She does. Always. It's a little intimidating.

To Susan Copeland: I can't imagine what my work would be like without your steady, empowering presence. That's a lie. Yes, I can. I just did. It was awful. Thank you thank you thank you thank you thank you!

To Michael N. Nayebi-Oskoui: You have become my analytical sounding board, chief of staff, and, quite unexpectedly, my fixer. While all of *Disunited Nations* is far stronger courtesy of your input, the Middle East chapters would have been a particular mess.

To all of you: Do. Not. Leave. Me.

Because we aren't yet done. Hope all y'all are ready to get moving on book number 4. . . .

INDEX

Page numbers followed by "*f*" indicate illustrations and maps. Page numbers followed by "n" or "nn" indicate notes.

A

ABOUT THE AUTHOR

PETER ZEIHAN is a geopolitical strategist and the founder of the consulting firm Zeihan on Geopolitics. His clients include energy corporations, financial institutions, business associations, agricultural interests, universities, and the US military. He is the author of *The Accidental Superpower* and *The Absent Superpower*. He lives in Austin, Texas.